Stem Cell Battles
Proposition 71 and Beyond

How Ordinary People Can Fight Back Against the Crushing Burden of Chronic Disease — with a Posthumous Foreword by Christopher Reeve

Stem Cell Battles
Proposition 71 and Beyond

How Ordinary People Can Fight Back Against the Crushing Burden of Chronic Disease — with a Posthumous Foreword by Christopher Reeve

Don C. Reed
Americans for Cures Foundation, USA

World Scientific

NEW JERSEY · LONDON · SINGAPORE · BEIJING · SHANGHAI · HONG KONG · TAIPEI · CHENNAI · TOKYO

Published by

World Scientific Publishing Co. Pte. Ltd.
5 Toh Tuck Link, Singapore 596224
USA office: 27 Warren Street, Suite 401-402, Hackensack, NJ 07601
UK office: 57 Shelton Street, Covent Garden, London WC2H 9HE

Library of Congress Cataloging-in-Publication Data
Reed, Don C.
   Stem cell battles : Proposition 71 and beyond : how ordinary people can fight back against the crushing burden of chronic disease : with a posthumous foreword by Christopher Reeve / Don C. Reed, Americans for Cures Foundation, USA.
     pages cm
   Includes bibliographical references and index.
   ISBN 978-9814644013 (hardcover : alk. paper) -- ISBN 9814644013 (hardcover : alk. paper) --
   ISBN 978-9814618274 (pbk. : alk. paper) -- ISBN 9814618276 (pbk. : alk. paper)
  1. Stem cells--Therapeutic use. 2. Regenerative medicine. 3. Embryonic stem cells--Research.
4. Embryonic stem cells--Research--Law and legislation--California. 5. Stem cells--Research.
6. Stem cells--Research--California. 7. Science and state--California. I. Reeve, Christopher, 1952–2004.
II. Title. III. Title: Proposition 71 and beyond.
  QH588.S83R44 2015
  616.02'774--dc23
                                                                                 2015023650

British Library Cataloguing-in-Publication Data
A catalogue record for this book is available from the British Library.

Copyright © 2016 by World Scientific Publishing Co. Pte. Ltd.

All rights reserved. This book, or parts thereof, may not be reproduced in any form or by any means, electronic or mechanical, including photocopying, recording or any information storage and retrieval system now known or to be invented, without written permission from the publisher.

For photocopying of material in this volume, please pay a copying fee through the Copyright Clearance Center, Inc., 222 Rosewood Drive, Danvers, MA 01923, USA. In this case permission to photocopy is not required from the publisher.

Printed in Singapore

CONTENTS

Foreword by Christopher Reeve — ix

Acknowledgements — xi

Chapter 1   The World's Deadliest Killers — 1

Chapter 2   The Naked Face of Hate — 11

Chapter 3   To Clone, or Not to Clone? — 21

Chapter 4   Season of Storms — 29

Chapter 5   The Trial(S) of the California Stem Cell Program — 41

Chapter 6   When Changing the World, What Do You Do First? — 49

Chapter 7   Battles With a Friend — 53

Chapter 8   Time-Lapse Scriptography — 61

Chapter 9   Ideology, Science, or Bible-Quoting Vampires? — 65

Chapter 10   Joan of Arc, and the Republican Stem Cell Reversal? — 71

Chapter 11   Deadly Definitions — 75

Chapter 12   What are We Fighting for? — 79

Chapter 13   Suffer Little Children — 83

Chapter 14   Klein Must Resign! — 89

Chapter 15   Last Day in the Wolverine State — 97

Chapter 16   Getting up in the Morning and Going to — Washington? — 105

Chapter 17   The Boy Who Loved Stanford Too Much — 113

Chapter 18   Stem Cell Summit, Stem Cell World — 117

Chapter 19   How Not to Have Sex in a Personhood State — 125

Chapter 20   Fighting the Killers: Leukemia and Cancer — 129

Chapter 21	The Thief of Memory	135
Chapter 22	The Anti-Science Society	139
Chapter 23	Stem Cell Thanksgiving	143
Chapter 24	Swimming from Alcatraz	147
Chapter 25	Bridge to a New Life	153
Chapter 26	Skidding on Ice	157
Chapter 27	Helen Keller and Stem Cell Research	161
Chapter 28	How to Mend a Broken Heart	167
Chapter 29	Why We Can't Afford *Not* to Cure Paralysis	171
Chapter 30	The War We Must Not Lose	175
Chapter 31	In Which I Get Cancer	179
Chapter 32	Sickle-Cell Anemia and the Politics of Pain	187
Chapter 33	Mowgli and the Matrix: A Year in the Life of the California Stem Cell Program	191
Chapter 34	To Whom Goes the Kingdom?	197
Chapter 35	The Liver List	201
Chapter 36	The Will of Connecticut	205
Chapter 37	Champions Find a Way	211
Chapter 38	Disenfranchise the Disabled?	215
Chapter 39	Adventures in Intellectual Property	219
Chapter 40	Spartacus Fights Back Against Stroke	223
Chapter 41	Diabetes Going Down?	227
Chapter 42	Jamie Thomson, or, How Do You Follow an Act of Genius?	233
Chapter 43	Sherley V. Sebelius	239
Chapter 44	The Woman Who Would Not be Silenced	243
Chapter 45	The Gorilla Gynecologist, or, the Pera-Chen Anti-Urinary-Incontinence Method	251
Chapter 46	Turning Over Rocks: The Battle for Paralysis Cure	255
Chapter 47	In Memory Still Green: The Passing of Three Giants	261
Chapter 48	Invitation to Mexico	265

Chapter 49	Of Presidents, and the Valley of Death	269
Chapter 50	Little Hoover and the Institute of Medicine	275
Chapter 51	Studying the Moon, Looking Through a Straw	283
Chapter 52	The Great Nebraska Compromise	289
Chapter 53	Stem Cell Tourism	295
Chapter 54	The Man Who Could Fly Without a Plane	299
Chapter 55	Thief of Lives	303
Chapter 56	Singapore, Biopolis, and the Power of the Small	307
Chapter 57	Singapore Scientists	313
Chapter 58	Fighters Against Parkinson's	321
Chapter 59	International Friends	327
Chapter 60	A Texas Miracle, or Thirty-Two	335
Chapter 61	The Stem Cell Musketeers of Brazil	339
Chapter 62	Adventures in China	343
Chapter 63	Professor Forever and the Giant Squid	349
Chapter 64	A Double Baker's Dozen of Disease Team Grants?	355
Chapter 65	The Greatest Speech You Never Heard	363
Chapter 66	A Stem Cell Mystery: The Resignation of Mahendra Rao	369
Chapter 67	Arthritis and the Fifty States	373
Chapter 68	Would You Drink from a Fountain of Youth?	377
Chapter 69	When Things Go Right	383
Chapter 70	Where Did the Money Go? (and a New Year's Delight at the End…)	387
Chapter 71	The End?	391
Appendix 1	Interview with Lim Chuan Poh, Chairman of A*Star and Biopolis	397
Appendix 2	Interview with Hans Keirstead	403
Appendix 3	Interview with Bob Klein	411
Name Index		415
Subject Index		419

FOREWORD
BY CHRISTOPHER REEVE

The late Christopher Reeve was kind enough to recommend this book, which would ultimately take me another ten years to complete.

— dcr

January 5, 2004

To Whom It May Concern:

I first heard of Don C. Reed several years ago, when he wrote a play, "A Night for No Mexican Tears: the Story of Juan Cortina" and his school put it on as a fund-raiser for our foundation.

I dictated a note to him then, saying in part, "One day, Roman and I will stand up from our wheelchairs, and walk away from them forever."

Since then, his involvement in the cause has deepened remarkably.

I was delighted to learn of the newly-established Roman Reed Laboratory in the Reeve-Irvine Research Center, under the leadership of Dr. Oswald Steward, funded by the Roman Reed Spinal Cord Injury Research Act. It was a pleasure for me to attend opening day of the lab via telephone conference.

Don Reed sees the fight for stem cell cure as a personal battle, in which victory for one will be a victory for all.

Like any parent, Don will do just about anything to protect his child, and he knows for Roman to be healed, political and financial barriers to medical research must come down. He is determined that nothing shall be allowed to stand in the way of that research.

I am glad he is on our side, and I look forward to reading his book.

Sincerely,

Christopher Reeve /TB

(signature by designated signer, TB)

Christopher Reeve

Bedford, New York, 10506

ACKNOWLEDGEMENTS

This book is dedicated to Gloria Jean Reed: beloved wife and best friend, the mother of my children, the color of my life. I also send family hugs to Roman, Terri, Desiree, Josh, Jackson, Roman Jr., Jason, and Katherine.

I owe appreciation to hundreds, only some of whom are recognized in these pages.

My friend in struggle is Karen Miner, paralyzed in body but never in spirit.

The California stem cell program would not exist without Bob Klein. Not just his endless energy, creativity and dedication to the cause of cure, but the sheer amount of donations he gave. He could not have made those donations without the hard-working people of his company, Klein Financial Enterprises. I would like to thank them for their endless efforts. They are, in no particular order:

Linda Menon, Mimi Gardner, Adrienne Brown, Alan Bogomilsky, Maati Benmbarek, Trent Paulsen, Chad Lewis, Danielle Guttman-Klein, Katelin Slifer, Ryan Ayers, Jessica Spinks, Jessica Anderson, Felicia Goodison, Aretha Hurt, Elizabeth Tafeen, and Annette Gettman.

David Bluestone, Jacqueline Hantgan, Mary Bass, Amy Daly, Susan DeLaurentis and many others gave strength to Americans for Cures Foundation, which sprang from the campaign to pass Proposition 71.

And to thousands of patients and advocates in America and around the world, who work without reward except to know they bring cure closer — let's make it happen.

I would also like to thank my editor Sook Cheng Lim of World Scientific Publishing for her patience and creativity; it has been a joy to work with her.

Joy from my family to yours
— Don C. Reed, 1/15/2015

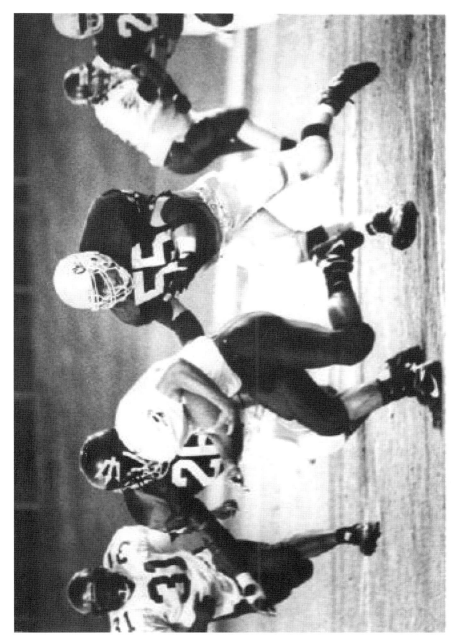

Football paralyzed Roman Reed, but he still loves the game.

Don and Roman Reed at the beginning of their journey.

A door opened.

"Bob can see you now," said the office assistant.

Wide-shouldered and craggy-featured, Robert N. Klein was seated at a long black table, scribbling something indecipherable on a yellow legal tab. He held up an index finger.

His handwriting is almost as bad as mine, I thought. My wedding proposal to Gloria had been in writing, but she could not read it, and threw my proposal away. We had been married 34 years so it worked out, but after that I typed everything important.

I looked around at the lovely office: carved wooden horses, a wall of books, a picture of young Bob Klein meeting young Bill Clinton, and a picture window above Palo Alto's green trees, swaying; a forest orchestra conducted by the winds.

"Ah!" said Bob Klein, putting down the pen and extending his hand. He has a great beaming smile, radiating warmth and energy. I sat down, expecting a mini-lecture on Bob's ideas, problems ahead, or chores that needed doing. But it was not like that.

"Tell me," he said. He leaned back in his chair, clasping hands behind his head and getting comfortable, as if this was the most important event in his day, and he had nothing but time on his hands. I knew it was an illusion. I had maybe ten minutes before his next appointment.

In a rush of words I told him about our family's most terrible night: September 10, 1994, a football game at Chabot College, Hayward, California.

Nineteen-year-old Roman Reed was the defense captain and middle linebacker. Under the floodlights he was playing his usual great game: 14 solo tackles, a bunch of assists, a diving one-arm interception, and a forced and recovered fumble.

Roman and a 341 pound giant blocker had been having an epic duel. Sometimes they crashed like trucks; other times Roman faded like smoke around the giant, reappearing

Lucy Fisher and Doug Wick, motion picture legends, and co-founders of Cures Now. With Jerry and Janet Zucker, they dreamed the dream of Proposition 71, and worked hard to make it become real.

on the other side, leaping on the runner, dragging him down. Once he dove and flicked out the ankle of the runner, breaking his balance like a cheetah tripping a gazelle.

Between plays he roared up and down the sidelines, exuding energy like flame, helmet off, challenging, inspiring — "Roman made you afraid to do less than your best," one player put it.

And then, in the surf-roar of shoulder pad collisions, on the third play of the fourth quarter — "Roman's down, it's bad!" said Gloria, "I saw it through the viewfinder!"

"Just the wind knocked out," I said, as the play was whistled dead.

The players trotted lightly to the sidelines, all but one.

Dear God, don't let it be Roman, I thought, looking frantically for him at the bench. He always liked to take his helmet off between plays, golden hair sweat-soaked, face red with exertion and competitive rage. But nobody had their helmet off, and a body lay still on the floodlit field.

I watched my feet trading places down the bleacher steps. This is a mistake, I thought, we just have to go back in time a few minutes and straighten it out.

"Don't worry, Mom," said Roman to his mother, "Nothing hurts. I don't feel a thing."

A shadow fell over us. I looked up. It was the giant blocker, number 22.

"Jesus God, Roman," he said, tears running down his face.

"Not your fault, two–two," said Roman, "Good hit," and he reached up to shake the other's hand.

I noticed something small, and terrible. My son's fingers did not move. Sparing the other's embarrassment, Roman sandwiched the giant's hand between both of his.

At the hospital, a doctor ordered Roman's shoulder pads to be sawed off. I started to object, this was expensive gear, designed to keep him safe from injury. But the doctor looked at me and I went silent, afraid he would say: what use will he have for them now?

After his helmet was also removed, a C-shaped metal tool was brought out. Roman's hazel eyes flicked briefly to me once, then straight ahead, blazing green, as shallow pits were drilled into bone, and smoke rose from the sides of his skull. The doctor secured a clamp in the holes and fastened it to weights on the gurney below, immobilizing our son's neck.

The X-rays came, and the diagnosis:

"Your son is paralyzed," the doctor said, "He will not walk again, nor close his fingers."

To make his point, the surgeon lifted Roman's hand, told him to hold it up. Before the accident, Roman could bench-press 430 pounds. But now? When he let go, Roman's hand fell like dead meat, so fast it slapped his own face. The brain-body connection was gone.

He could breathe, which was a blessing. But if he needed to cough we had to do that for him, shoving his stomach in hard, like a punch to the gut. It hurt him; the nerves inside still worked.

A flicker of hope. As I was sitting in the hallway, a book landed in my lap: *Rise and Walk: The Dennis Byrd Story*, about another football player who had been paralyzed. Byrd's millionaire coach provided him with an experimental drug called Sygen, made from dried cow brains. The Sygen had been injected right after the accident, and Dennis Byrd walked again.

"Get that for my brother," snapped Desiree, Roman's sister, just in from college in Arizona.

More good news! There were FDA-approved clinical trials of Sygen going on right now, a nurse told me. If Roman was involved in the trials, we could get the drug free...

The medicine had to be injected no more than 72 hours after the accident. But it was raining and midnight when I called the hospital. Bring him in the morning, I was told, it will be all right. But it was not all right. We missed the cutoff deadline by one hour, and were denied.

I located the drug's inventor, Dr. Fred Geisler, at the University of Chicago. He said the cutoff time was only for the clinical trials, and that the drug might work after the deadline — if we could get FDA approval. I called U.S. Representative Pete Stark, who contacted the FDA, helping us get "compassionate use" permission for the medication, though it was not yet officially approved.

Finding a drug source in Switzerland, I ordered the medication. It was not cheap. We borrowed. No doctor wanted to write a prescription for an experimental medicine.

But a kindly Chinese-American doctor, Chi-Chen Mao signed the prescription slip.

But what about side effects? The documents only mentioned minor stomach upsets, but still things can always go wrong. The first injections needed to be done in the hospital.

When the boxes of Sygen arrived, a nurse set up Roman with an intravenous needle in his arm. "Ready?" she asked, and turned the valve.

Roman's eyes rolled back; he stopped breathing.

"Oh my god, mouth to mouth resuscitation!" I said.

Our son's eyes went back to normal. He grinned and said: "Gotcha, Pops!"

Rehabilitation was wonderful; caring therapists, great exercise. We learned about the all-important "transfers", helping him shift bodyweight from bed to shower chair. But 39 days of rehabilitation was all our insurance would provide, and Roman was still paralyzed.

When our insurance quit, we found another "rehab center" (about which I have nothing good to say), and enrolled Roman, borrowing more money.

Roman's girlfriend Terri was the bright spot in the nightmare. She and I put Roman through seven months of massive rehabilitation, both at the "gym" all day, and at our apartment in the evening. She stayed by him almost constantly, in a chair beside the hospital bed or curled up beside him on top of the covers.

The money burned away. I begged systematically; anything to get money for the rehabilitation expenses. Senator Dianne Feinstein and Representative Pete Stark donated $500 apiece, and author Beverly Cleary sent us $100. The kids at my school did two penny drives for us. A friend bicycled nearly to Los Angeles on a fundraiser, but his calf muscle went out and he had to stop.

Our local church? My wife and children are Catholic, though I am not, and it seemed reasonable to ask the priest's help with a fund raiser. To my astonishment, the priest got angry, saying that I was "*going against God*" by not accepting His will that my son be paralyzed!

But another church, a Presbyterian one, came through for us without even being asked. They read a newspaper article about our struggle and donated $5,000. Also, a Mormon temple sent members to our house to help with meals and errands, kindnesses I will not soon forget.

And one bright day, Roman bench-pressed — a *broomstick*. Maybe it only weighed one pound, but his triceps were working again, long past the time the doctors said recovery was possible.

The restored arm function let him learn to drive an adapted van instead of needing an attendant. It meant, in time, he would be able to get himself out of bed in the morning without a mechanical hoist or needing me to pick up all his 235 pounds, as I had been doing.

When we visited one of our doctors, Ilyas Chaudry, Roman saluted him like a soldier. Dr. Chaudry was astonished; the recovered motion was pure triceps power, which was supposedly gone.

But he was still paralyzed.

Gloria adapted our home for wheelchair use: borrowing still more money to add on three rooms that met disability standards. The whole back wall of the house was taken off and at night the wind whistled through the plastic sheeting. But at last it was done. Roman would live in the adapted rooms of our house with Gloria and I — forever? Terri and I would share the care giver chores.

But cure? The time for that was over, I thought. We had recovered some motion in his arms, more than the doctors thought we would get. I was sure Roman would be reasonable about this when I explained it to him.

Unfortunately, when Roman was built, they forgot the "Quit" button.

Sooo…. I read science books about nerves and spinal cord injury, or more accurately, tried to. I was no doctor. How could I make sense out of these gigantic Latin words?

But there was one book I treasured. Written by Sam Maddox, *Quest for Cure* was published by the Paralyzed Veterans Association. It asked the question, "Could there be cure?" and answered "Maybe." Compared to the hopelessness I had been hearing, that was like a candle in a cave.

I attended a convention on spinal cord injury research in the seaside town of Asilomar. The words the scientists used were so big I kept falling asleep from brain overload. It was like the famous cartoon by Gary Larson of *Far Side* about a dog named Rex who digs up the roses, and the owner is yelling at Rex, who only knows one word of human — namely, Rex. All he hears in the long lecture is "Rex, blah blah Rex, blah blah…" I knew the words "spinal cord injury" and would wait for them. But I approached one scientist afterward and asked him:

"I don't understand what you guys are saying, except I know it is important. I can't help you on the science, but is there anything I can do, something you need?"

"Money," he said. "If I work six months on a grant request and it is rejected, I cannot put food on the table for my family — and the research won't happen."

Money… okay, how hard could that be?

Ever see the movie where kids are trying to raise money and somebody says, "Let's put on a show!"? With the help of my 8th grade student club True Colors, I wrote, directed, and produced *A Night for No Mexican Tears* about Mexican revolutionary Juan Cortina. It featured a live horse ridden across the stage and a 40-shooter cap gun battle. The Parent Teacher Association (PTA) counted our proceeds on stage, and we sent $4,000 to Christopher "Superman" Reeve.

An act of kindness helped us. When Christopher Reeve's riding mount threw him, the "Superman" actor broke his neck and became paralyzed from the neck down. His brothers wanted to shoot the horse. But Reeve said no, it was an accident, and would not let them do it. Joan Irvine-Smith is a lady who loves horses. When she heard about the paralyzed Superman's kindness, she gave him a million dollars — and the Reeve-Irvine Research Center was born.

Christopher Reeve sent the cast of our show a wonderful letter, saying, in part:

"One day, Roman and I will stand up from our wheelchairs and walk away from them forever."

But the money we raised was chicken feed, compared to the need. What to do, what to do?

I read about a New York policeman, Paul Richter, who was shot while in the line of duty. A gun runner blasted him in the chest, shotgun pellets grazing the spine. Richter was firing as he went down, and the perpetrator fled. (He was caught and sent to prison, where he later had a heart attack and died on the toilet.) Richter was not paralyzed, but had extensive nerve damage and had to retire.

"Uncle" Paul wanted to fight paralysis. If the number one cause of spinal cord injury is car crash, why not charge reckless drivers a fine? Fifteen dollars a ticket sounded great, but Republican Governor George Pataki said he would only allow the bill if it was approved by every legislator, Republican and Democrat alike. Paul Richter made it happen. The bill passed unanimously.

Maybe we could do something similar? I sent letters (not emails) to all the California Assembly-folk and Senators. Most said nothing; the few who answered said no.

But one day in my 8th grade English classroom, the telephone rang. Would I mind, asked Assembly Member John Dutra of Fremont, if they made a program called the Roman Reed Spinal Cord Injury Research Act?

Paul Richter (on *right*) with friends: shot in the spine on duty, "Uncle" Paul established a New York spinal cord injury program, raising millions of dollars for research.

Would I mind? I screamed and pounded the wall, then ran to the office, taking over the PA system to announce the good news. Nobody understood what I was talking about, but they gathered I was happy.

Well, that was that, I thought, knowing nothing about what was to come next. The bill could only be introduced by its sponsor; after that, it had to be pushed through multiple committee hearings, passed by the California Assembly and Senate, and lastly approved by the Governor.

Paralyzed friend Karen Miner turned out to be a natural organizer, and we began. Many people helped, in large ways and small. Bob Yant, Shelly and Marco Sorani, Susan Rotchy, Fran Lopes, and many more lent the power of their groups. World champion karate star Bill "Superfoot" Wallace (he fought Jackie Chan in *The Protector*) endorsed us, as did General "Stormin' Norman" Schwartzkopf, who said, "I have led troops into many battles, but never one so important as the fight against paralysis."

There was unexpected opposition: the American Automobile Association (AAA) came out against us. They did not want their drivers (even bad ones!) interfered with. But we mounted a letter-writing campaign, targeting their Sacramento lobbyist. So many letters were pushed through the mail slot on her door that she had a hard time opening it. They changed their stance from "oppose" to "neutral".

Republicans disliked the traffic fine and came on strong against the bill at the Senate budget committee hearing. But Senate President Pro Tem John Burton held sway. As Roman and Karen and friends watched from the audience, I was astonished to hear:

"How about $19 million a year, paid for by the General Fund?" asked Senator Burton, and the Republicans agreed to it. Nineteen million dollars? Wow.

But then came the energy crisis. Every bill which had a funding cost went into the "Suspense" file (meaning it had to be reconsidered), and ours did not come out. So, we started over again. This time, we only asked for one million dollars a year. "It's a feel-good law that doesn't cost much," one Republican aide said. He was right. One million dollars was the approximate cost of taking care of *one* severely paralyzed person in their first post-accident year.

But the Roman Reed Spinal Cord Injury Research Act of 1999,[3] or "Roman's Law," was alive. On March 1, 2002, the opening day of the Roman Reed Lab at the University of California Irvine, I held in my hands a laboratory rat named Fighter, which had been paralyzed but walked again. It had been given human embryonic stem cells.

On the speaker-phone, Christopher "Superman" Reeve said, "Oh, to be a rat this day!" Gloria was so excited, she photographed the phone!

Back in Bob Klein's office, the telephone rang. Bob frowned, then nodded:

"I have to take this," he said, "It's my son." He picked up the receiver, from which a burst of rapid talk emerged from the phone.

"Okay," said Bob at last, "Call Doctor — and tell him... Okay? Then call me back, and tell me what he said. Right..? Love you."

In that instant, I recognized Bob Klein. He was a parent trying to help his son, like me and millions of others. The difference was that if he succeeded, everyone on earth would benefit.

He looked at his watch. I braced myself, sure he would say something about how he really appreciated my stopping by, but —

"Let's go meet the others," said Bob Klein.

---

[3] Roman Reed Spinal Cord Injury Research Act of 1999, AB 750 [statute on the Internet]. 1999 [cited 2015 Jan 30]. Available from: http://www.leginfo.ca.gov/pub/99-00/bill/asm/ab_0701-0750/ab_750_bill_20000927_chaptered.html

2 THE NAKED FACE OF HATE

The "Three Amys": Amy DuRoss, Amy Daly, Amy Lewis, with Christina Olson, friendly security guard, and Kirk Kleinschmidt.

Autographed posters of Pixar's *Toy Story* beamed down from the walls of the office hallway. The unmistakable smell of pizza wafted from a closed door, as did a hum of conversation.

We stepped into cheerful chaos, like lunchtime at a college cafeteria. Everybody seemed young, and there were open pizza boxes on one counter top.

A giant cardboard thermometer towered over a lady with a telephone snugged to one ear. Her fingers were typing madly, but she smiled like Popeye's girlfriend, a beaming Olive Oyl.

"Amy Lewis," said Bob, as if those two words took care of everything, then added, "She is in charge of fundraising for the campaign." She smiled again and kept on with the call.

Amy DuRoss worked side by side with Bob in formation of Prop 71.

The ten-foot thermometer had markers on the side — 5, 10, 15, 20 — marking points for *millions of dollars*. At the top was $30 million, the amount needed for a successful campaign. But the thermometer was blank: empty, except for a little green smear at the bottom of the bulb — $100,000, a donation from Bob. Where would all the millions come from?

"We don't take money from biomedical corporations or Big Pharma, the drug outfits," said Bob, "People need to know this is a patient advocate community effort."

Heads turned, as when the lion shows up at the water hole. People remembered things they wanted Bob's opinion on; a queue began to form.

"I'm Amy Daly," said a strongly built woman, turning to me after showing Bob a sample flier, "But since there are three Amys, we just go by our last names. Call me Daly," she said, "I work with patient advocate groups. That was Lewis you just met."

"Bob, your ten o'clock!" snapped a woman with green eyes. Bob said, "Ah!" and was gone.

"That was DuRoss, the third Amy," said Daly, "Want some pizza?"

I had never been on a campaign with an actual budget before. Everything before had been on the Zero Budget Option: be creative as you wished, as long as you paid for it yourself. This was different: one campaign to raise the money — before we could win the main campaign.

"Just to get on the ballot," said Daly, "we need five hundred and fifty thousand signatures, plus maybe another hundred thousand extra, to make sure there are enough valid addresses."

Two-thirds of a million signatures? And after that we had to persuade California (in the midst of a recession) to risk six billion dollars (three for the cost, three for interest) on medical research?

Jaclyn Hantgan, Christina Olson, Lorraine Stiehl — volunteers extraordinaire.

Each signature sheet held a brief description of the program, plus space for names and addresses.

"Take some with you when you go," said Daly. A pause.

"Is there anything we can do — right now?"

"Well…." She pointed to a pile of empty cardboard boxes, "The clipboards for the signature gatherers did not arrive. But we have these boxes. If somebody could take a razor blade and cut them up into squares, we could use them for backing…"

I went home with a blood blister on my thumb. The pile of boxes was gone. So was the pizza.

When Gloria told me she was going to help me gather signatures, I did not really think she would be too helpful. I had studied the information sheet back and forth, and Gloria is more of a get-up-and-do-it kind of person. But I do not argue with my wife a whole lot. It is not safe.

Let's see, did we have everything? Two folding chairs, blank petitions, a sheet of message points[1], ballpoint pens, and a card table with a sign that read "STEM CELLS ON THE BALLOT?" We were ready!

Several managers shooed us away when we tried to set up shop in front of their stores. "Too controversial," one said. But another had a sister with diabetes, and he said, "Absolutely!"

---

[1] Key talking points about Proposition 71: the California stem cell research and cures initiative [Internet] [cited 2015 Feb 2]. Available from: http://digital.library.ucla.edu/websites/2004_996_027/documents/KeyTalkingPoints.pdf

Amy Lewis with Amy Daly and friend Erin Robbins of JDRF.

Seated comfortably, I glanced over my message points, waiting for someone to stop by and visit. I envisioned a leisurely conversation, sharing thoughts on the possibilities of stem cells.

A stream of people passed. One or two looked our way. But nobody stopped, and nobody signed.

"This is not working," said Gloria, and jumped off her chair. She did not actually tackle the nearest passerby, but she had him by the sleeve, and his face went through several changes of expression before she maneuvered him to the table.

"This gentleman wants to help get stem cells on the ballot," she said.

I went into my memorized lecture from the message point sheet.

"Stem cells are 'unspecialized' cells that can generate healthy new tissues and organs," I told him, "Possible treatments and cures for many diseases and injuries include…"

Gloria had sent two more people over. A line was forming. The first man shifted foot to foot.

"… cancer, heart disease, diabetes, Alzheimer's, Parkinson's, HIV and AIDS, multiple sclerosis, Lou Gehrig's disease, spinal cord — "

"He just wants to sign the petition, Hon!" called Gloria.

"Yes, dear," I sighed, and handed the man a pen.

We watched a professional signature gatherer once and it was a revelation. He had the "ask" down to four words — "Support stem cell research?" — and offered them the clipboard. If they said no, he just moved on.

In terms of signature gathering, it seemed, there were four kinds of people.

Category one: YES — put your name here, please.

Category two: MAYBE — give a brief run-down (2–3 short sentences), but not much more; no sense lecturing a "maybe" while two or three "yes" voters walk by.

Category Three: NO — if someone was against the research, we did not waste their time or ours.

And the fourth category?

I was "tabling" alone at the corner of University and Telegraph Avenues, in front of the Sather Gate entrance to University of California Berkeley. If there was ever a safe place to gather signatures for a progressive initiative, this had to be it, I thought: perhaps the most liberal spot on the planet, and Berkeley was the town where I was born.

I was working with one man who had a couple of questions when the flint-eyed elderly woman approached, she and a little bald man. They read my stem cell sign, looked at each other, then stood by, listening intently. The woman's lips were compressed to a thin line. She must be irritated by having to wait, I thought, and I hurried the gentleman to sign his name.

I got to her fast as I could. "Support stem cell research?" I asked cheerfully.

She *spat* on me. I heard the noise, saw the twist of white foam, and watched it land on the toe of my shoe.

I had wondered what it would be like to see the naked face of hate. Would it be terrifying, a person devoid of reason, or infuriating, someone who would deny my son his chance to walk?

What I felt, but did not say, was: "Lady, you are a little bit crazy."

The bald man guided her away. She went calmly now, apparently having delivered her message. The spittle dried on my shoe. I went on gathering signatures.

But I was glad she was not carrying a gun.

A couple weeks later, when I showed up at the campaign office, a security guard was blocking the door. What was this about, I asked as the guard looked me up on a list.

"Didn't you hear?" he said, "They blew up a stem cell lab."

"Police Confirm Pipe Bomb Blast at Stem Cell Lab[2]

"BOSTON (Reuters) — An explosion that blew out a number of windows at a Boston-area laboratory specializing in stem cell research was caused by a pipe bomb, local police said.

"[…] No one was wounded in [the] blast at Watertown, Massachusetts-based Amaranth Bio, which is working on cures for diabetes and liver disorders."

The opponents of research disclaimed responsibility. Nothing to do with them, they said, probably just some disgruntled former employee.

I was not so sure, because where do people like my little old lady friend get their information? They were being pumped full of poisonous misinformation.

---

[2] Police confirm pipe bomb blast at (Boston) stem cell lab. *Yahoo News* [Internet]. 2004 Aug 28 [cited 2015 Feb 2]. Available from: http://www.freerepublic.com/focus/f-news/1201313/posts

Imagine someone emotionally on the edge, who goes around mumbling to themselves. Now think how they feel when someone tells them stem cell research is... *murder*?

One lady, a friend of the family, asked, "Where do you keep the babies?"

"What babies?"

"The ones you get the spare parts from, like they told us about in church," she said.

She had been hearing statements like the following:

"It's indisputably killing," says Douglas Johnson, legislative director for the National Right to Life Committee in Washington. "Living human beings should not be used for harmful research without their consent."[3]

That is close to an allegation of murder and it must have an effect on impressionable minds, even though it is utter nonsense. For how can there be "killing" when there is no "living human being" involved? Living *tissue*, yes; like a growing hair or a microscopic heart cell, each of which contains the full DNA makeup of a person — but living tissue is not a life.

Embryonic stem cells for research come from materials left over from the In Vitro Fertilization (IVF) process. Consider how the stem cells are really gathered in the actual process. A childless couple goes to the IVF clinic to try to make a baby. This has been done successfully by more than five million families using the standard IVF treatment.[4]

The man's part is easy; he goes into a room by himself and donates his biological materials. That's it, he's done.

The woman's part is far more serious. She first takes hormones to increase her egg production, and later has the eggs removed by a surgery called oocyte removal.[5]

As with any medicine, there can be side effects. While millions of women have done the procedure safely, a small percentage do have reactions to the hormones involved, and some have actually died.[6] The donor deserves full information, including every negative possibility.

The gathered eggs and sperm are then mixed in a dish, making 15 to 20 fertilized eggs, called blastocysts. The healthiest one or two are put inside the woman's womb. Hopefully, this implants in the wall of the uterus, becomes an embryo, and a baby begins. We wish them well.

But what happens to the other fertilized eggs, which are NOT placed in the womb? If not implanted, it is biologically impossible for any of these to become a child. These "leftover" blastocysts can be frozen and stored (for a monthly fee), or given to another

---

[3] Belsie L. Should U.S. fund embryonic-cell research? *The Christian Science Monitor* [Internet]. 2001 Mar 13 [cited 2015 Feb 2]. Available from: http://www.csmonitor.com/2001/0313/p2s1.html/(page)/3

[4] Bryner J. Five million babies born from IVF, other reproductive technologies. *NBC News* [Internet]. 2012 Jul 3 [cited 2015 Feb 2]. Available from: http://www.nbcnews.com/id/48060498/ns/technology_and_science-science/t/million-babies-born-ivf-other-reproductive-technologies/

[5] In vitro fertilization for infertility [Internet]. 2013 Nov 14 [cited 2015 Feb 2]. Available from: http://www.webmd.com/infertility-and-reproduction/in-vitro-fertilization-for-infertility

[6] Ovarian hyperstimulation syndrome. *Wikipedia* [Internet] [updated 2014 Nov 22; cited 2015 Feb 2]. Available from: http://en.wikipedia.org/wiki/Ovarian_hyperstimulation_syndrome

couple. Embryo donation can be done through a company for five or ten thousand dollars. Not many IVF donors choose this option, apparently fearing that the potential child may be adopted by a bad family.[7] Or, as typically happens, the fertilized eggs are simply discarded.

How many blastocysts are frozen and stored? Roughly half a million of such fertilized eggs are "cryo-preserved" right now — test tubes in tanks of nitrogen at minus 190°C. No one knows for sure how many millions are flushed away, incinerated, or tossed in the landfill.

If something is going to be discarded, might it not be better to use it to try and help people?

When the decision to donate has been made, the blastocyst (so small it could literally rest on the point of a pin) is taken apart under a microscope. It feels no pain; it cannot, having no nerves.

The stem cells are extracted and put in a dish of feeder gel.[8]

I have seen the petri dish, the orange gel, and the infinitesimal tiny specks; the stem cells.

That's it, the whole process. Where is the "killing" of the "living human being"? There is none. But wait; the argument is made, if one of those blastocysts was implanted in a woman's uterus, a baby could begin. The blastocyst has the potential for life.

That is perfectly true. But are potential and reality the same?

Consider a drop of semen. Inside it are multitudes of wriggling spermatozoa, each of which (if put inside a woman's uterus) could potentially join an egg, implant in the wall of the womb, become an embryo, and maybe a baby. If a male produces 250,000,000 sperm in a single ejaculation,[9] must every sperm be preserved? Each one has the potential for life. One ejaculation, if every sperm was preserved, could almost equal the population of America!

Remember the silly song in the Monty Python movie, *The Meaning of Life*? "Every sperm is sacred…" If every sperm must be regarded as an infant, a teenage male's wet dream would be accidental manslaughter of cataclysmic proportions, and masturbation would be mass murder.

As the name implies, stem cell research is microscopic cells, cells, nothing but cells — and then the possibilities of eventual cures. Such possibilities! There seems almost no limits to what might be achieved.

Need a new liver? Maybe we can grow one.[10]

---

[7] Kapralos K. Evangelicals embryo adoption: devout Christians seek a future for thousands of frozen embryos. *Huffington Post* [Internet]. 2012 Sep 10 [updated 2012 Sep 13; cited 2015 Feb 2]. Available from: http://www.huffingtonpost.com/2012/09/10/evangelicals-embryo-adopt_n_1871832.html

[8] Myths and misconceptions about stem cell research [Internet] [cited 2015 Feb 2]. Available from: http://www.cirm.ca.gov/our-progress/myths-and-misconceptions-about-stem-cell-research

[9] Olson ER. Why are 250 million sperm cells released during sex? *LiveScience* [internet]. 2013 Jan 24 [cited 2015 Feb 2]. Available from: http://www.livescience.com/32437-why-are-250-million-sperm-cells-released-during-sex.html

[10] Scientists create human liver from stem cells. *NBC News Chicago* [Internet]. 2013 Jul 3 [cited 2015 Feb 2]. Available from: http://www.nbcchicago.com/news/national-international/214206791.html

Heart attack scars? Stem cells might turn these into functioning cardiac muscle.[11]

Lose your sight? Perhaps we can regrow your retina.[12]

Paralyzed? Growing nerves may reconnect body and brain — and eliminate the need for a wheelchair.[13]

"False hope!" the opposition cries, "Peddling empty dreams!"

But does not everything great begin with a dream? What improvements could there ever be, if we did not dare to dream? "Without a vision, the people perish," the Bible says, Proverbs 29:18.

Across the state, excitement grew. People gave up weekends and evenings to gather signatures and help their friends and family members who were suffering.

How many times had well-meaning friends said, "I want to help, is there anything I can do?" Now there was something they could do. Fill a signature sheet, turn it in, and get another.

If Roman in his wheelchair could harass passersby for signatures, what able-bodied person could refuse to help — for what might be the greatest advance in medical history?

Our friends, the Kaplan family, showed what could be done. Young Ben had cerebral palsy; his twin brother Oliver did not. It must have been terribly frustrating sometimes for Ben to see his brother with all the physical abilities and freedoms that were denied him. But the Kaplans did not sit around groaning. They unified in effort, battling for every signature, giving signature sheets to friends, family, and neighbors — going door to door, asking, asking, and asking. Gloria, Roman, and I collected around 2,000 signatures. The Kaplans organized their extended family and brought in five times that much. (To see the Kaplan twins, visit https://www.youtube.com/watch?v=uzpEccZP6JM)

We had memorable moments. Once, our chartered bus lost power going up a steep San Francisco hill and started rolling backwards. I was sitting behind the driver. As we careened out of control, it occurred to me to whisper, "We have confidence in you." He just laughed and said, "Happens all the time," and steered the bus onto somebody's lawn.

Sometimes there was beauty, as when we gathered in thousands and walked across the Golden Gate Bridge. Behind us was the sweep of green hills, and on both sides the blue of San Francisco Bay; ahead of us the Embarcadero awaited, a signature-gathering heaven with thousands of passersby, great food from Fisherman's Wharf, and sea lions barking from the piers nearby.

---

[11] Malliaras K, Zhang Y, Seinfeld J, et al. (2013). Cardiomyocyte proliferation and progenitor cell recruitment underlie therapeutic regeneration after myocardial infarction in the adult mouse heart. *EMBO Mol Med* 5 (2): 191–209.

[12] Stem cells and the future of eye treatment. *AOA News* [internet]. 2013 Nov 4 [cited 2015 Feb 2]. Available from: http://www.aoa.org/news/clinical-eye-care/stem-cells-and-the-future-of-eye-treatment?sso=y

[13] Woodbury MA. Hans Keirstead can make mice walk again (and humans, too?). *Esquire* [Internet]. 2009 Nov 17 [cited 2015 Feb 2]. Available from: http://www.esquire.com/features/best-and-brightest-2009/human-embryonic-stem-cell-research-1209

Sometimes it was surprising. Gathering signatures in something called a Pride Parade, I turned around to see a naked man about six feet away; somehow I forgot to ask for a signature from the person with the all-over tan.

But mostly it was just work: like the caregiving chores for a loved one, and done for the very same reason. And when the goal seemed far away, it helped to remember Bob Klein's words:

"Let's send a message," he said, cheerful and confident, "We need 550,000 signatures to get on the ballot, 650,000 to be on the safe side. But if we can get a million…"

It was impossible, of course. But we did it anyway.

We turned in 1.1 million verified signatures, twice as many as required. We had passed the first hurdle. We had a number, Proposition 71, and a name, the California Stem Cells for Research and Cures Initiative, and we were on the ballot.

Now we would be judged by the voters.

3 TO CLONE, OR NOT TO CLONE?

Shoukhrat Mitalipov with Reed family and friend: Mitalipov was the first to achieve SCNT (therapeutic cloning) of human cells.

Just before the baptism of my grandson, Jackson Reed-Francois, he and I were sitting on the edge of a fountain, in front of the Church of Carmel. Jackson wore a white baby tuxedo and an infant car-seat. I had my gray Sunday-go-to-meeting suit. Both of us were stiff and uncomfortable. Jackson kicked and fussed; I was not allowed the privilege.

"I'll take him, Dad," said my daughter Desiree. She unfastened the little safety buckles and carried Jackson, still struggling mightily, into the shadows of the church.

I followed, to be immediately confronted by a table stacked with glossy four-color brochures. Automatically I took one, something to read during Mass, if I could get away with it. I was not Catholic myself, but Gloria was. I had to promise to raise my children in the faith, or she could not have married me in the Church.

The brochure was titled *Voter's Guide for Serious Catholics*. (Later I would see another version, almost identical, but with the title altered: *Voter's Guide for the Serious Christian*.) It was, I believe, aimed at any faith-oriented person, with the intent of persuading them to vote on ideological grounds. An estimated 70 million "voter guides" were made.

"This voter's guide helps you cast your vote [...] consistent with Catholic moral teaching [...] [and to] avoid choosing candidates who endorse five actions that are *intrinsically evil*."[1] (emphasis added)

What were these five unforgivable sins?

1. Abortion
2. Euthanasia
3. Homosexual marriage
4. Embryonic stem cell research
5. Cloning.

Of the five "non-negotiable" sins, two were about stem cell research.

Cloning. The word conjures up science fiction movie images of soul-less duplicate "people", alien invasions, *Star Wars* cartoons, and mad scientists.

In real life, of course, cloning is a tool with many uses. When a gardener "cuts a slip" from a plant and grows it, that is cloning.[2]

A Crime Scene Investigator uses cloning in a DNA test to determine guilt and innocence.[3]

On the farm, cloning breeds better milk-producing cows, faster horses, and sheep with superior wool.[4]

But the science fiction fantasy of duplicating people? That would be human reproductive cloning, which no one has done and which we should never attempt.

First, why would we want to clone more people? We can make plenty the old-fashioned way. To replace a lost loved one? A personality is comprised of memories and experiences, none of which are "clone-able."

Above all, attempting to clone children would endanger both mother and child. Dolly the sheep was cloned by Ian Wilmut. She was experiment number 278 — *the first 277 died*.[5] This is one of many reasons why human reproductive cloning is illegal in California.

---

[1] Voter's guide for serious Catholics [Internet] 2004 [cited 2015 Feb 3]. Available from: http://www.politicalresponsibility.com/voterguide.htm

[2] How to clone plants. *Wikihow* [Internet] [cited 2015 Feb 3]. Available from: http://www.wikihow.com/Clone-Plants

[3] Woodrow M, Mardigian R. Crime scene investigator PCR basics™ kit [online notes] [cited 2015 Feb 3]. Available from: http://www.cpet.ufl.edu/wp-content/uploads/2012/10/Crime-Scene-Investigator-PCR-Basics-manual.pdf

[4] List of animals that have been cloned. *Wikipedia* [Internet] [updated 2015 Feb 2; cited 2015 Feb 3]. Available from: http://en.wikipedia.org/wiki/List_of_animals_that_have_been_cloned

[5] Whitfield J. Obituary: Dolly the sheep. *Nature* [Internet] 2003 Feb 18 [cited 2015 Feb 3]. Available from: http://www.nature.com/news/2003/030217/full/news030217-6.html

As Wisconsin bioethicist R. Alta Charo said in an interview with CNN,[6] "[It would be] an act of medical malpractice to do human reproductive cloning […]. Animal data shows how very risky it is, [and] how high the frequency is of miscarriage, stillbirth, [and] birth defects."

To the best of my knowledge, no scientific organization in the world supports reproductive cloning of humans.

But there is another kind of cloning which ought to be considered: therapeutic cloning of cells, or nuclear transfer, which is short for somatic cell nuclear transfer (SCNT). It is easier to understand if we change one word in the SCNT phrase. Let's call it *Skin* Cell Nuclear Transfer instead.

Here is how it works. First, swab a Q-tip inside the mouth of a patient. That will gather skin cells. Then, under a microscope, remove the nucleus from an unfertilized human egg, and transfer the nucleus of a skin cell into it. That completes the skin (i.e., somatic) cell nuclear transfer. But what do we call this new biological entity? It is not an embryo — there is no fertilization from sperm involved, nor any implantation in a womb.

Dr. Ann Kiessling calls it an *ovasome*.[7] That sounds reasonable.

Keep the ovasome in a dish of salt water, shock it gently with electricity to start the cells multiplying, put in feeder cells, and wait 5–7 days. Then, take it apart under a microscope and extract the stem cells.

Therapeutic cloning may one day be a source of body repair cells. Think how hard it is to transplant a human heart, or a liver. The body's immune system may reject it, so the patient dies. With therapeutic cloning, it may be possible to regrow one's own heart or liver or nerves for a damaged spine, with cells that might not be rejected. They come from you; they would *be* you.

Again, this is therapeutic cloning, the copying of cells; human reproductive cloning is against the law in California.

As this is written, an Oregon scientist friend, Shoukhrat Metalipov, has succeeded in making a human stem cell line via SCNT. As far as I know, he is the first in the world to successfully accomplish this important step in regenerative medicine.[8]

Is SCNT legal? In some states yes, some states no, while most have no laws pertaining to it.[9]

But because of the dedication of one woman, therapeutic cloning is legal in California.

When her mother was diagnosed with cervical cancer, Deborah Ortiz haunted the internet, seeking a way to defeat this dreaded killer of women. Sadly, she was not successful. But she continued the struggle: to honor her mother's memory and help save the lives of others.

When she became State Senator, Ortiz decided to make California the nation's first "stem cell friendly" state. She authored Senate Bill 253[10] to specifically legalize therapeutic cloning and embryonic stem cell research.

---

[6] Zahn P. To clone or not to clone? *CNN* [internet]. 2002 Dec 30 [cited 2015 Feb 3]. Available from: http://edition.cnn.com/TRANSCRIPTS/0212/30/ltm.18.html
[7] Kiessling AA (2004) What is an embryo? *Conn L Rev* **36**: 1051–1092.
[8] Cyranoski D (2013) Human stem cells created by cloning. *Nature* **497**: 295–296.
[9] Stem cell laws and policy in the United States. *Wikipedia* [Internet] [updated 2014 Nov 23; cited 2015 Feb 3]. Available from: http://en.wikipedia.org/wiki/StemcelllawsandpolicyintheUnitedStates
[10] Embryonic Stem Cell Research, SB 253 [statute on the Internet]. 2002 [cited 2015 Feb 3]. Available from: ftp://leginfo.ca.gov/pub/01-02/bill/sen/sb0251-0300/sb253bill20020922chaptered.html

Bernie Siegel organized a United Nations hearing to educate world leaders about somatic cell nuclear transfer (SCNT).

As grassroots organizer for the bill, I testified at numerous committee hearings and rallied support for it through all the various steps of the legislative process.

What a joy it was to stand on a platform beside Senator Ortiz, when Governor Gray Davis signed America's first stem cell permissions bill into law.[11] Pride compels me to add that the Governor said (after a whispered aside from Ortiz): "Don Reed's involvement was integral to the passage of Senate Bill 253."

The next part may sound like I am making it up.

Bernard "Bernie" Siegel was a Florida trial lawyer with no connection to the stem cell world, until the day he read some astonishing news. A quasi-religious group called the Raëlians claimed that the world was begun by outer space aliens who cloned the population.[12]

The Raëlians also claimed to have *cloned a child*, called "Baby Eve."[13]

Now, Florida has a very sensible law to prevent child abuse. If neighbors suspect a child is being harmed, they can sue for a court-appointed guardian to look after the rights of the minor.

---

[11] Davis signs legislation to bolster embryonic stem cell research, therapeutic cloning. *California Healthline* [Internet]. 2002 Sep 23 [cited 2015 Feb 3]. Available from: http://www.californiahealthline.org/articles/2002/9/23/davis-signs-legislation-to-bolster-embryonic-stem-cell-research-therapeutic-cloning

[12] Raëlism. *Wikipedia* [Internet] [updated 2015 Jan 30; cited 2015 Feb 3]. Available from: http://en.wikipedia.org/wiki/Ra%C3%ABlism

[13] Begala P, Novak R. Rael defends claims of cloned baby. *CNN* [Internet]. 2003 Jan 3 [cited 2015 Feb 3]. Available from: http://edition.cnn.com/2003/ALLPOLITICS/01/03/cf.opinion.rael/

Siegel, as a lawyer, figured a cloned child would have medical legal problems and worried that he/she might be "treated like a lab rat". He sued the court to appoint a guardian for Baby Eve.[14]

That night, he and his wife were watching television before going to sleep. CNN ran a headline ribbon across the bottom of the screen reading, "Florida attorney sues for custody of cloned child".

Cheryl Siegel turned to her husband and inquired, "Bernie, what have you done?"

It swiftly became clear the Raëlians were not about to produce any cloned Baby Eve, or any serious evidence thereof.[15] The farce should have ended right there.

But the world was watching and the press attention was immense. No one loves a headline more than a politician. Florida Representative David Weldon (R-FL) and Kansas Senator Sam Brownback (R-KS) perked up their ears.[16] To them, the exploitation of cloning terrors must have sounded like the opportunity of a lifetime. Promoting the fear of human reproductive cloning might advance their carreers — and it did not seem to matter that it might also block some very needed medical research.

If they had wanted a law criminalizing reproductive cloning, they could have had that easily. No government official has ever supported human reproductive cloning, so a bill banning that would probably have passed without disagreement.

But it would also have ended the headlines. By going against good cloning as well as bad, Brownback and Weldon insured a lengthy fight, whereby the issue (and their names) would stay in the headlines.

The Cloning Prohibition Act (HR 2505 in the House, and S.790 in the Senate), was invented to defend against a non-existent threat.

Will Rogers famously said, "When Congress makes a law, it's a joke." The consequences here were, however, not humorous. Under this law, both reproductive and therapeutic cloning were felonies: crimes punishable by ten years in jail and/or a million dollar fine.

If Roman went overseas and had a therapeutic cloning operation done in England (where it is legal) he might actually be arrested on returning to America.

HR 2505[17] was rushed through the Republican-controlled House of Representatives, and President George W. Bush promised to "enthusiastically" (his word) sign it into law.

A virtually identical anti-science law was offered to the Senate.[18]

But the patient advocacy community rallied, led by our friends at the Coalition for the Advancement of Medical Research (CAMR). CAMR never had much of a budget (at its peak they had a grand total of one-and -a-half paid employees), but their value and contribution to the history of stem cell research cannot be overestimated.

---

[14] Philipkoski K. Clone newcomer bends UN's ear. *Wired* [Internet]. 2004 Jan 6 [cited 2015 Feb 3]. Available from: http://archive.wired.com/medtech/health/news/2004/06/63636?currentPage=all

[15] Siegel B. (2007). Reflections on the cloning case. *Cloning Stem Cells* **9** (1): 40–46.

[16] Cloning ban bill reintroduced in Congress. *Jewish World Review* [Internet]. 2002 [cited 2015 Feb 3]. Available from: http://www.jewishworldreview.com/0103/cloning_ban.asp

[17] Wright MV. Cloning: a select chronology, 1997–2003. Congressional Research Service Report for Congress (U.S.); 2003 Aug 19. 15 p. Report No.: RL31211.

[18] Human Cloning Prohibition Act of 2001, S.790. 107th Congress (2001).

CAMR leaders like Larry Soler, Dan Perry, Tim Leshan, Tricia Brooks, and Michael Manganiello should have their names carved in granite for their endless hours of volunteer labor. CAMR led the fight in this and many other regenerative medicine issues and really deserves to have its own book.

On both sides of the political spectrum, informed voices made themselves heard.

Voices such as the late Republican President Gerald Ford[19]:

"[…] Therapeutic cloning — more precisely known as somatic cell nuclear transfer […] — holds limitless potential to improve or extend life for 130 million Americans now suffering from some chronic or debilitating condition. [A ban on all cloning would mean] slamming the door to lifesaving cures and treatments merely because they are new.

"[…] During my Presidency, similar questions were raised about research into recombinant DNA. […] A quarter century later, would anyone turn back the clock? Would anyone discard vaccines traceable to recombinant DNA research?"

Former President Jimmy Carter[20]:

"One of the great scientific accomplishments of our time, therapeutic cloning, or nuclear transplantation, presents promising new opportunities for the treatment of many serious illnesses and injuries [including] heart disease, Parkinson's, and spinal cord injury […] to name a few.

"[…] I support banning reproductive cloning, […] but oppose restrictions on therapeutic cloning."

And First Lady Nancy Reagan, in an open letter to Senator Orrin Hatch of Utah[21]:

"Dear Orrin,

"As you may know, Ronnie will observe his 92nd birthday soon. In earlier times, we would have been able to celebrate that day with great joy and wonderful memories of our life together. Now, while I can draw strength from these memories, I do it alone, as Ronnie struggles in a world unknown to me or the scientists who devote their lives to Alzheimer's research. Because of this, I am determined to do what I can to save other families from this pain.

"I'm writing, therefore, to offer my support for stem cell research and to tell you I'm in favor of new legislation to allow the ethical use of therapeutic cloning.

"[…] Orrin, there are so many diseases that can be cured, or at least helped, that we can't turn our back on this. We've lost so much time already. I can't bear to lose any more."

In the House, Republicans ruled; they passed the anti-research bill. In the Senate, the Democrats had the numbers — and they blocked it.[22]

---

[19] Ford GR. Curing, not cloning. *The Washington Post* [Internet]. 2002 Jun 5 [cited 2015 Feb 3]. Available from: http://www.fordlibrarymuseum.gov/library/speeches/20020605.asp

[20] Carter, Jimmy. Message to: Bush, George W. Available from: http://lobby.la.psu.edu/107th/121HumanCloning/OrganizationalStatements/CAMR/CAMRPresidentCarter.htm

[21] Reagan, Nancy. Message to: Hatch, Orrin. 2003 Jan 29.

[22] Human cloning: must we sacrifice medical research in the name of a total ban?: hearing before the Committee on the Judiciary of the United States Senate, 107th Congress, 2nd Session (2002).

But the story deepens. President Bush had another card to play. If the United Nations would ban therapeutic cloning all across the world…

America's UN representatives, accompanied by robed priests, were sent to the representatives of small countries urging them to ban cloning. Costa Rica, for example, had no stem cell research program in its country — but it did receive U.S. foreign aid — and Costa Rica led the charge to ban therapeutic cloning everywhere in the world.

But Bernie Siegel organized a meeting in the United Nations, bringing top scientists together to discuss therapeutic cloning. He led a patient advocate movement to contact the United Nations.

I found a fax address which would reach every member of the United Nations. But each fax had to be separately transmitted; somebody had to physically stand by a facsimile machine and send off each one, lest it would be thought to be mass-produced.

On the West Coast, Bob Klein's office was headquarters for the effort. Matt Jordan stepped into the stem cell history books when he volunteered to spend a three-day weekend in the Klein Financial offices, manning the FAX machines day and night, so that approximately 70,000 individual faxes were generated in the days before the United Nations vote.

We gained a year for more serious consideration of the issue. President Bush tried again, but though in the end he won a paper victory, there were no teeth to it. Any country that wanted to do the research was free to do so. Nationally and internationally, the patient advocate community had protected the nation's — and the world's — freedom to develop stem cells through a variety of methods.[23]

But research freedom doesn't mean much, if there is no money to pay for it.

---

[23] Philipkowski K. United Nations balks at cloning ban [Internet]. 2003 Dec 9 [cited 2015 Feb 3]. Available from: http://www.grg.org/UNStopCloningBan.htm

4 SEASON OF STORMS

"Cyclone: a storm with a mass of thunderstorms, [around] an eye of calm... the smaller the eye, the more fierce the storm..."

— Dahloan Hembree

The days between April, 2003 and November, 2004 were a cyclone of stem cell storms: crises by the dozen, large and small, and they all came together in Proposition 71.

Some of it was pure fun. One afternoon, Gloria, Roman, and I found ourselves being filmed for a commercial — directed by Hollywood legend Jerry Zucker of *Ghost and Airplane!* fame. Jerry would ask questions about our hopes for stem cell research. We would answer, and most of it landed on the cutting room floor.

"Arnold and I had a four hour conversation about stem cells," said Zucker later, his angular face like a gentle hawk's. Out his front window we could see the Schwarzenegger complex, fronted by a gigantic boulder and a waterfall. The Republican Governor/bodybuilder/movie star had not yet declared himself on the issue.

Brad Pitt spoke out strong for Proposition 71, as did Michael J. Fox, Dustin Hoffman, Rhea Perlman, and other champions of the entertainment world. Hollywood folk often support progressive causes. Perhaps, it is because they understand so well the fragility of life, toiling in an industry built on near-impossible standards of youth and beauty (and a reported 96% <u>un</u>employment rate!), or maybe because actors must have empathy to understand the suffering of others.[1]

The only major Hollywood star who came out *against* Proposition 71 was *Lethal Weapon* star Mel Gibson, who did a commercial attacking the initiative. He talked very fast, sounded angry and (to my ears) unconvincing.

The professionals hired to manage the campaign, Chuck Winner, Paul Mandabach, and Bob Deis, were a delight to work with. Fiona Hutton of Red Gate was in charge of message control. We also had guidance from Fairbanks, Maslin, Maullin and Associates, Porter Novelli, and more folks than there is room to appreciate.

Our primary scientific advisors? The best of the best.

---

[1] Yes on 71 Coalition Members [Internet] [cited 2015 Feb 2]. Available from: http://digital.library.ucla.edu/websites/2004_996_027/coalition.php.htm

David Ames with children Isabel and Joaquin and parents John and Genevieve —
ALS stole a father from his children, and a son from his parents.

Larry Goldstein, the great scientist and explainer, author of *Stem Cells for Dummies*; Irv Weissman, pioneering champion of stem cell research, there from the beginning of the field and responsible for many of its major developments; Paul Berg, Nobel Laureate for his work on DNA; and many more leaders of the field.

But it was the patient advocates on whom the campaign rose and fell.

Genevieve Ames was Senior Scientist at the Pacific Institute, University of California Berkeley. She was also the mother of David Ames, who had amyotrophic lateral sclerosis (ALS), or Lou Gehrig's disease, and who was himself a fighter for change. So was his father John. When their son was first diagnosed with ALS, the family took him

VICTORY FOR PROP 71. Tessa Wick, *left*, who has diabetes, gets a hug from her mother Lucy Fisher at the Yes on Proposition 71 stem cell research initiative election night reception, Nov. 2, 2004. (AP Photo/Chris Carlson.)

all around the world, trying to find a cure. They were unsuccessful, but their battle went on, day after day. At night, John would physically lift his son, transferring him from wheelchair to bed, and John is not a large man, except in spirit. When the Ames family talked, people listened.

Joan Samuelson was a lawyer and the founder of Parkinson's Action Network, as well as suffering Parkinson's disease herself. She had to balance her medications so that her limbs would neither "freeze" nor spasm out of control, while somehow mustering the strength to fight.

At one Southern California meeting, Attorney David Carmel, paralysis advocate, was confronted by a non-handicap-accessible steep flight of stairs. Without missing a beat, the advocate organized bystanders: telling me to pick him up by the left leg and shoulder, nominating another to take his right side — we walked sideways upstairs, he chose a chair — the meeting went on.

Every advocate I met shared a common conviction: Proposition 71 was the most important thing we would ever do in our lives; it must succeed.

The odds against Proposition 71 were steep. Progressive measures typically start strong with high approval ratings, which go down as the cost is counted, and the opposition gets into gear. Without a wide margin of support from the beginning, most will not survive.

We began with a very narrow margin: 45% approve — 42% oppose, according to California's respected Field Poll.

Melissa King and Ben Kaplan: Hard-work heroes of stem cell effort.

Democrats were in favor, two to one: 57% Yes, 27% No;
Republicans were against it, two to one: 59% No, 31% Yes.
Non-partisan folks were slightly more inclined to support: 48% Yes, 43% No.
We could not afford to lose a single vote.[2]

Opposition came from conservative organizations like the California Pro-Life Council, a branch of the National Right to Life Committee.[3] Their rhetoric was polished, professional, as if they had waited all their lives for this chance to attack. In a way they had. Their arguments had indeed been rehearsed for decades — against abortion. Since our research had nothing to do with abortion, the assault was like a beautiful map of the wrong planet. But they could not be faulted for lack of enthusiasm. They came roaring at us like a mudslide of misinformation: even calling our initiative the "loan to clone and kill" bill.

If you looked at the contest in terms of *money declared*, it would seem we had the contest won. The opposition declared about a million dollars total; our side ended up with between $30–$34 million.[4]

---

[2] DiCamillo M, Field M. Voters sharply divided on stem cell research bond measure: Favor two other health-related propositions but oppose a fourth. The Field Poll (US); 2004 Aug 15. 6 p. Report No.: 2130.

[3] California Proposition 71, Stem Cell Research (2004). *Ballotpedia* [Internet] [updated 2014 Dec 11; cited 2015 Feb 2]. Available from: http://ballotpedia.org/California_Proposition_71,_Stem_Cell_Research_(2004)

[4] Proposition 71 [Internet] [cited 2015 Feb 2]. Available from: http://www.followthemoney.org/entity-details?eid=10246406

T-shirts spread the message: Yes on 71!

But in terms of actual resources? Leading the opposition was the Catholic Church, and *one in four Americans is Catholic*.[5]

There are Catholic Churches in 1,358 California cities, and there are often multiple churches per city. San Francisco alone has 52 Catholic churches, meaning 52 free meeting places to organize against Proposition 71, spreading the message from the California Council of Catholic Bishops.[6]

The Catholic Church is the world's largest property owner. The American branch of the Church *spends* roughly $150 billion a year[7] — not even counting the parishes, on which it spends an additional $11 billion — that is power.

Our side had to raise the money for ads to run on radio and TV stations — the Catholic Church *owns* TV and radio stations, as does the Religious Right. If you ever drive across the country with your radio on, run up and down the channels, and count the number of religious channels.

---

[5] Catholic Church in the United States. *Wikipedia* [Internet] [updated 2015 Jan 30; cited 2015 Feb 2]. Available from: http://en.wikipedia.org/wiki/Catholic_Church_in_the_United_States#Catholicism_by_state

[6] California Cities [Internet] [cited 2015 Feb 2]. Available from: http://www.thecatholicdirectory.com/directory.cfm?fuseaction=show_state&country=US&state=CA

[7] Yglesias M. How rich is the Catholic Church? *Slate* [Internet]. 2013 Mar 14 [cited 2015 Feb 2]. Available from: http://www.slate.com/articles/business/moneybox/2013/03/catholic_church_and_pope_francis_religious_institutions_are_exempted_from.html

Don Reed, Gloria Reed, and Amy Lewis, celebrating that magic night Prop 71 became law in California.

But aren't churches supposed to be non-political, since there is a Constitutional separation of Church and State, and since they mostly don't pay taxes on their wealth?

In theory, yes. But reality?

Specially trained priests visited Catholic schools, lecturing children about the alleged evils of embryonic stem cell research. One of the schools visited was attended by my grandson, Roman Junior. His regular classroom teacher, a kindly woman, actually apologized to "Little Man", saying she knew how difficult it was for his father to be paralyzed, and how hard he worked on the research — but the priest went ahead with his lecture just the same.

As advocates, we made an effort not to argue. Whatever the opposition said, we tried to respond with information: staying on message, saying what we came to say.

During one radio/telephone interview, a caller hissed at Bob, "You are like Josef Mengele, the Nazi Angel of Death!" …for supporting embryonic stem cells in research…

Mengele, of course, was the Nazi who tortured Jewish people in concentration camps, injecting blue dye into their eyeballs or removing their organs, all without anesthesia. A more despicable insult is hard to imagine, especially since Klein (though Congregationalist by faith) is of Jewish ancestry. I wanted to punch the radio.

But Bob just said, "Thank you for raising that important point." He spoke as if the man had done him a favor. Then Bob explained how embryonic stem cells come from biological materials left over from the in vitro fertilization (IVF) procedure, which could only be used after being scheduled to be thrown away.

When Bob was done, the caller said, "Well, I never thought of it like that before!"

We could not win every time. But there were undecided people listening, and those were the friends we hoped to make, as well as encouraging our supporters to keep working.

We came prepared to say, "Nearly one in two Americans has a chronic disease, an incurable condition. These are not empty statistics, but our loved ones, members of your family and mine. Proposition 71 is a fight to save lives and ease suffering."

My favorite line was: "American families deserve the best medicine science can provide." This came from a visit Bob had with the great linguist, Professor George Lakoff, from the University of California at Berkeley.

(For a look at the official ballot arguments, check out: http://vote2004.sos.ca.gov/propositions/prop71-arguments.htm)

Surprisingly, this was not a religion-versus-science debate. I was raised in two religions: Southern Baptists, who opposed the research, and Presbyterians, who supported it.

To our delight, many communities of faith supported us.[8]

An especially important endorsement came from IMPACT, the one and a half million member action arm of the California Council of Churches.[9]

And if Catholic leadership was against us, what about regular-folk Catholics themselves? In a poll by Russell, Beldonello, and Stewart (George Bush's pollsters) **72% of Catholics** were *in support* of embryonic stem cell research.[10]

The opposition produced a lot of angry words against the research and our program, but when you looked closely, there was emptiness behind the rage.

For example, one opposition website was called "A List of Scientists Against Proposition 71".[11] At first I thought, uh-oh, what was this? But the actual list had a grand total of *two* scientists.

Scientists? We had 22 *Nobel Prize winners* in support of Proposition 71.

If it seems like I mention Bob Klein a lot, it is because it would be false not to do so. Many helped, laboring heroically. When complimented, Bob would always turn the praise around and say (rightly) that it was a team effort all the way. But it cannot be denied that he was the irreplaceable man.

For him personally, Proposition 71 was a financial loss. He sold part of his company, Klein Financial, and took out loans on his house, contributing roughly four million dollars. But it cost him more than that. His involvement in Proposition 71 had meant time away from his company, and the business inevitably suffered. Bob estimated that the financial value of his firm was cut by as much as half due to time spent on the fight for stem cell research.

---

[8] Yes on 71 Coalition Members.
[9] California Church IMPACT endorses Proposition 71. *PR Newswire* [Internet]. 2004 Jul 8 [cited 2015 Feb 2]. Available from: http://www.prnewswire.com/news-releases/california-church-impact-endorses-proposition-71-71217887.html
[10] Catholics support stem cell research. *Catholics for Choice* [Internet]. 2004 Jun 30 [cited 2015 Feb 2]. Available from: http://www.catholicsforchoice.org/news/pr/2004/6.30.2004PressReleaseCatholicsSupportStemCellResearch.asp
[11] A List of Scientists Against Proposition 71 [Internet] [cited 2015 Feb 2]. Available from: http://noon71.godandscience.org/scientists.html

Beyond patient advocates, Bob found support among Republicans like George Schultz, economic adviser to Ronald Reagan; and from the Chambers of Commerce he contacted all across the state. Not only were Chamber members eager to protect their families from chronic disease and disability, but they understood the economy; it was their world. They recognized biomedicine as the wave of the future. If provided with facts, Chambers usually would support us, and many did — including the California Chamber of Commerce, and a majority of Chambers branches, up and down the state.

Elected officials? *One hundred and seventy-five* endorsed us, including State Controller Steve Westly and State Treasurer Phil Angelides.

For me as a writer, Proposition 71 was a series of small accomplishable chores. In the paralysis struggle, I had developed a fat phone book. I used that binder within an inch of its life, contacting every advocate or group I had met or could find, asking for support, requesting letters, and referring new friends to the www.yeson71.com website. I used what public relations people call the "F U" strategy — follow up, follow up, follow up! Few people act on just one notification; they need to be contacted multiple times.

Everywhere I looked people struggled with chronic disease — everybody with a story deserving of being told. I wrote perhaps a hundred letters to the editors of newspapers across the state. Most never saw print, but still they were part of the wave; we must educate the editors.

Far more important was to help other supporters reach their own local papers. It was a natural match. Every advocate has a story to tell and every paper depends on local input. From my days as a reporter, I knew that 70% of all newspaper content is planted. If presented professionally, they would consider almost any rational copy.

How is it done? How do you get "in the paper"?

Just write a one page letter telling your story. Make it a good one, spend a week on that one page, work on it every day, carry it around with you, and keep thinking about it. Your story is the struggle of a family against incurable disease — in the course of which you find a way to mention your support for Proposition 71. Once the letter is ready, call up the paper and tell them what your letter says. Read it aloud if you want to. At the end, ask if you can send them information — they will almost certainly say yes — and you send them the letter. In a few days, check back with them, ask if they will send a reporter to interview you, or should you come to their office. Make it convenient for them and you are practically guaranteed to succeed. Once you have that story, photocopy it (or use the URL) and send it everywhere.

Share information in as many languages as possible. For instance, California is blessed with a strong Chinese population. So I wrote a long article, had a friend translate it into Mandarin, and a Chinese newspaper ran it on the front page.

The three Amy's reached out to the patient advocate community, which responded with a roar. No fewer than 80 major groups endorsed us, and some were groups *of* groups, like the Coalition for Advancement of Medical Research (a group combining nearly 100 additional groups).[12]

---

[12] Yes on 71 Coalition Members.

Governor Arnold Schwarzenegger's endorsement was of course eagerly sought. But Republicans had hopes of running him for President — there was a distinct element of political risk for him. Would he be willing to go against the conservative establishment? We waited and waited, suspense mounting, until, with only three weeks left in the campaign —

"I am, of course, a supporter of stem cell research. Research that we do now holds the promise of cures for tomorrow. California has always been a pioneer. We daringly led the way for the high-tech industry and now voters can help ensure we lead the way for the biotech industry. The creativity and resources are right here in California. We are the world's biotech leader and Proposition 71 will help ensure that we maintain that position while saving lives in the process. I encourage Californians to join me in voting for Proposition 71."

— Governor Arnold Schwarzenegger, October 18, 2004

"The Terminator" had come through for California, and the world.

But for me as the father of a paralyzed young man, there was another actor whose endorsement was even more meaningful.

Do you remember where you were when Christopher Reeve became paralyzed? Roman and I were in that nameless Southern California rehab center when the receptionist came running back with a boombox radio.

"You've got to hear this!" she said.

In shocked horror we learned how Christopher Reeve had been thrown from his horse. He was terribly injured: his neck broken in the worst possible place, the first cervical vertebra, just below the brain. He was paralyzed from the head down, unable even to breathe unassisted. Had not someone given him mouth to mouth resuscitation, he would have died on the spot.

A different sound: a soft thud… thud… thud… Roman was banging his wrists together.

"What are you clapping for," snapped someone, "Don't you know what this means to Christopher Reeve?"

"Don't you know what this means to us?" he answered, "We just got a face."

Christopher Reeve did indeed become "a face" for the paralysis community, and for anyone with an "incurable" condition, who had been told there is no hope. He embodied our defiance.

I met the man millions knew as "our Christopher" at a fund raiser, the "Night of a Thousand Stars". It did not begin well.

I was babbling the usual inanities, trying to find words to express appreciation for his commitment to cure, when he interrupted me, saying something I did not understand. He repeated himself, but was plainly exhausted, and the words came out garbled.

"What?" I said, after his second attempt. He made a frustrated noise, between a hiss and a snarl.

Fortunately, his wife Dana was there, and she translated, "He said, he wants to talk to Roman."

That I understood, and was off.

"Christopher wants you!" I said to Roman when I found him. He turned his chair to ramming speed, and woe for the shins of anyone in his way.

Christopher wanted to know how "Roman's Law" was doing, and of course Roman filled him in on the latest, like Hans Keirstead's work with embryonic stem cells, and Director Oswald Steward's long term efforts as leader of the program.

And then Roman expressed what everyone was feeling. "You are more like Superman in real life than in the movies," he said.

I did not attend Christopher Reeve's funeral, which I have regretted many times. There were money problems, etc., but still I wish I could have stood in that crowded church to honor him.

But at home I was interviewed, and one television reporter's unbelievably mean-spirited question gave me a chance to speak my mind. The man with the microphone asked, "Was it not maybe a *good thing* that Christopher Reeve had died — because it would 'rally the troops'?"

Had I heard him correctly? Surely it was a mistake. But no, during the interview, the reporter kept coming back to it. Three times he asked that same miserable question, and three times I gave him an answer.

"Our champion has fallen, but the flame of his faith still lights our way. We will 'go forward', as Christopher always said — and we will prevail — because California has picked up the torch".

The final crisis might sound small at first, but it was huge in its consequences — the First International Stem Cell Action Conference.

We needed a place to hold the event, and it had to be cheap. Advocates like Idelle Detlaf, a hard-driving multiple sclerosis survivor, Dr. Raymond Barglow, a college professor, Bernie Siegel, Joe Riggs, founder of the Students for Stem Cell Research organization, Matt Jordan, and others worked together. But when soft-spoken Ray Barglow told me he thought the University of California would let us *borrow* a conference hall, I thought he was kidding. Free? That was not going to happen. Raymond smiled and nodded and said I was probably right. But he did it.

For one day, the Hall by Telegraph Avenue was ours. Bob Klein volunteered to speak, as did Alan Trounson (the Australian scientist who had helped develop the IVF procedure), and naturally I was always ready to babble about how the Roman Reed Spinal Cord Injury Research Act provided funding for the first state-funded embryonic stem cell research in the world. Suddenly, we had a program.

Our speaker accommodations might have been less than lavish (like Professor Barglow's living room couch), but admission was free, and we packed the house.

And when a beaming Bernie Siegel took over the podium, he was obviously where he belonged, the consummate master of ceremonies.

In the middle of his speech, someone handed him a note. Bernie frowned, paused, and said, "I have just received notice that Ronald Reagan, former President of the United States, has died. We will now observe a moment of silence."

In the hush of respect for a fallen President, I think everyone remembered also Nancy Reagan, who had turned sorrow over her husband's Alzheimer's disease into loving concern and action. A friend of Lucy Fisher and Doug Wick of CuresNow, Ms. Reagan had supported research which might help my son. I wished I could hug her and say, thank you, so much.

At last the day came: Tuesday, November 2, 2004. Election day. California's Proposition 71 (not to mention the Presidency of the United States) would be decided tonight.

We gathered at the beautiful Omni hotel in Los Angeles. Bob Klein, ever confident, had rented the hotel in advance to celebrate the successful passage of Proposition 71.

But have you ever noticed how life's most important moments often have a ludicrous side? At the moment when all the speechifying began, I was on a mission to find a canine restroom.

Karen Miner had become paralyzed in a car accident, same year as Roman, and we had met on the web. She wanted to help and I was very willing to let her. Bone-thin and fragile, she was nonetheless iron-willed and unstoppable. For example, I can write, but know nothing about setting up websites. So, Karen was my webmaster for about ten years. When she *broke her shoulder*, I figured that meant she could no longer help with my weblog (*Stem Cell Battles*) for a while. "Oh no," said Karen, "I will just use my left arm."

I took the leash of the beautiful golden Labrador receiver, one of the best assistance animals in the world, and we went outside.

But this was downtown Los Angeles, shopping malls and hotels. In mounting frustration I hunted — nothing, nothing — until we found a park. Perfect! Dog heaven! Unfortunately, the park was defended by an economy-sized policewoman, who said, pointing at my canine companion, "Don't you be pooping in my park!"

I looked at my watch. The ceremonies would have just begun. Bob Klein would be stepping to the podium, to give the latest numbers on the exit polls…

At last we found some bushes beyond the dragon's view. Baxter did his duty (it took forever).

We dashed back to the hotel, through the slow-rotating doors, across the golden lobby —

Too late. Bob was already speaking, standing on a raised platform jammed with advocates and campaign staff. I could only give Baxter back to Karen, and settle in to watch from the audience.

But no, one of the three Amys, green-eyed DuRoss, gestured from the stage: hurry! I ran around behind the platform, clambered ten feet up onto the stage, made it, *and fell*, landing on my back, crosswise on a support beam. For one terrible moment, *I could not move*. What bitter irony I thought, if I should become paralyzed on this night of nights… But no, the breath was just knocked out. I clambered back up.

Senator Deborah Ortiz, whom I once referred to as "the mother of California stem cell research", was being deservedly lauded for her contributions.

Remember the early Field Poll showing Proposition 71 and its opposition essentially tied, with "Yes" at 43% and "No" at 41%? Instead of losing ground, Proposition 71 had increased its margin of approval, receiving 59.1% support, essentially a 60–40 victory.

Seven million voters had said yes to fighting chronic disease with stem cell research.

5 THE TRIAL(S) OF THE CALIFORNIA STEM CELL PROGRAM

James Harrison: Super-lawyer for progressive causes, especially the California stem cell program.

"Uh-oh," said Roman, as we sat in the Hayward courtroom, waiting for Judge Bonnie L. Sabraw.

"Uh-oh, what do you mean, uh-oh?!" As Mark Twain once said, I was already nervous as a long-tailed cat in a room full of rocking chairs.

"That's an unusual last name Judge Sabraw has," said Roman, "I think I dated her daughter. It did not work out, and the mother may not like me a whole lot."

"Go home right now," I said, "We don't need any more complic — "

"All rise," said the bailiff, as, in a rustle of black robes, Judge Bonnie L. Sabraw entered her courtroom.

"Is it her…?" I whispered to Roman. He nodded.

The California stem cell program was under attack. We (I regard the people of the program as family) were enduring attempts to shut us down. Lawsuits came at us like armies.

The National Tax Limitation Foundation (NTLF) was the toughest anti-tax group around. Originally the Howard Jarvis Taxpayers Association, it had organized Proposition 13. By making it all but impossible to ever again raise taxes on the rich, Proposition 13 essentially gutted California public school funding. My father, Dr. Charles H. Reed, was Assistant Superintendent of Schools for a local school district the night Proposition 13 passed, and he said, "This is the end of California's leadership in education," and so it has proved. This fabulously wealthy state now has the highest number of students per teacher in the nation — 24.9 to one — and that was the average; some of the classes I taught had as many as 36 students apiece.[1]

We also spent the least on the education of our children of any state.[2] Some studies said the lowest, others said 44th, just ahead of Texas. http://ed100.org/support/californiaskimps/

The NTLF and a group called People's Advocate were represented by Life Legal Defense Fund (LLDF). The LLDF was led by Dana Cody, reportedly the person who persuaded President George Bush to do a press conference in his pajamas to dramatize the Terry Schiavo case where a woman apparently brain-dead was kept on life support for months in a right-to-life legal battle.[3]

The California Family Bioethics Council was a just-formed group from the James Dobson "Focus on the Family" evangelical empire. Dobson was a preacher who rose to fame by authoring a book, *Dare to Discipline*, which advocated corporal punishment for children. He had parlayed that into a string of religious radio stations (more than 60 in America alone) and was a force behind the Religious Right wing of the Republican party.[4]

Finally, a fictitious embryo, "Mary Scott Doe", was represented in a separate lawsuit by a group called the National Association for the Advancement of Pre-Born Children.[5]

After due consideration, the group representing the embryo was denied standing, and the other lawsuits were combined into one. Both sides applied for an early resolution of the case: they claimed our program was unconstitutional, while we claimed their lawsuit was nonsense.

But Judge Sabraw said no, the case was important, and should be heard. Both sides called that a victory. Bob Klein called it "not a home run but an inside the park triple

---

[1] Rankings of the States 2013 and Estimates of School Statistics 2014. National Education Association (US); 2014 Mar. 129 p.
[2] California schools rank low-again-in Education Week report. *The Sacramento Bee* [Internet]. 2014 Jan 10 [cited 2015 Feb 2]. Available from: http://blogs.sacbee.com/capitolalertlatest/2014/01/california-schools-rank-low---again---in-education-week-report.html
[3] Meyerson H. Target of opportunism. *The Washington Post* [internet]. 2005 Mar 23 [cited 2015 Feb 2]. Available from: http://www.washingtonpost.com/wp-dyn/articles/A58466-2005Mar22.html
[4] James Dobson. *Wikipedia* [Internet] [updated 2015 Jan 21; cited 2015 Feb 2]. Available from: http://en.wikipedia.org/wiki/James_Dobson
[5] Uttley L. The politics of the embryo. *Center for American Progress* [Internet]. 2005 Jul 19 [cited 2015 Feb 2]. Available from: https://www.americanprogress.org/issues/women/news/2005/07/19/1556/the-politics-of-the-embryo/

Tamar Pachter, lawyer who brilliantly defended Prop 71 in court.

play," noting that many of the plaintiff's complaints had been thrown out. David Llewellyn said the judge's ruling proved the substance of their case.[6]

On the opposing side, lawyers included Robert Taylor. Big, white-haired, and dangerous, he had the veteran trial lawyer's ability to think on his feet and change directions quickly. David Llewellyn was a scrapper, happiest when the battle was joined. Terry Thompson had a Southern accent, was very polite, always inquired after my family, and said "God bless you" very naturally; essentially the sort of person you would want for a neighbor.

And then there was Dana Cody, Executive Director of the Life Legal Defense Foundation, she was apparently the real power behind the lawsuit, paying the bills for the court case. She rarely spoke, perhaps because she was so clearly identified with anti-abortion causes, and this case was structured from a different angle. She was there every minute, making sure all was in order.

An attractive woman, she was overwhelmingly determined but unfailingly courteous. She definitely knew people. One day, a cigarette lighter was left behind overnight. The next day the bailiff announced it and held it up. I do not smoke, but it was a very cool lighter, a 1960's Zippo lighter with Led Zeppelin inscriptions — a college kid might have used it to light an illicit cigarette. Ms. Cody turned to a reporter covering the story and said, "I bet that's yours."

That's all it was, nothing the slightest bit untoward or improper, but the force and warmth of her personality just poured out, enjoying the bit of fun, and the reporter got flustered, no, no, it wasn't his, really it wasn't, he didn't even smoke — and they both shared a laugh.

Lawyers associated with her group often contributed work on a pro-bono basis, i.e. free. Considering the costs of lawyers, that is power. She got things done. For example,

---

[6] Hall CT. Stem cell program wins key court ruling, poised to issue grants. *SFGate* [Internet]. 2005 Nov 30 [cited 2015 Feb 2]. Available from: http://www.sfgate.com/news/article/Stem-cell-program-wins-key-court-ruling-poised-2558997.php

the judge once requested a particular document, which would have required extra effort to obtain in the short time period allowed. One of their lawyers started to object, but Ms. Cody interrupted his objection, saying of course the document would be provided. The next morning, she entered the courtroom early, and handed the bailiff a folder. In the four days of the trial, I was usually the first person there, but she would be the second.

I knew she was against our research. She would come at us with everything she had, from every legal avenue. If her side won, our research would be blocked.

On impulse I got up and approached her. I told her who I was, adding quickly, "Nothing negative," so she would not feel threatened. I told her about Roman's Law and how we funded some of the research she opposed, but that while she and I were 100% in opposite corners, I respected the strength of her convictions.

Then I felt stupid from the impulsive action. All she had to do was turn away, and I would look idiotic. She could lift a finger and say, "Bailiff, this person is annoying me."

But she just said, "I hope you find a cure for your son, I really do," and stuck out her hand.

We shook hands, like boxers touching gloves, and retired to our corners.

Judge Sabraw? She maintained perfect courtroom control while still allowing everyone their due.

Once, when an attorney jumped up and thrust his arm almost violently into the air, eager to be heard, the Judge smiled and said: "I see that Mr. _____ is champing at the bit." Her tone was in no way insulting, but she immediately softened the words' impact anyway, saying: "But I do not wish to make light of it; what did you want to say?"

Another time a lawyer argued a little bit too long.

"I believe that objection was overruled," she said softly. There was a sudden hint of frost in the air and the attorney did not comment further.

The plaintiffs went first, voicing their arguments for a solid 90 minutes, and ending with an astonishing request for 14,000 pages of information and documents from the California stem cell program — which was provided.

They had many objections to the California stem cell program, including:

1. They called our program unconstitutional. State money could not go to institutions not under state control — they made the program sound like some sort of rogue agency.
2. Initiatives can legally cover only one subject. Proposition 71 (according to them) had several.
3. They felt our board of directors had automatic conflicts of interest, and members might abuse their positions to get research grants for their home organization.

When they talked, things looked grim; a flood of big words and lawyer-ese.

But we would have our turn.

If James Harrison had been in charge of the maiden cruise of the great ship Titanic, it would have made a boring movie — because nobody would have died. The co-writer of Proposition 71 would have foreseen the danger, avoided the iceberg, and brought everybody safe back home.

Don Reed speechifying before Governor Arnold Schwarzenegger, Roman Reed, ICOC board members, and super-advocate Karen Miner, on day the Governor approved a $150 million loan to the California stem cell program. (Notice the Governor's crutch — a skiing injury — not even "The Terminator" is safe from accidents.)

Of the law firm Remcho, Johansen & Purcell, Harrison was tall, soft spoken, and seemingly incapable of becoming flustered. It was said that he and Bob Klein had worked through 200 drafts of Proposition 71's language, trying to anticipate political attacks like these. Now their work would be put to the test; confronting it were lawyers who would go over the language like killers with small knives, to weaken, cut, slash, and destroy.

In charge of defending the new government agency was Bill Lockyer, Attorney General of California. But his work, like James Harrison's, was mainly behind the scenes.

The person who would actually speak for us was Tamar Pachter, Deputy Attorney General.

When I first saw her in the courtroom, I thought she was too young to be defending all the hopes and dreams of millions of suffering people and their families.

But once she began to talk, I mentally kicked off my shoes and leaned back. Tamar Pachter was clear, well-informed, and very tough.

Her introduction was cold fire.

"We are here to make sure California voters get what the Constitution guarantees them […] and what they voted for, by overwhelming majority.

"I don't doubt the sincerity of plaintiffs who are opposing Proposition 71 here. They have strongly held beliefs that merit consideration. But those beliefs did not prevail in the most important forum, the ballot box.

"[…] The Constitution is the bedrock foundation of our democratic system and it must be preserved. The court's role, we submit, is not to use the Constitution to interfere with the initiative process, but to uphold it by giving effect to the legitimate exercise of the voters' power.

"[…] The implementation of Proposition 71, like the language of the law itself, is fully compliant with all Constitutional and statutory requirements."

After that she went to work on the charges against us.

I did not understand everything she said, but the opposition did. It got very quiet in the room.

From what I grasped of her arguments:

1. Far from being an institution out of control, the California stem cell program was itself a part of the State government, and subject to numerous state laws and controls.
2. "Single subject"? Every word of Proposition 71 had to do with stem cell research therapies and the hunt for cures; it was absolutely "single subject".
3. There was no "conflict of interest" in our structure. The voters had decided (with clear information before them, including on the ballot) that a board of directors made of experts was desirable. All board members were subject to conflict of interest regulation and supervision in numerous forms including routine public audits. If a grant might affect their institution, they were required to recuse themselves from voting, or even discussing it.

The plaintiff's complaint took an hour and a half to deliver.

Ms. Pachter took 15 minutes to gut their arguments.

When she was through, the opposition looked at each other, and there was much reaching for the water pitcher. They would regroup to fight again, of course. These were warriors, but for this day they seemed to struggle to think of something new to hit us with.

Surprisingly, they did not call a stream of witnesses. They originally intended to bring 73 witnesses to oppose us. Instead, they made an attempt to interrogate Bob Klein. But Bob is a lawyer too. It was not a profitable use of their time, though enjoyable to watch.

I thought we were crushing their arguments, but then, I was not a lawyer. And I was absolutely biased. I felt about CIRM the way Biblical Ruth felt about her new husband; CIRM's friends were my friends, and her enemies were mine. I would be polite, of course, but this was about the protection of my family, and yours. We had to win.

I asked James Harrison what if we lost on **one** of the charges against us — was that it, game over?

"Could be," he said, "But we did write a severability clause into Proposition 71, meaning if part of it was found unconstitutional, we hoped to have just that part removed."

"Oh, good!"

"Of course," he added, "They filed to overthrow the severability clause…."

Win or lose, the case would be appealed to the next court and the next, perhaps all the way to the State Supreme Court. How long would that take? How many years?

If the lawsuits continued, our money supply was cutoff, and with it, the research. All we had for sure was the $3 million the state had guaranteed to hire staff — nothing like the $300 million a year intended for the research.

If their funding is cutoff, most government programs just die. But in an obscure English law book, Bob Klein found something called a Bond Anticipation Note, or BAN, and figured out a possible way to use it. People would be asked to loan the program money, and *maybe* they would get their money back! Maybe. If we lost the lawsuit, their loan became a donation.

Imagine going to a bank and telling them you wanted to borrow money, and you hoped you would be able pay them back, but if not, they should regard the loan as a donation?

That was Bob Klein's plan.

Thanks to the kindness of the Dolby Family Foundation, we had $5 million to start us off.

The Bond Anticipation Notes? Bob and friends managed to put together $45 million.

And then Governor Arnold Schwarzenegger came through for California, authorizing a loan to CIRM of $150 million.

At the press conference, I had the honor of giving the patient advocate response. I spoke for five minutes, but the only important thing was the last sentence:

"The California stem cell program is the glory of a state, the pride of a nation, and a friend to all the world."

Meanwhile, back at the courtroom… At last, we would have the Judge's verdict.

"The court finds that the plaintiffs have not shown that [Proposition 71] is clearly, positively, and unmistakably unconstitutional. The Act [is] valid."

So was it over? Well, no.

The plaintiffs appealed.

The trial dragged on more than another two years.

Here is the Appeals Court Conclusion:

"After careful consideration of all […] legal objections, we have no hesitation in concluding […] that Proposition 71 suffers from no constitutional or other legal infirmity. Accordingly, we shall affirm the […] decision of the trial court upholding the […] initiative.

"DISPOSITION

"The judgment is affirmed."[7,8]

With these words, the trial was essentially over. The plaintiffs did indeed appeal their case to the California State Supreme Court, which affirmed the decision.

And Judge Sabraw? We met her in the hallway of the building after the first trial, and she said of course she remembered Roman. "I recognized you immediately," she said. They shook hands and chatted briefly, and it made no difference either way.

Now it was up to California.

---

[7] *California Family Bioethics Council v. Independent Citizens' Oversight Committee* (2007), 147 Cal. App., 4th 1319.

[8] *People's Advocate v. Independent Citizens' Oversight Committee* (2007), Cal. Ct. App., A114282, 6 MRLR 143.

P.S.:

"It was, perhaps, a blessing in disguise that ideological opposition kept us in the court system for the first 2.5 years of the agency's existence, unable to use bond funds, so we had enough time to create gold standard procedures and regulations before awarding our first grant. The tireless efforts of the Board in those first years, combined with those of key outside legal counsel, patient advocates and other members of the public, were essential to laying the foundation for CIRM's success."
— Melissa King, Executive Director, CIRM, 2004–2011

6 WHEN CHANGING THE WORLD, WHAT DO YOU DO FIRST?

Dr. Kathryn Ivey, Ph.D., received a training grant from the California Stem Cell Program. Today, she is fighting heart disease, the number one killer of women and men around the globe. Young women, take note — science is for you!

The first official meeting of the California stem cell program, December 17, 2004, lasted 15 minutes. A critic had complained about us to the District Attorney's office. He noted that California state law required ten days' notice before public meetings, and only three days' notice had been posted on the website. (The program was so new at this point that it did not even have a secretary.)

Accordingly, the meeting of the Independent Citizens Oversight Committee (ICOC) was cut short, allowing just the election of the chair and vice chair, which was necessary for operations to begin.

The four most powerful state officials, the Governor, Lieutenant Governor, Treasurer, and Controller, all recommended one person for the job of Chairperson — Bob Klein. The ICOC voted unanimously to approve that choice.

For Vice President, the biomedicine business pioneer, Ed Penhoet, was one of three candidates, and he was also voted into office by the board. Some of the critics were not happy about him, calling Penhoet "the ultimate insider" as if that was an insult. We wanted an expert, we got one, and his value would soon become clear.

Kathy Ivey and Deepak Srivastava of Gladstone Institute on Mission Bay: Texas's loss is California's gain.

But at last the delays were all dealt with, and the full meetings could begin, a bunch of them. The board generally met once a month, but in the early years (when we were setting up procedures) it was sometimes two or three meetings a week. I tried to attend them all.

A firm was hired to hunt for our first official President. In the meantime that position was filled by Salk Institute's CEO Richard Murphy. How wonderfully appropriate: Jonas Salk developed the vaccine that prevented polio, and our first President worked at the institute Dr. Salk founded.

Housekeeping chores had to be done, much of it quite boring, which I will try to skip.

The very first grants? **Training grants**. If there was to be research, there must first be student researchers, and that meant helping both scholars and schools. Research Training Grants would go first to 16 colleges or institutions, supporting their most outstanding students. These were significant grants — one to three years long, some as high as $75,000 per year, plus benefits.

Was the money well-spent? Let's follow one person.

The Gladstone Institutes are housed on the University of California San Francisco's Mission Bay Campus. It is a large building, but everybody in it knows everyone else. "This is where the stem cell lab used to be," said athletic and cheerful Kathryn Ivey, PhD. We were standing in a closet-sized room, with plainly very limited equipment.

"But right through here..." she opened a door into stem cell modernity.

This floor was big as a factory, but with a buzz of conversation and no interior walls. There were cabinets, test tubes, and bio-safety "hoods" with their perpetual gush of cold and sterile air.

Kathy Ivey was brought up in Texas, earning her bachelor's degree at Texas A&M University. In 2004, she received her PhD from the University of Texas Southwestern Medical Center.

Just one year later, her scientific mentor, Dr. Deepak Srivastava, was recruited to California for a job directing the Cardiovascular Disease Institute at Gladstone. He asked Dr. Ivey to come along.

I had to convince my husband we should relocate to California," she said, shaking her head about San Francisco's housing costs.

But when she arrived, Gladstone had just been approved for the training grants. Dr. Ivey applied for and received one of the Institution's ten grants.

Kathy Ivey was hard-working, self-starting, and talented — the kind of young scientist whom California should support, and did. For three years, she immersed herself in stem cells.

Her goal? To attack heart disease, the number one killer of women in the world. In America alone, 600,000 people die of heart disease every year[1] — one in every four deaths is due to heart disease — and just one kind of heart disease (coronary) costs our nation $108.9 billion a year. She would draw the genomic blueprint of the developing heart, learn from it what happens when things go wrong, and predict how stem cells might help.

For instance, a heart attack leaves a scar — could stem cells turn that scar into new and healthy tissue? Questions led to more questions. For instance, one piece of the puzzle might be microRNAs, little-known blueprinting cells which might activate heart regeneration and restore its vital function. Could they signal an injured heart to heal itself?

Working with other Gladstone scientists, she performed experiments leading to important research papers, such as "MicroRNA regulation of cell lineages in mouse and human embryonic stem cells", published in the journal *Cell Stem Cell*.[2]

The grant "made it possible for her to take on more big-risk, big-reward science," said Gladstone Senior Investigator Bruce Conklin. "It was a game-changer, and for Gladstone as well."

"[…] Today, the Gladstone Institute is known for its fantastic stem cell work. Nobel Prize winner Dr. Shinya Yamanaka works part-time with us. But before the training grants, we had very little.

"Of all the grants Gladstone ever received, this was the most transformative. It seeded the beginnings of new ideas. You can draw a direct line from that to multiple projects […] like reprogramming to make cardiac cells; potential drugs to cure Huntington's disease and ALS; drug-screening processes and others; these began only because the Research Training Grants allowed new scientists financial assurance," said Bruce Conklin in an interview.

As Dr. Ivey's ability became clear, she was trusted with more responsibilities, moving up the ladder. Today she is Director of the Gladstone Institute Stem Cell Core. In addition to her own research, she and her assistants John Chunko and Uma Lakshmanan look after

---

[1] Heart disease facts [Internet]. 2015 Feb 1 [updated 2015 Jan 28; cited 2015 Feb 2]. Available from: http://www.cdc.gov/heartdisease/facts.htm
[2] Ivey KN, Muth A, Arnold J, *et al.* (2008). MicroRNA regulation of cell lineages in mouse and human embryonic stem cells. *Cell Stem Cell* **2** (3): 219–229.

the stem cell needs of some 40 scientists. She is an Investigator for the Roddenberry Center for Stem Cell Biology and Medicine, and an Adjunct Assistant Professor at the University of California San Francisco.

Since the program's inception, 690 trainees like Dr. Ivey have benefited. As stated by Gil Sambrano, Associate Director of the California Institute of Regenerative Medicine, these are comprised of "clinical fellows, post-doctoral fellows, and pre-doctoral students, participating in research projects across more than 400 California labs." They include:

- Louise Laurent, University of California San Diego;
- Matthew Blurton-Jones, University of California Irvine;
- Nil Emre, BD Biosciences;
- Laura Perrin, Children's Hospital, Los Angeles;
- Chris Schaumberg, Allergan;
- Duncan Lieu, University of California Riverside;
- Megan Hall, PLoS Biology;
- Laura Elias, Boston Consulting Group;
- Edward Kavalerchik, Scripps Translational Science Institute;
- Laura Saunders, StemCentRx;
- Derrick Rossi, Harvard Stem Cell Institute;
- Chiro Nakano, Ohio State University Medical Center;
- Ann Foley, Weill Cornell Medical College, New York.

Here was a new generation, fighting to turn the hope of cure into reality. But what if there had been no California stem cell program, and no Research Training Grants to change their lives?

"If I had stayed in Texas, I could never have done the work I have done here," says Dr. Kathryn Ivey, now Director of the Gladstone Institute Stem Cell Core.[3]

---

[3] CIRM funding commitments [Internet] [cited 2015 Feb 2]. Available from: http://www.cirm.ca.gov/our-progress/summary-cirm-rounds-funding

7 BATTLES WITH A FRIEND

At a campaign hearing for Proposition 71, a critic seemingly tried to pull the wool over the eyes of Senator Deborah Ortiz (D-Sacramento). As if referring to established fact, the woman sniffed, "As you know, all the major women's groups oppose Proposition 71".

"Really? Which groups are those?" answered the Senator. "It can't be NOW, the National Organization for Women, or Planned Parenthood, or the Feminist Majority Foundation, because they are all in strong support. So exactly which 'major women's groups' oppose Proposition 71?"

There was a painful silence. The critic fumbled in her briefcase as if there might be some opposition groups inside, and allowed as how she might not have the actual names handy.

I had helped Senator Ortiz pass her three pro-stem cell bills, the first such laws in the country. Senate Bill (SB) 253 legalized embryonic stem cell research and therapeutic cloning in California, SB 322 established research guidelines, and SB 771 set up a registry for embryo donations with procedures for informed consent.

Imagine how I felt when she planned laws AGAINST the California stem cell program! Her Senate Constitutional Amendment 13 (SCA-13) would change the stem cell program in ways large and small, and chisel each alteration permanently into the State Constitution.

Why was she doing this?

"I introduced Senate Constitutional Amendment 13 with one purpose: to strengthen Proposition 71 regarding conflicts of interest and financial disclosures, ensuring public accountability and open participation in the $3 billion taxpayer-funded California stem cell research program."

These were positive words, like motherhood and apple pie. But the actual practical effects? Those might not be so pleasant. SCA-13 was a minefield of lawsuits waiting to happen.

"Good fences make good neighbors," said Robert Frost. But lawyers wanting lawsuits hope for bad fences; they feed on ambiguity. Those who sued us had not gone away. They did not attack us right now, because our legal structure had proved solid with no weak spots to exploit. But any slightest change in the law — let alone something major like a Constitutional alteration — would be studied intensely. If their lawyers found even a hint of violation against the new standards, they would sue us again and again, wasting our precious research dollars on court costs.

Senator Art Torres: Sharing a laugh with Roman and Don Reed. The Senator has been a champion of progressive values since early days with legendary farm labor union organizer Cesar Chavez.

Read just one sentence from SCA-13. It potentially opens floodgates of trouble, threatening researchers wanting a grant and patients hoping for a cure.[1]

"Any clinical treatments […] resulting from the biomedical research are made available, *at the costs of producing them,* to California residents who are eligible to

---

[1] California Stem Cell Research and Cures Act, SCA 13 [statute on the Internet]. 2005 [cited 2015 Feb 2]. Available from: http://www.leginfo.ca.gov/pub/05-06/bill/sen/sb_0001-0050/sca_13_bill_20050523_amended_sen.pdf

receive assistance through state and county healthcare and preventive health programs" (italics mine).

In a personal conversation, Bob Klein stated that: "The Board had already adopted a commitment to provide these price discounts. Furthermore, the Board had incorporated into their bylaws a provision that they would notify the President of the California Senate and the Speaker of the Assembly if they ever changed this provision.

Access to affordable treatment is important to the state and the nation — every one of us deserves good medical care at an affordable cost. The stem cell board was already working on ways to insure low-income Californians would have access to stem cell therapies.

> "But did that mean new inventions were automatically cheap? Do not most inventions start off expensive, and then become more affordable as market economies come into play? Think of computers; they were once huge as houses and cost millions; now they can fit in cereal boxes, and are within the grasp of most."[2]
>
> ...Personal Statement, Bob Klein

There was a very practical reason why the Board could not go beyond the changes it had already made, namely: price–fixing for medical research had been tried before — and it did not work. Grants called Cooperative Research and Development Agreements (CRADAs) were available in the late 1980's, but they came with strings attached. If a researcher wanted the money, he or she had to agree that any products developed would be available — at reasonable prices — to the public. Sound good? The CRADA grants were a total flop: with those requirements, nobody wanted them. Only when the "reasonable pricing" requirement was removed, did the grants begin to be picked up.

But perhaps I was worried for nothing? You can tell a lot about a bill by its sponsors. Her co-author on SCA-13? Senator George Runner, perhaps the most dedicated enemy of embryonic stem cell research in Sacramento.

"Senator Ortiz has [...] teamed up with virulently anti-embryonic stem cell Republican George Runner (R-Antelope Valley) to propose [...] a constitutional amendment," reported Idan Ivri of the *Los Angeles CityBeat*.[3]

Above all, fighting Ortiz-backed legislation was not easy because she had a great reputation. For years she had been the go-to person on stem cell issues. Again and again I would hear, "If it is about stem cells, and Ortiz approves it, it must be good." I understood that viewpoint, because it had been my own — until the Senator began to mash Sacramento's thumb down on us.

---

[2] Why we oppose Senate Bill 1565 [Internet] [cited 2015 Feb 2]. Available from: http://www.americansforcures.org/article.php?uid=4766

[3] Ivri I. Cell out: management issues plague distribution of $3 billion in state stem cell research funds. *Los Angeles City Beat* [Internet]. 2005 Mar 24 [cited 2015 Feb 2]. Available from: http://www.lacitybeat.com/article.php?id=1837&IssueNum=94

Gloria and I drove to Sacramento for a meeting with the Senator and her chief legislative aide, Peter Hansel. Gloria generally does not mix in my stem cell politics, but she liked the Senator, and thought may be she could help. But probably nothing could have helped that conversation.

The first thing Senator Ortiz said was to ask *why I had called her a Nazi*. I was befuddled, without an idea of what she was talking about. She pointed to an editorial in the *Sacramento Bee* (the most influential newspaper in the State Capitol, and a long-term conservative opponent of Proposition 71).

The editorial read, "The institute's oversight committee and its supporters have launched a campaign to demonize [Senator Ortiz] [...]. At Monday's meeting, oversight committee member Joan Samuelson, an advocate for people with Parkinson's disease, said Ortiz's measures 'will be measured in extra suffering and death.' [...] Even more disturbing were the comments of Don Reed, an audience member and patient advocate. He compared critics of Proposition 71 to 'Nazis' and left the impression that Ortiz was in that camp."[4]

Fortunately the transcript of ICOC meetings was a matter of public record. My actual words?

"My name is Don Reed. Like everyone in America, I have a reason to defend the California Institute of Regenerative Medicine. My sister Barbara has cancer. [...] My son Roman is paralyzed. [...] How do we feel about attacks on our own California Institute of Regenerative Medicine? We must feel somewhat like the way the English people felt in World War II when Nazis were raining bombs on London."

— ICOC transcript, May 23, 2005.

There had been no name calling. The Senator and I disagreed, but with hopefully mutual respect.

But if someone challenged the California stem cell program, the patient advocate community would definitely respond. Should we sit idly by and watch our best hope for cure being micro-managed into defeat, a death of a thousand cuts? Not likely! If lawsuits came, or religious opposition, or legislative attacks, whatever — we would defend what we had helped to build.

Just weeks ago the Senator and I had been fighting side by side, urging Washington to support HR 810, the Stem Cell Research Enhancement Act (Castle/DeGette), to do for the nation essentially what SB 253 had done for California — legalizing embryonic stem cell research.

But now? Emotions ran high. At one point I said, "Hey, this is Don you are talking to, remember — I worked free for you for two years?"

"Oh, you just did that for your son," she snapped back.

Which was both true and not. Of course I was inspired by my son. Most advocates do begin for personal reasons, but gradually realize the only way we can win is if the entire field advances.

---

[4] Stem cell compromise. *The Sacramento Bee* [Internet]. 2005 May 26 [cited 2015 Feb 2]. Available from: http://www.sacbee.com/content/opinion/editorials/story/12956334p-13803706c.html

We argued for two hours, and neither one of us were likely to run dry of words.

Finally she whispered in Peter Hansel's ear. Her aide left the room. When he returned a moment later, it was to mention another appointment: our cue to leave.

Everybody hugged, but it was a formality.

Instead of organizing support in favor of an Ortiz bill, patient advocates had to rally against it — or rather, against them, because there were several. At the committee hearings in Sacramento, we were there, in opposition. I wrote editorials and emails, opposing her legislation.

We did what we had to do to defend our program. But it was pure pain, fighting with a friend.

I worried that the program would not be willing to defend itself. They had the right to do so; every government agency must lobby to protect its funding — look at the Defense Department, sending Generals and the Admirals to testify before Congress, lobbying vigorously.

But sometimes scientists are shy to stand up for themselves. In their labs, they are at ease. To step out of their comfort zone is not pleasurable. But it must be done.

Quotes below are taken from an official ICOC Transcript[5], by recorder Beth C. Drain.

First to speak on SCA-13 was Nobel Laureate Dr. Paul Berg. His work on DNA led to artificial insulin which keeps millions of diabetics alive today.

PAUL BERG: "If passed, SCA-13 would cripple CIRM's ability to operate, and imperil the progress of stem cell research in California."

BOB KLEIN: "The current language in SCA-13 would destroy Proposition 71; it's time to work together and get down to the business of finding treatments and cures for chronic diseases. The Assembly's legislation, ACR 252 and this year's ACR 24, both introduced by Gene Mullin, along with ACR 1 by Gloria Negrete McLeod, represent extremely thoughtful legislative initiatives providing direction through the Assembly's taskforce on Intellectual Property and constructive guidance and reporting requirements on conflicts of interest. We want to work with the Legislature."

We had an outstanding interim President, Zach Hall, formerly Head of the National Institute of Neurological Disorders and Stroke (NINDS), one of the branches of the National Institutes of Health (NIH) and a former leader of the Roman Reed Spinal Cord injury Research Act. Tall, thin, and be-whiskered, Zach was cautious and careful — and in a dignified way, furious:

ZACH HALL: "We are doing our best at CIRM to start our scientific program in stem cell research. If enacted, SCA-13 will stop us in our tracks. […] As a neuroscientist and former Director at the National Institutes of Health, I can say with assurance that SCA-13 will cripple our efforts, making it impossible to put California at the forefront of stem cell research."

---

[5] Meeting before the Independent Citizens' Oversight Committee to the California Institute for Regenerative Medicine, organized pursuant to the California Stem Cell Research and Cures Act. (2005). BRS-72141 [cited 2015 Feb 2]. Available from: http://www.cirm.ca.gov/sites/default/files/files/agenda/transcripts/05-23-05.pdf

JOAN SAMUELSON (who has Parkinson's disease, and was the founder of the Parkinson's Action Network): "I am looking at this series of provisions that are covered in the Ortiz bill, and everyone of these tasks were assigned to us by the voters. [...] The voters gave it to us, and we're endeavoring to work as hard and as fast as we can. [...] [SCA-13] very likely can cause harm that will be measured [...] in extra suffering and death."

JEFF SHEEHY: "This bill is a disaster."

Bob Klein expressed the board's eagerness to work with Sacramento, as for example, ACR 252 and 24 by Gene Mullin, as well as ACR 1, and Gloria Negrete-McLeod's bill, ACR, all of which had been supported both by CIRM and the patient advocate community.

It was decided to write a public letter stating opposition. The final version read:

"The Independent Citizens oversight committee opposes SCA-13.

"We are committed to working with the legislature to advance stem cell research, to ensure transparency, to prevent conflicts of interest, to provide an outstanding peer review system, to provide a strong and effective intellectual property program to protect the interests of the state of California and its citizens, and we believe we have put standards and policies in place to achieve these objectives. The ICOC also shares the goal that therapies and cures developed through research funded by Proposition 71 be made available to all members of the California public.

"As currently drafted, however, SCA-13 will make it extremely difficult, if not impossible, for scientists to do their jobs, and it will delay critically needed medical therapies."

The letter was approved unanimously. The word got out and the message spread.

California Senate Pro Tem Don Perata suggested the CIRM be allowed to do what it had said it would do all along, and what it was already doing — develop its own regulations.

Was it over now? Not even close. Senator Ortiz threw bill after bill at us — SB 18, SB 401, SB 1565, and more, attempts at legislation which continued even *after she was gone*.

How was that possible? California has term limits, and her time in the Senate was over. But before she left, Senator Ortiz called a meeting, and announced that Senator Sheila Kuehl would be carrying on related legislation in her behalf. This was bad news for us; not only was Senator Kuehl an effective lawmaker, she was also genuinely beloved.

It would be five years' grueling exertion before the last Ortiz-inspired law was dealt with.

SB 1064 (Senator Elaine Alquist's bill) was a reasonable compromise, and it became law with the cooperation and compromise of the ICOC, with the specific involvement of Bob Klein, Senator Art Torres, James Harrison, Senate Pro Tem Don Perata, and others.

As Bob Klein put it: "In the spirit of Senate Bill 1064, it is important to note that the agency continued to work with the legislature in Sacramento to establish affordability standards, including discounted prices for publicly-funded hospitals, clinics and programs. An important goal for CIRM in the future is to continue to work the state to implement Cal RX pricing standards." — Personal Communication.

Every issue Ortiz raised had been answered by the board. But perhaps, I must grudgingly admit, the issues were dealt with more thoroughly than if she had not spoken up.

As ICOC member David Serrano Sewell said, "[Senator Ortiz] has had a profound influence on the committee. […] We were going to make these changes, but if we have this [Ortiz proposal] out there, we're more motivated to do it."

Why did Senator Ortiz come at us so hard? No one can know another's heart.

But was she sincere in wanting a better deal for low-income Californians? Of that there can be no doubt, because look at what she has done with her life since then.

As Vice-President for Governmental Affairs for the California Primary Care Association, she is fighting full-time for decent medical care for every Californian.[6]

I met Senator Ortiz just once more, when she was speaking at a public hearing for something or other, I don't remember what. The announcer called for questions, and I stood up.

As soon as she saw me, Senator Ortiz found it necessary to turn around and give me her back while going through some papers. I waited for a moment, but it became clear that she was not going to turn around anytime soon.

The room went still. By now I had forgotten what I had stood up to say. The words that came out were halting, clumsy:

"Proposition 71 was too huge an endeavor not to have friction. There were hard feelings on both sides, no question. But you were so important in its development. Whatever our differences were, for me at least, you will always be the Mother of California stem cell research."

The audience agreed, rocking the room with the biggest applause of the day.

And though she may have instantly regretted it, Senator Deborah Ortiz turned around and blew me a kiss.

---

[6] Deborah Ortiz. *Wikipedia* [Internet] [updated 2014 Dec 6; cited 2015 Feb 2]. Available from: http://en.wikipedia.org/wiki/Deborah_Ortiz

8 TIME-LAPSE SCRIPTOGRAPHY

As Mayor of San Francisco, Gavin Newsom brought the California Institute for Regenerative Medicine (CIRM) to the city by the Bay.

In the midst of all this drama, a cheerful competition arose.

Where should we put "Stem Cell Central", the headquarters of the California stem cell program?

Presently the meetings were held in Emeryville, a lovely city, but the facilities were too small.

Many cities could do an excellent job of housing the California Institute for Regenerative Medicine (CIRM). Los Angeles and Berkeley had magnificent universities, San Diego held an abundance of biomedical companies and research institutions, Sacramento was the hub of politics, San Francisco was the home of pioneering biomedical research — but only one could be chosen.

The Mayors made special presentations to the board; these were public meetings on buses.

One presenter stood out: San Francisco Mayor Gavin Newsom.

Gloria and I had met him before. It was a fundraiser, and we were standing around eating the free mini-foods, when Gavin Newsom entered the crowded room. He had indefinable star quality, like there was a Technicolor spotlight on him. He also had the gift of memory. The first time we met, we chatted briefly. The second time, he remembered me and asked about Roman.

Most Mayors rely on staff for key statistics. But Newsom sat himself down behind the driver, and just started talking. He used no notes, but the enthusiasm for San Francisco shone through.

A reporter asked me, as someone who came to all the meetings, which city I liked best for the HQ. I responded that the board had a tough decision to make, because all were terrific.

But for me, well, as the song lyrics go, "I left my heart in San Francisco."

But on the last day, the ICOC would make its decision.

Everybody who had something to say spoke. Someone said, as a Hispanic, they felt the headquarters should be in Sacramento — whereupon Gloria stood up and said that she was also Hispanic and she liked San Francisco!

To our delight, the city by the Golden Gate Bridge became Stem Cell Central.

That was the easy part.

Far more difficult were the decisions on the new stem cell laboratories…

In time-lapse photography, a building can leap to the sky in 30 seconds. In real life, of course, that same building may take decades to dream, design, fund, get permits for, and construct… and at any step of the way the project may fail.

Lab space was key to the California stem cell program. Scientists need a well-equipped laboratory and also a safe place to work, with no worry about politics shutting them down.

"Bricks and mortar" are expensive. Construction costs can devour funds and short-change research. But not here. Proposition 71 put an absolute limit on how many dollars could be spent on construction: 300 million dollars, ten percent of the total project. We wanted laboratory buildings that were solid and well-equipped, but not wasteful, and we wanted them now. We had no idea of the incredible surprise waiting for us.

The Facilities Working Group took on the challenge. They would develop proposals to put before the governing board, which would decide.

October 28, 2005 was their first major meeting; let's listen in for a couple of minutes…

The official transcript, at 140 pages, is quite long. I cut it down to three. If someone spoke twice on the same subject, I sometimes combined his or her remarks. My edits or comments are in parentheses; other than that, the words are exactly as transcribed by ace recorder Beth C. Drain.

Members of the Facilities committee were Chairman Rusty Doms, David Serrano-Sewell, Marcy Feit, Deborah Hysen, Ed Kashian, Robert Klein, Sherry Lansing, David Lichtenger, Jeff Sheehy, Janet Wright, and Joan Samuelson

BOB KLEIN: "Joan Samuelson was going to be able to attend. […] But Joan is — "

JOAN SAMUELSON: "Present!"

BOB KLEIN: "We are excited that Joan made it."

COMMITTEE CHAIRMAN RUSTY DOMS: "We have on the agenda, the pledge of allegiance, but since we don't have a flag yet, I think we will unfortunately have to dispense with that today. […] On this committee we have four members from the real estate sector, six patient advocates, and Bob Klein, [...] our ICOC Chairman. I'm from Southern California, born and raised there. I've been in the real estate business basically all my life, [including] the development of two very large hospitals."

DAVID SERRANO-SEWELL: "I have the honor of representing the Multiple Sclerosis and the ALS community as a patient advocate."

BOB KLEIN: "The initiative provides [...] $300 million for construction, hopefully with at least a one-to-one match" (matching funds — new money from a different source).

DAVID LICHTENGER: "I've been in construction [...] for over 20 years. [...] In the last five, six years, [I have been involved] heavily on the life science side, doing a lot of labs, technical facilities, clean rooms... also some hospital and healthcare work as well."

ED KASHIAN: "My background is in real estate, universities, [and] state colleges."

JANET WRIGHT: "I am a cardiologist (heart surgeon), one of the patient advocates on the ICOC. This group is about *getting the science a home* — I am excited to participate."

JEFF SHEEHY: "I'm a patient advocate representing people with HIV and AIDS."

ZACH HALL: "I'm the President of CIRM right now. Most of my career was at UCSF. I was a neurologist once upon a time. Now I've become a science administrator."

JAMES HARRISON: "I'm James Harrison, outside counsel to the CIRM, and it's my job to help keep you in line" [laughter].

JANET WRIGHT: "A full-time job!"

BOB KLEIN: "James Harrison was one of the key attorneys who helped me with the drafting of Proposition 71, and is extraordinarily well-versed in the legal issues behind Proposition 71, its structure and implementation. So we're very privileged to have his services."

CHAIRMAN RUSTY DOMS: "This is an extraordinary opportunity to turn hope into reality. [...] People around the table have all been affected in some way or another, either personally or with family with terrible diseases. I have a daughter that has serious learning disabilities."

ZACH HALL: "As you know, an important part of our mandate [...] is to provide facilities that can be used particularly for human embryonic stem cell research.

"Unfortunately, [...] United States' scientists have not been able to participate fully, because the major supporter of biomedical research in this country, the NIH [National Institutes of Health], has been restricted by Federal policies. Because of this uncertainty, [...] most of the research [buildings] that institutions have [...] are not suitable for this very important work.

"What we envisage would be having [lab] space with necessary basic equipment; that is, incubators, hoods, sterile hoods, freezers, microscopes.

"$100 a square foot [might be the price]. But how much equipment do you put in these [...]? If you isolate a protein from a human embryonic stem cell, you run it out on a gel and now you want to do mass spectroscopy on it, do you have to have a separate mass spectroscopy facility [...]?

"You might decide [...] to set up a core facility [...] in which you would have a very expensive piece of equipment that would be shared, used by a lot of people. [...] There are *many* questions to answer. What space and facilities do they presently have for human stem cell research [...]? How large would the proposed new facility be [...]? Do they plan to expand an existing facility [...]? What is the budget and timetable for completion [...]? Is this a reasonable cost? Are the timelines and milestones satisfactory? Is there adequate oversight [for] construction work [...]? What is the level of institutional commitment [...]?"

BOB KLEIN: "What we have before us is a tremendous challenge in timing [...] facilities that can be built *within two* years of a grant award. [...] We'll probably need to have a partnership with cities to accomplish that [...] (such as fast-tracking projects through permit processes, etc.). It is imperative we protect our institutions and our researchers from [political] pressures that may [...] come upon us in a blink of an eye. [...] But it is with great optimism that I look at the group assembled here. I know they are up to the task" (italics mine).

DON REED: "On behalf of people who are suffering, thank you so much for the incredible amounts of work that will be demanded of you for almost no money.

"[...] I have to tell a story about [committee member] Sherry Lansing since she's not here to defend herself. [...] I am movie crazy. So Sherry Lansing was the President of Paramount Motion Picture Studios. I had to ask her: How can you, the head of this powerful organization (she did the movie *Titanic*), [...] justify the time to work on this particular assignment? And she said, 'It is the most important thing in my life.' I believe you all feel that way. Thank you."

CHAIRMAN DOMS: "Thank you. [...] Too bad there are not a lot more people like you who support us and help us with what we're trying to do here."

DON REED: "There *are* a lot. I am just noisy!"

End of transcript.

And now, thanks to the miracle of time-lapse scriptography (i.e., print), let's jump ahead and see what actually happened. We will skip multiple meetings and thousands of hours of many people's hard work, and go right to a press release from the California stem cell agency.

On May 7, 2008, the governing board of CIRM voted to distribute $271 million to 12 institutions to build stem cell research facilities throughout California.

As stated on a CIRM report[1] published that day, "The institutions themselves committed an additional $560 million, bringing the total statewide investment in new research space to $831 million, [...] plus additional institutional commitments. [...] **In total, the state funding will have leveraged $1.1 billion in new resources** to accelerate the pace toward therapies for patients with chronic and debilitating disease and injury."

Did you see what just happened? The places which won new research centers had also chipped in their own dollars as well, multiplying the money, in some cases two and three times as much.

Total return on investment?

California's $271 million was leveraged into $1.15 billion to build its laboratories with.

---

[1] California Institute for Regenerative Medicine Report [Internet]. 2008 May 7 [cited 2015 Feb 2]. Available from: http://www.cirm.ca.gov/print/3195

9 IDEOLOGY, SCIENCE, OR BIBLE-QUOTING VAMPIRES?

Right about this time, as California's program was getting started, and America struggled to understand the new science, I had a family reunion — with my super-religious relatives.

For them, faith was no spectator sport, practiced on Sundays and forgotten the rest of the week; it touched every corner of their lives. Grace at dinner ran long because everybody added their own commentary. But if the food was sometimes cold, the hearts were unquestionably warm.

I kept waiting for someone to bring up the subject of stem cells. The dinner was peaceful with plenty of family catch-up. Afterward, when the table was cleared and everyone was gathered in the living room, the oldest woman said:

"All right, Don. We've heard a lot about this — stem cell stuff?"

Her voice was polite but firm. It got very quiet in the room; this was no private conversation. They all knew about Roman's struggle with paralysis and that we were trying to find a cure. That they approved. But there were people in the room who regarded every word in the Bible as literal and inerrant truth. Was stem cell research something evil?

To general puzzlement, I acted out a scene from the 1931 Bela Lugosi movie *Dracula*.

The Count is speaking with doomed Renfield. Wolves howl in the distance, and Dracula says, "Listen to them, children of the night — what music they make…"

Then he notices a cobweb on the wall, and remarks, "The spider, spinning its web for the unwary fly… the blood is the life, Mr. Renfield…"

He offers his prospective victim a glass. "This… is very old… wine."

"It's delicious!" says Renfield, "But aren't you drinking?"

Dracula smiles.

"I never drink… wine."

I paused. My relatives looked at each other.

"The blood is the life…" I repeated, suddenly nervous. I had figured they would pick up on it right away. But if they thought I was trivializing the issue, or insulting their faith…

"It is a quote from the Bible, Leviticus, Chapter 17, Verse 11!" I said. "The blood is the life! The movie is a joke, but the subject is not. If there is no blood, there is no human being! Only in the womb does the fertilized egg develop blood — but outside the body, in a dish of salt water? There is no heart, no veins, no blood! Embryonic stem cell research is living tissue — not a person!"

As the vampire in the classic 1931 film DRACULA, Bela Lugosi said: " The blood is the life," which helped the author explain stem cell research.

I babbled on, but they were not listening to me now. They were checking in with a higher source.

Flip-flip-flip. Pages turned.

And then, exoneration. The Biblical fact-checker said, "The quote is 'the life is in the blood'." Apparently that was close enough. Everyone relaxed, and I survived the evening without further incident. Once it was established that the Bible was not actively against stem cell research, and we were not sacrificing one person to benefit another, we could talk.

Why does this matter? Because the only serious objections to embryonic stem cell research are ideological: religious opinion — and religious people often disagree, even with each other.

Let me prove that. Remember I said only 17 groups opposed the Stem Cell Research Enhancement Act, and everyone of those was a religious or ideological group? Here they are.

The list was obtained from the Chairman of the Republican Study Committee, Representative Jeb Hensarling (R-Texas). It refers to HR 3, the second attempt to pass the Stem Cell Research Enhancement Act.[1] — http://www.dailykos.com/story/2007/02/02/297483/-WHO-OPPOSES-EMBRYONIC-STEM-CELL-RESEARCH#

1. National Right to Life Committee
   http://www.nrlc.org/Missionstatement.htm
2. US Conference of Catholic Bishops
   http://www.usccb.org/index.shtml

---

[1] Reed DC, Sarah Palin and Christopher Reeve: a special needs parent speaks out. *Daily Kos* [Internet]. 2008 Sep 19 [cited 2015 Feb 3]. Available from: http://www.dailykos.com/story/2008/9/19/64022/9921/599/603501#

3. Family Research Council
   http://www.frc.org/get.cfm?i=PGO3FO6@V=PRINT
4. Christian Coalition
   http://www.cc.org/about.cf
5. Concerned Women for America-- http://www.cwfa.org/about/issues/sanctity-of-life/
   http://cwfa.org/about.asp
6. Focus on the Family-- http://www.focusonthefamily.com/socialissues/social-issues/stem-cell-research/stem-cell-research-issue
   http://www.focusonthefamily.com/aboutus/A000000408.cfm
7. Christian Medical Association:
   http://www.cmdahome.org/index.cgi?BISKIT=3982036250&CONTEXT=cat&cat=100011
8. Eagle Forum
   http://www.eagleforum.org/misc/descript.html
9. Traditional Values Coalition
   http://www.traditionalvalues.org/about.php
10. Southern Baptist Convention
    http://www.sbc.net/aboutus/default.asp
11. Susan B. Anthony List
    http://www.sba-list.org/aboutSBA.htm
12. Republican National Committee for Life
    http://www.rnclife.org/about/
13. Cornerstone Policy Research
    http://www.nhcornerstone.org
14. Culture of Life Foundation
    http://thefactis.org/default.aspx?control=Contentmaster&pageid=13
15. Religious Freedom Coalition
    http://rfcnet.org/about.htm
16. Coral Ridge Ministries
    http://www.coralridge.org/about_cirm.htm
17. Center for Reclaiming America for Christ
    http://www.reclaimamericaforchrist.org/history.htm

One group was of particular interest to me. In terms of stem cell research, the Family Research Council was more dangerous than all the others put together — because it is the home of Dr. David Prentice.

Question: What is the connection between Erik Prince, owner of Blackwater/Xe, perhaps the world's largest private army, and David Prentice, opponent of embryonic stem cell research?

Money.

Prentice is Senior Fellow for Life Sciences on the Family Research Council (FRC), a right-wing religious lobbying group, founded and funded by the family of Erik Prince.

The FRC mission statement reads, in part:

"The Family Research Council promotes the Judeo-Christian worldview as the basis for a just, free, and stable society."

Their version of the "Judeo-Christian worldview" apparently includes opposition to gay rights, abortion, evolution, environmentalism, public healthcare, the Department of Education, taxes — especially taxes — and embryonic stem cell research.

Erik's father, the late Edgar Prince, financially began the FRC, including construction of its six-story building in Washington DC. Others led it: evangelist James Dobson of Focus on the Family (one of his groups, the Family Bioethics Council, sued the California stem cell program), and conservative icon Gary Bauer; but Prince donated the money to make it happen. As Bauer said in a eulogy for the elder Prince, "Without Ed and Elsa and their wonderful children, there simply would not be a Family Research Council."

Son Erik used inheritance money ($1.35 billion was realized from the sale of his father's auto-parts business) to make the private military company Blackwater, recently renamed Xe, several of whose employees were recently found guilty of the unprovoked massacre of Iraqi civilians.[2]

Erik's sister Betsy famously proclaimed, "My family is the largest single contributor of soft money to the national Republican party." Married to Dick DeVos of the Amway fortune, she is an active funder of conservative causes, including such gifts as, according to an article in *Salon*, "$670,000 to the Family Research Council and $531,000 to Focus on the Family," between July, 2003 and July, 2006.[3]

And David Prentice, full-time employee of the FRC?

No individual has had more of an impact on Republican stem cell research policy.

"Karl Rove, head of the White House's Office of Political Affairs, has declared that embryonic stem cells aren't required because there is 'far more promise from adult stem cells.' [...] It seems that the White House received this idea from David Prentice, a senior fellow for life sciences at the Family Research Council and an adviser to Republican members of Congress.

"In a report of the President's Council on Bioethics, Prentice claimed that adult stem cells can effectively treat more than 65 diseases. Not only is this assertion patently false, but the information purveyed on the Family Research Council's website is pure hokum," wrote Robert Schwartz in the *New England Journal of Medicine* in 2006.[4] Prentice's views may be popular in ultra-conservative circles, but are not mainstream science.

For example, the Stem Cell Research Enhancement Act, House Resolution (HR) 8 (Castle/DeGette), would have legalized the funding of new embryonic stem cell lines,

---

[2] Apuzzo M. Blackwater guards found guilty in 2007 Iraq killings. *The New York Times* [Internet]. 2014 Oct 22 [cited 2015 Feb 3]. Available from: http://www.nytimes.com/2014/10/23/us/blackwater-verdict.html?_r=1

[3] Van Heuvelen B. The Bush administration's ties to Blackwater. *Salon* [Internet]. 2007 Oct 3 [cited 2015 Feb 3]. Available from: http://www.salon.com/2007/10/02/blackwater_bush/

[4] Schwartz RS (2006). The politics and promise of stem-cell research. *N Engl J Med* **355**:1189–1191.

made from blastocysts otherwise scheduled to be thrown away. Prentice opposes such research, calling the microscopic joinings of sperm and egg "young human life" and the research "immoral". His position was echoed by the above-named 17 ideological groups, publicly opposing HR 8.

And on the other side, in favor of expanding the research? *Five hundred and ninety-one patient advocate, science, and medical groups* — not to mention majorities in both House and Senate.

Apparently influenced by the Prentice prescription, then-President George W. Bush vetoed the Stem Cell Research Enhancement Act not once, but twice, and funded adult stem cell research far more heavily than embryonic. In the fiscal year of 2008, human embryonic stem cell research received $88 million from NIH compared to $381 million for adult stem cell research.

Prentice's list of alleged adult stem cell treatments was publicly discredited[5]:

"Adult stem cell treatments fully tested in all required phases of clinical trials and approved by the U.S. Food and Drug Administration are available to treat only nine of the conditions on the Prentice list, not 65. [...] By promoting the falsehood that adult stem cell treatments are already in general use for 65 diseases and injuries, Prentice and those who repeat his claims [...] cruelly deceive patients."

Restrict cure research to adult stem cells only? According to Peter Donovan, Director of the Sue and Bill Gross Stem Cell Research Center, "It's like someone in the early 1900's saying, 'Why develop the motor car, when we already have the bicycle [...]? Patients deserve more, a lot more."

Dr. Sean Morrison, then Director of the University of Michigan Center for Stem Cell Biology, also said, "David Prentice has been one of the most prolific sources of misinformation on stem cell research. I don't know any leading stem cell biologists who agree with the claims he has made."

But now here is something puzzling.

Every year, the Family Research Council puts on an ultra-conservative gathering, the Values Voters Convention. All the most conservative Republican Presidential candidates attend and speak, hoping for the endorsement of the Religious Right. The nearly 6,000 attendees are arguably the most conservative Religious Right voters in the nation.

In 2007, at this Values Voters convention, a survey was taken by 5,775 attendees.

The poll was on issues of importance. What bothered them the most? It had astonishing results.

Issue number one was abortion, with 41.5% voting it their top concern.

Now, if embryonic stem cell research is supposedly a form of abortion, it should be issue number two in importance, shouldn't it? But it wasn't. Number two in importance

---

[5] Reed DC, Eric Prince and David Prentice: the Blackwater/anti-stem cell connection. *Huffington Post* [Internet]. 2010 Apr 4 [cited 2015 Feb 3]. Available from: http://www.huffingtonpost.com/don-c-reed/eric-prince-and-david-pre_b_446709.html

to the Values Voters was Same-Sex Marriage (19.7%). After that the list of concerns included:

3. Tax cuts (10.7%).
4. Permanent tax relief for families (9.7%).
5. Federal "hate crimes" legislation (5.7%).
6. No vote on this question (3.1%).
7. Tax payer funding for abortions (2.6%).
8. Prayer in schools (1.6%).
9. Reinstatement of the "Fairness Doctrine" (1.5%).
10. Public display of the Ten Commandments (1.0%).
11. Enforced obscenity laws (0.94%); and finally, way down on the list.
12. Embryonic stem cell experiments (0.83%).

In this ultra-conservative religious community, less than one percent listed embryonic stem cell research as an area of concern.[6]

My impression was that the leadership was trying to get the membership all worked up about the research, but the members were not buying it. Like the Catholic voters, they were perhaps over-propagandized, and had listened to the other side as well, and made up their own minds.

My quarrel with the Religious Right has never been about their faith. What a person believes in their heart and how he or she chooses to worship is none of my business. But if a group attempts to impose their religion on me, especially when it comes to protecting my family from disease and disability — then we have a difficulty.

Will Prentice's narrow view prevail? According to Past President Dr. Alan Trounson, President of the California Institute for Regenerative Medicine:

"The enemy is the disease or injury that is destroying our friends and families' quality of life, and that is what we must defeat. The mission is to bring the benefits of pluripotential and progenitor cell therapy to regenerative medicine. That is what we are doing, and neither Dr. Prentice nor any other critic shall deter us from fulfilling this vision."

---

[6] Bridges J. Straw poll on the issues. *Family Research Council Blog* [Internet]. 2007 Oct 23 [cited 2015 Feb 3]. Available from: http://www.frcblog.com/2007/10/straw-poll-on-the-issues/

10 JOAN OF ARC, AND THE REPUBLICAN STEM CELL REVERSAL?

Joan of Arc's courage and faith inspires us still. But did you know her mother sued the Pope? Artist unknown.

We all know the story of Joan of Arc: how a 12-year-old girl led French soldiers to war against the occupying English and beat their armies, again and again.[1]

After many victories, Joan was dragged off her horse by English soldiers, and put on trial by a collaborating French Church. She stood alone, defying 62 religious lawyers and

---

[1] See Mark Twain's *Personal Recollections of Joan of Arc*, a two-volume novel drawn from court documents of the trial. Twain considered the book his most important work.

priests, defeating their arguments by faith and logic. But the verdict was set before the trial began.

Her "crime" was heresy: she claimed to have been guided by the voices of Saints — and she wore men's clothes. In battle she dressed as a soldier; in prison she wore trousers, tightly knotted with ropes to defend her virtue against male soldiers, who were actually put into her cell at night. (She was also chained to a log of wood, ostensibly to prevent her flying away.)

Bishop Pierre Cauchon eagerly condemned Joan to the stake. "I die through you, Bishop," Joan said. Cauchon hated Charles VII — to whom Joan's victories had given power.

Even Shakespeare lowered himself by attacking Joan, calling her a "witch" and a "harlot." In the play Henry VI, Part I, he falsely portrayed her begging the court for mercy, claiming one moment she was a virgin, the next that she was pregnant — by a long list of men! (Joan's virginity was physically verified during the trial; the prosecution hoped to prove her a witch who had intercourse with Satan; if Joan had not been a virgin, that would have been used to discredit her.)

The only act of kindness shown to her was at the execution. As she stood at the stake, she asked for a cross. An English soldier broke a piece of wood and tied it together. She thanked him, and placed the cross in her prison shirt, to be with her as she died.

Her last words were, "Jesu, Jesu, Jesu!" Then there was silence from the flames.

Her body was burnt, her ashes scattered, in an attempt to wipe out the memory of Joan. But here is a part of the story you may not know.

Joan's mother sued the Catholic Church. For more than 20 years, Isabel Romee fought for a new trial for her daughter. She persuaded Pope Nicholas V to reopen the case. When that Pope unfortunately died, Isabel convinced the next Pope to continue.

And she won. In 1456, Pope Callixtus III reversed the original court's judgment, and in 1920, Joan of Arc was officially declared not a witch, but a saint.[2]

Joan's murder enraged the French; with her name as their standard, they won battle after battle, and they drove the English out.

On July 18, 2006, a crucial bill was finally allowed a vote in the U.S. Senate. It was HR 810: the Stem Cell Research Enhancement Act.

One side said, "Illegal, immoral, unnecessary!" and that was (mainly) the Republicans.

The other side said, "Ease suffering, save lives!" and that was (almost entirely) the Democrats.

The issue was federal funding of new embryonic stem cell lines. Under the Bush doctrine, only a few outdated, mouse-contaminated, racially limited, and expensive cell lines were fundable.

It is hard to imagine a more carefully written and moderate bill than HR 810. There was not even a money element; no funding was guaranteed at all, just that it might be allowed. President Bush's policy allowed stem cell lines to be made from fertilized eggs

---

[2] Robo E. The holiness of Saint Joan of Arc [Internet] [cited 2015 Feb 3]. Available from: http://www.ewtn.com/library/mary/joan1.htm

which would otherwise be thrown away — but no new lines could be made after 2001. HR 810 allowed research to be funded on stem cell lines made after that date. It was very nearly the Bush guidelines — with the cutoff date removed.

Here is the bill's key sentence: the line in the sand.

"Notwithstanding any other provision of law [...], the Secretary shall conduct and support research that utilizes human embryonic stem cells."[3]

The Republican-controlled Senate blocked the bill for more than a year, not allowing a vote.

But a new voice was being heard, what Dan Perry called "patient's voices: the powerful sound in the stem cell debate." Perry, President of the Coalition for the Advancement of Medical Research, had earlier described why patients should involve themselves in a patients' coalition.

"When medicine cannot relieve their suffering," Perry said, "Patients are the most compelling witnesses to the value of research that can quite literally save their lives."[4]

Across the country, patients and their families reached out to Congress and the White House, becoming a political force. We are millions, we families affected by chronic disease, and there will come a time when we can no longer be safely ignored.

The bill passed in both houses, but look at the vote.

In the House of Representatives, 50 Republicans supported the research while 180 voted against it. Among Democrats, 187 supported the research and 14 opposed it.[5]

In the Senate, 18 Republicans said "yes" and 36 said "no." Democratic Senators were 44 in favor, with only one against.[6]

The two Independents voted yes: both in favor of the Stem Cell Research Enhancement Act.

As expected, the bill was vetoed by President Bush, the first veto in his two terms in office.

But if just a few more Republicans had voted with us, we could have overridden the Bush veto.

A second attempt was made to pass the same legislation, with identical results. House and Senate passed the bill, but with only a few Republicans — and once again the Bush veto killed the bill.

So what is the stem cell connection to Joan of Arc?

The Catholic Church had the courage to admit its mistake of burning Joan of Arc at the stake. Can the Republican Party change its mind on the question of stem cell research?

---

[3] Stem Cell Research Enhancement Act of 2005, HR 810. House veto and override, 109th Congress (2006).
[4] Perry D (2000). Patients' voices: the powerful sound in the stem cell debate. *Science* **287** (5457): 1423.
[5] Stem Cell Research Enhancement Act of 2005, HR 810. House vote, 109th Congress (2005).
[6] Stem Cell Research Enhancement Act of 2005, HR 810. Senate vote, 109th Congress (2006).

11 DEADLY DEFINITIONS

Health food and weightlifting pioneer Bob Hoffman's misunderstanding of a word's definition led to comical consequences; other linguistic errors can have far more serious results.

Definitions of words make a difference: the words "intercourse" (sex) and "conversation" (talk) had opposite meanings in earlier times. Imagine the possibilities! "Oh, I just had intercourse with your wife!" "That's nice, what did you talk about?" But if you said you had "conversation" with her — that risked a punch in the nose!

Medical terms spread Latin across the page, and with it, endless misunderstandings.

For instance, in the late 1960's, I was Associate Editor for a weightlifting magazine called *Strength and Health*, owned by the late Bob Hoffman.

Hoffman not only began the health food industry in America (combining dried milk, soy, and egg into protein powder, the basis of health food bars), but also developed the sport of Olympic weightlifting. His York Barbell Club won the American national championships 38 years in a row.

But the 70-year-old health foods pioneer seemed puzzled one day when I told him why two of our weightlifters were not at work in the office.

"They're at the hospital," I said, "Having vasectomies."

"Vasectomies?" The word did not seem to register.

"You know, so they won't have children any more?"

His face went pale and he took a step backward.

I thought no more about it until a few days later at the annual *Strength and Health* picnic. This was a wonderful event, and lifters, body builders, and fitness friends from around the world gathered.

Hoffman was at the microphone when the aforementioned two great lifters stepped onto the stage, chalking their hands before approaching the bar.

Bob Bednarski had just set two world records in official competition. Bill Starr had a build like Conan the Barbarian — but could move like a cat. They slung weights around and made it look easy. The crowd was delighted and let them know it.

But Bob Hoffman turned his head on an angle, looking at the two lifters oddly, and said, "That is remarkable lifting, considering they have both just been *castrated*!"

And here the joke ends; for one of the most dangerous stem cell battles ever fought was about the definition of a word.

Proposition 73, or the "Waiting Period and Parental Notification Before Termination of Minor's Pregnancy Initiative Constitutional Amendment", was an anti-abortion initiative.

Proposition 73 required parents to have a 48-hour notification before their minor daughter could have an abortion. That sounded reasonable, at first. But what if the parents were violent and abusive? A family member might even be the rapist responsible for the pregnancy. Or — and this could happen in the best of families — what if the girl was too ashamed to talk about it?

Rather than tell her parents, a frightened teenager might try to "take care of the situation" on her own and have an abortion done under back-alley conditions, endangering her life.

So I knew I would vote against Proposition 73, but did not plan to take part in the battle against it. There are only so many hours in the day. Besides, it wasn't an attack on stem cells, or so I thought. But my threat-alert antennae were twitching, so to speak.

Then a red-headed woman named Susan Fogel approached me. Ms. Fogel was one of Proposition 71's most enthusiastic critics. Whenever she stood up in meetings to make a public comment, I would think, uh-oh, here comes trouble! But she always had rational comments, and I listened when she talked. Some of her suggestions had affected the program's policy framework.

"Proposition 73 has language," she said, "Which might endanger stem cell research."

Proposition 73 had a new definition for abortion, calling it "the death of *the unborn child — a child conceived but not born.*" If it became law, that phrase would be part of the State Constitution.

"Conceived"? That was the joining of sperm and egg, invisibly small. If a fertilized egg (the microscopic blastocyst) could now be legally defined as a child — even before it was implanted in the womb? … what did that mean?

Stem cell scientists might be arrested for murder — for trying to cure disease?

Those opposed to Proposition 73 included such heavy hitters as the American Medical Association, the League of Women Voters, Planned Parenthood, California Medical Association, Juvenile Court Judges of California, California Nurses Association, NARAL Pro-Choice California, and others. Visiting their websites helped clarify the issues.

I reached out to friends, asking them to look closely at Proposition 73.

Rayilyn Lee of Arizona, a Parkinson's fighter, argued, "If microscopic cells can be legally redefined as a child, then scientists doing embryonic stem cell research could be said to be murdering a person. For ideologues to whom invisible dots in a Petri dish are more important than an actual person, passing Proposition 73 may be a way to stop the research."

The more people heard about the bill, the less they liked it.

Margaret Crosby, of the American Civil Liberties Union, said, "To equate an embryo with a child would raise serious issues about legality of many forms of biotechnology."

And in another interview, Crosby further said, "Putting that language into the State Constitution could lay legal grounds for those wishing to ban stem cell research and IVF. […] Because embryonic stem cell research and IVF involve unused embryos, the practices could be vulnerable to the legal challenge that they are terminating human life."

I produced sample letters and editorials, and shared them widely. But it could not just be the usual advocates talking; everybody had to get involved, including the researchers. If scientists shrink from the battles hoping others will defend their rights and funding, that will amount to political suicide.

If the educated do not take part, the ignorant will be happy to make the decisions.

Fortunately, modern scientists are beginning to pay attention.

Director of the Burnham Institute stem cell program Evan Snyder admitted, "Our [research] program would probably not be able to derive therapies and cures, [or] even to understand how diseases may unfold," if Proposition 73 passed.

Dr. Snyder and I did a radio commercial on the issue, trying to break through the reasonable-sounding language the opposition offered, to the dangerous reality beneath.

Conservative religion was behind the initiative. The *Los Angeles Times* did a major article[1] detailing the massive effort put on by Catholic and evangelical churches.

http://articles.latimes.com/2005/nov/07/local/me-nuabortion7

The article read:

"Millions of voters began hearing about the initiative [...] in their neighborhood church: [...] glossy 'Yes on 73' fliers slipped into church bulletins, [...] information tables set up behind the pews, [...] a two minute DVD featuring teenage actresses.

"The California Catholic Conference distributed [materials] for priests to read at the state's 1,100 parishes, which serve 11 million people.

"[...] San Diego publisher James Holman, an anti-abortion activist and the measure's top financial backer printed petitions in a string of Catholic newspapers he owns.

"[...] The initiative [...] had a narrow lead in a recent Times poll."

But we were fighting too. In the last days before the vote, an anonymous donor (no, not Bob Klein!) provided $2,000 to put my editorials in newspapers in areas where the vote was close.

Proposition 73 was ahead, but its lead had begun to narrow. The night before the election, a November 7, 2005 article in *SurveyUSA Elections* wrote: "Proposition 73 [...] led by 11 points one week ago, and leads today by 4 points."[2]

The morning after the election, there was a frightening headline in the *Los Angeles Daily News*: "PROPOSITION 73: ABORTION MEASURE FINDING FAVOR WITH VOTERS," followed by an editorial suggesting that Proposition 73 had majority support.

In San Diego, early results showed 309,629 voters approving Proposition 73 (52.3%) and only 282,848 voters against it (47.7%).

But then the tide turned. The final tally of the voters was:

YES: 3,676,592 (47.2%)

NO: 4,109,430 (52.8%)[3]

We had won by five percent and kept the research safe for another day. But another and far more threatening wave was building toward a tsunami, and this from a single word...

Personhood.

---

[1] Rivenburg, R. Abortion proposition finds its forum in the churches. *Los Angeles Times*, 2005 Nov 7.
[2] Results of Survey USA Elections Poll #7443 [Internet] [cited 2015 Feb 3]. Available from: http://www.surveyusa.com/client/PollReport.aspx?g=6dc431a5-8970-4caa-8c41-8480863dfe28
[3] California Proposition 73 (2005). *Wikipedia* [Internet] [updated 2012 Nov 10; cited 2015 Feb 3]. Available from: http://en.wikipedia.org/wiki/California_Proposition_73_%282005%29

12 WHAT ARE WE FIGHTING FOR?

Frank Capra's Award-Winning film series, "Why We Fight" made clear the reasons for American involvement in World War Two. (Wikipedia photo.)

In World War II, famed director Frank Capra and Walt Disney put together a series of motion picture shorts, titled WHY WE FIGHT.

Patient advocates also need to know why their efforts matter.

Here are a dozen embryonic stem cell projects. Remember, these could all be illegal if anti-science elements get their way.

I wrote the list (widely circulated but seldom credited) in 2006, so it will be interesting to see what happened in the months and years to come. Watch for the research in the chapters ahead.

They are in alphabetical order by disease. Is your "favorite" chronic condition among them?

Amyotrophic lateral sclerosis (ALS; also known as Lou Gehrig's disease)

At the University of Wisconsin at Madison, scientists turned embryonic stem cells into motor neurons (nerves which carry messages between brain and body), offering possibilities for repairing damage caused by ALS, spinal cord injury, and other nerve-related disorders.[1]

Alzheimer's disease

Until now, it had been impossible to study the progression of this horrific disease, which first robs sufferers of their memory, before it steals their life. We know so little about it: not even exactly when it begins. With human embryonic stem cells, we maybe able to isolate the disease and observe its progress on human tissue cells, instead of only watching human beings as they lose their minds and die. The special cells may offer a new way to design better Alzheimer's medicines. Dr. Goldstein, from the Howard Hughes Medical Institute at University of California San Diego, is using human embryonic stem cells to test new ideas of how Alzheimer's disease develops, and how it might be treated.[2]

Blindness

The major cause of blindness in Americans over age 60 is macular degeneration, which refers to the loss of retinal cells in the eye. Dr. Robert Lanza and Dr. Irina Klimanskaya of Advanced Cell Technology in New Jersey used human embryonic stem cells (hESCs) to make replacement cells, which may one day offer the return of vision to millions.

Cancer

The speed at which cancer develops is a major obstacle in curing this devastating disease. At Kumamoto University in Japan and Cambridge University in England, surface proteins were developed that could mark cancer stem cells, laying ground work for new drugs that may one day slow, or even turn off, tumor formation.[3]

Cystic fibrosis

Cystic fibrosis inflames the lungs, strangling sufferers in slimy mucous. Using hESCs, Dr. Stephen Minger of King's College developed a stem cell line of cystic fibrosis. The

---

[1] Minkel JR (2005). One small step. *Sci Am* **292** (4): 34.
[2] Goldstein L. Message to: Don Reed. 2005 Mar 26.
[3] Stemline licenses high throughput drug discovery technology from University of Cambridge. *Stemline* [Internet]. 2005 Jan 17. [cited 2015 Feb 4]. Available from: http://www.stemline.com/newsArticleDetails.asp?id=9

disease can thereafter be studied in a human cell line that has genetic mutations like those seen in cystic fibrosis patients.[4]

Deafness

The death of tiny hair cells inside the ear contributes to deafness for an estimated 28 million Americans. These cells do not naturally regrow. However, using hESC techniques, Dr. Stefan Heller of Boston's Eye and Ear Infirmary is capable of generating these inner-ear hair cells, raising the possibility that this technique may lead to new treatments for the deaf.[5]

Diabetes

At Stanford University, researchers have made insulin-producing cells from mouse embryonic cells. When transplanted into diabetic mice, these cells reduced blood sugar fluctuations and increased the animals' lifespan.[6]

Growing human tissue

At the Massachusetts Institute of Technology, Dr. Robert Langer used embryonic stem cells to grow liver, cartilage, nerve tissue, and blood vessels, all of which appeared to function normally when transplanted into mice.[7]

Hemophilia

At the University of North Carolina, Chapel Hill, Dr. Jeffrey Fair and Dr. Oliver Smithies used embryonic stem cells to reverse hemophilia, which is a blood clotting disorder, in mice.[8]

Immune system disease

In the battle against HIV/AIDS, a certain gene, the RAG-2, is apparently needed for a healthy immune system. At Cambridge University, mice were purposely bred without the

---

[4] Rincon P. Cystic fibrosis stem cells made. *BBC News* [Internet]. 2004 Sep 9 [cited 2015 Feb 4]. Available from: http://news.bbc.co.uk/2/hi/science/nature/3639126.stm
[5] Li HW, Roblin G, Liu H, Heller S (2003). Generation of hair cells by stepwise differentiation of embryonic stem cells. *Proc Natl Acad Sci USA* **100** (23): 13495–13500.
[6] Hori Y, Gu X, Xie X, Kim SK (2005). Differentiation of insulin-producing cells from human neural progenitor cells. *PLoS Med* **2**: e103.
[7] Levenberg S, Golub JS, Amit M, Itskovitz-Eldor J, Langer R (2002). Endothelial cells derived from human embryonic stem cells. *Proc Natl Acad Sci USA* **99** (7): 4391–4396.
[8] Lang LH. Embryonic stem cells treated with growth factor reverse hemophilia in mice: UNC researchers. *The University of North Carolina at Chapel Hill* [Internet]. 2005 Feb 15 [cited 2015 Feb 4]. Available from: http://www.unc.edu/news/archives/feb05/es021505.html

gene. Then, using somatic cell nuclear transfer (or therapeutic cloning) to make the cells, RAG-2 was given to the mice, partially restoring their missing immune system.[9]

Parkinson's disease

Israel's Dr. Benjamin Reubinoff transplanted hESCs into the brains of rats that did not have dopamine-producing nerve cells. Dopamine in a healthy body controls motion; loss of dopamine production in the brain is associated with the Parkinson's disease symptom of uncontrollable shaking. Implanted stem cells produced dopamine and brought significant improvements.[10]

Spinal cord injury paralysis

Using hESCs, Dr. Hans Keirstead in the Roman Reed Laboratory at University of California Irvine restored myelin insulation around damaged nerves, returning motion to partially paralyzed rats. This was the famous rat that walked again, thanks to research paid for by Roman's Law.[11]

---

[9] Rideout WM, Hochedlinger K, Kyba M, Daley GQ, Jaenisch R (2002). Correction of a genetic defect by nuclear transplantation and combined cell and gene therapy. *Cell* **109**: 17–27.
[10] Ryan C. Stem cell therapy for Parkinson's. *BBC News* [Internet]. 2004 Jun 30 [cited 2015 Feb 4]. Available from: http://news.bbc.co.uk/2/hi/health/3853791.stm
[11] Keirstead HS, Nistor G, Bernal G, et al. (2005). Human embryonic stem cell-derived oligodendrocyte progenitor cell transplants remyelinate and restore locomotion after spinal cord injury. *J Neurosci* **25** (19): 4694–4705.

13 SUFFER LITTLE CHILDREN

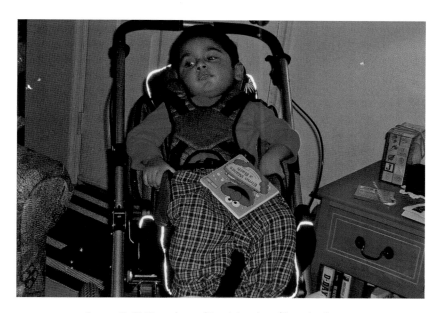

Pranav B: SMA took our friend, but he will not be forgotten.

"Similar to ALS/Lou Gehrig's disease, but in babies, Spinal Muscular Atrophy (SMA) eventually impacts every muscle in the body, hindering the ability to walk, sit, stand, eat, breathe, and swallow. SMA is degenerative: SMA is brutal, and, as of today, SMA is terminal."

— Bill Strong

"He is right here," said Pranav's mother Kavitha, in her lilting Indian accent, "Would you like to speak to him?" I realized she was talking from beside his bed.

For a moment I was actually afraid; I don't know why. Two-year-old Pranav had SMA, a disease like a slow spinal cord injury. Like the paralysis which afflicts my son Roman, SMA is not "catching," and in any event we were on the phone.

"Pranav, say hello to the nice gentleman," she said.

"How do you do?" came a surprisingly deep voice, very dignified.

Slight awkward pause.

"Um — Pranav, do you like sharks and dolphins? I used to swim with them in an aquarium called Marine World!"

Gwendolyn, Victoria and Bill Strong: Challenging SMA, terrifying disease which paralyzes and kills children, often before the age of 2.

I could feel him processing the information.

"*Finding Nemo*, very good," he said. We discussed Disneyland, and Sesame Street's Elmo, both of which he approved. But the conversation was tiring for him and we soon said goodbye.

Because he could no longer swallow, his food had to go into a tube in his stomach.

"We cannot bear to eat in front of him," his mother said, "He used to eat well until he was 14-month-old, and now he cannot."

Because his lungs choked frequently, Pranav had to be suctioned, many times a day. He knew it helped him and would request it. He slept with a mask puffing air into his lungs. He could not turn over in the bed; his mom and dad did that for him, every two hours all night.

Most of their friends were gone. "Maybe my fault," Kavitha said, "I find it hard to see kids running around when mine cannot. Also, some people don't feel comfortable being around a handicapped child. He is so bright and cheerful," she said, "He speaks three languages!"

Pranav wanted to become President someday, and to walk on the moon (in the year 2035, his mom said he always had a date for his goals). He got straight A's in his home school.

An experimental medicine, *hydroxyuria*, became available, and the family traveled from their home in New Jersey to Stanford University in California, four days a month — until Pranav could not travel any more. He caught an infection, which put him in the hospital for two months.

"Maybe the medicine helped a little," Kavitha said, "But it was not a cure."

Aside from that, and to go to the store, she had not left the house for 18 months. She gave up her career as a Certified Public Accountant to stay home to fight for her son's life. When Pranav had a cold, she would sit beside his bed all night long, keeping him breathing.

"As a mom, I do not object," she said, "But who wants to see your child suffer like that, before your very eyes, and you not be able to help?"

She sent me a picture of him, a handsome little boy resembling Rico Rodriguez of the television Series *Modern Family*. But whereas "Manny" on the hit show is always active, running around, even fencing with a sword, Pranav was paralyzed, confined to a black-framed wheelchair with a head brace, and a hospital bed with oxygen tanks underneath. Pranav's picture is on the wall beside my computer at home, where it will remain.

Time rushed by, years of constant effort, to raise funds, and protect scientific freedom.

One day I wrote Pranav's mother. In the past, when I needed a letter written to fight for research, Kavitha was always willing. Her letters were strong, full of fire and honesty.

This time there was no answer. She must be exhausted from the work of tending her son, I figured... or maybe just tired of hearing from me! The latter would be understandable. Sometimes I get sick of the sound of my own voice, endlessly exhorting people to do what they don't want to do, interfering with their too-busy lives, asking them to write just one more letter of support.

But this was not a chore, just something nice. Roman had given a speech about children affected by spinal cord maladies, and naturally he talked about Pranav. So I wrote her again.

Kavitha responded with the following, reprinted here with her permission.

"Hard day for us — one month since Pranav passed away. But very glad Roman mentioned Pranav and it had an impact.

"I knew Pranav was getting weaker; he wanted to spend more time on the computer than going out. But he was so cheerful and happy.

"On March 23, he started coming down with a cold — just a minor one; we had seen this a million times, managed it at home. His numbers were okay — good O_2, heart rate slightly high. We just did more chest physical therapy, lots of rest, more *bipap* (oxygen mask) time — nothing unusual.

"Night came and he was not comfortable, asking to be turned many times. Again not unusual. Numbers were fine, so we were not concerned.

"March 24 was better. Had school at home, did painting project as he was studying about DaVinci and Monet.

"March 25, he woke up in the morning and asked me to cancel school, unusual for him. For some reason Dad wanted to stay home, saying he would work from home that day.

"I stayed with Pranav at his bedside and he told me he was tired. I told him to relax and go to sleep. His numbers were okay, so no big worry.

"Suddenly at 12:03 he said, 'Mom, I can't breathe.' That was the last time he spoke. It happened in seconds. We did everything possible. The EMT (emergency medical technician) was there in five minutes. Pranav still had a pulse, so was taken to ER and flown to Children's Hospital in Philadelphia. He had gone into a coma. They did everything, but there was no response.

"We had to say goodbye on March 29. His eyes, kidneys, and tissue were donated.

"I cannot believe yet what happened. He still lives here with us, I can feel his presence.

"No child should be taken away so mercilessly by disease. He was so looking forward to doing great things in life. He made me a better person, a better mom.

"I wish I had just one more chance to make everything alright."

Pranav was gone.

He had wanted to make a movie starring Barack Obama. He reasoned that when he was old enough to do the film, Obama would no longer need to be President and would have the time.

Kavitha and her husband are shining examples of what parents can be. They gave everything they had to protect their son, but the medical answers were just not there.

How do we fight this evil thing, a disease that paralyzes children and slowly kills them?

One key technique for scientists may be to find common grounds among diseases, so attacking one may help defeat another. This is vital for "orphan diseases", meaning the number of people affected is not enough to justify Big Pharma risking millions developing a therapy.

At the University of California San Francisco, Jeremy Reiter was doing just that. Using the power of embryonic stem cells, he was attacking not only ALS, but also SMA in children and other disorders — *as well as challenging the biggest single reason medicines are expensive.*

Here is my unscientific summary of what he is up against, followed by his own words. The information is taken from his research grant at the CIRM website and also a telephone interview.

The disease does its damage in a three-step process.

1. Diseased genes (a mutation) affect the body's good stem cells, making them bad.
2. The now-diseased stem cells pass the sickness onto the motor nerves (neurons).

3. As motor nerves break down, the body loses basic abilities — a sick child "forgets" how to move, or breathe, or swallow. When enough motor neurons die, so does the child.

As Dr. Reiter puts it, "Diseases […] affect motor neurons that control muscle movement […]. Genetic […] mutations cause these diseases. We study motor neurons with these mutations […] to understand how the diseases [kill cells. To make healthy] motor neurons, we use embryonic stem cells."

Next, by developing a cell-based model of the disease (called a "disease in a dish"), Reiter can follow the progress of ALS or SMA, seeking ways to disrupt the deadly process. Also, he says, "Diseased motor neurons […] can be quickly screened with potential drugs to discover [what may] slow, halt, or reverse the […] damage."

Studying the life cycle of a disease in a petri dish (rather than in a patient) and being able to test drugs quickly could lower the costs of medications, quickly eliminating ones that don't work. This brings us quicker to cure by not wasting time going down the wrong paths. This "disease in a dish" approach could lower the cost of drug testing, potentially saving hundreds of millions of dollars for just one medication.

Like all CIRM researchers, Dr. Reiter shares what he discovers in both published papers and resources he develops: such as his process to make and compare motor neurons, both healthy and diseased, known as the Floxin system.

"We have made the resource of Floxin vectors and [24,000 cell lines] available to the research community. Application of the Floxin technology [will allow] modification of more than 4,500 genes in [embryonic stem cells]."

The California stem cell program is helping Dr. Reiter with a CIRM New Faculty II grant of $2,259,092. Will his work add another rung to the ladder that must be climbed to reach cure?

Only time will tell — and time, unfortunately, runs out all too quickly for patients with SMA. SMA is the number one genetic killer of children today.

Roman has a little friend, Gwendolyn Strong, a beautiful child with vivid blue eyes. She communicates by blinking, the only voluntary motion left to her. But her spirit shines through.

Her parents, Bill and Victoria Strong, are fighters too. As Bill said:

"At six-month-old, Gwendolyn was diagnosed with a terminal, degenerative genetic disease called spinal muscular atrophy, or SMA. We started the Gwendolyn Strong Foundation, a non-profit public charity, in 2009 to fight alongside her, increase global awareness of SMA, accelerate research, and support families impacted. To date, we've raised over $1,500,000 and formed 'smart partnerships' which have enabled us to advance advocacy efforts on Capitol Hill, provide 200 life-changing iPad grants to those with SMA, develop iPad communication app technology focused on those with limited dexterity.

We are also attempting to move the needle on critical SMA research through targeted, strategic funding of promising research programs including: Dr. Hans Keirstead's stem cell-derived motor neuron transplantation therapy currently awaiting FDA approval to commence the Phase I clinical trial process."

14 KLEIN MUST RESIGN!

Years ago in the Chicago Zoo, the wolf pack did something inexplicable. In the dead of winter, the wolves would suddenly run in a circle for 15 minutes or so — and then stop and go back to normal. It was as if they were practicing something, but what? For years, zoologists offered different reasons for this odd behavior — fitness exercise, warmth-making, group-think, hunting strategy — but nothing really made sense, until an old Russian wolf-hunter came to town.

It was a survival mechanism, he said. Sometimes the winter was too fierce, and their normal prey had burrowed under the snow, fast asleep. There was no food. When nothing moved across the frozen surface and death by starvation was very near, then the wolves would make their circle and begin to run. Faster and faster they would go, around and around, panting for breath — until one wolf tripped and fell, *and the other wolves ate him alive*.

With similar enthusiasm, the critics of the California stem cell program tracked our every move, eager for any slightest error they could use against us.

Mostly, they came at us sideways. Few would admit: "We were against Proposition 71 from the start, and intend to harass you every step of the way." Some had an irritating sense of self-righteousness, as if the program was something evil which should be destroyed for the good of the world. For some, it was just a job; they earned a salary tracking and attacking the stem cell program. Others were good government folks, who felt no tax dollar should be spent without rigorous inspection. I often felt like suggesting they investigate the military, which might have a little bit of waste in its Godzilla-sized budget. But that was perhaps too formidable a target.

But critics are useful, like the mosquitos that bite the caribou herds and keep them on the move. Denied a lengthy stay, the deer do not eat the vegetation completely to its roots.

Not to mention they had a right to be there.

The king of the critics was unquestionably David Jensen. His weblog, *California Stem Cell Report*, was (and remains) the single most widely read source of information about the program — even though it has a negative slant. Jensen worked 22 years for the *Sacramento Bee*, an opponent of the California stem cell program. His blog will run virtually any reasonable attack on the program, followed by CIRM's response, and of course the repetition of the charges. The good stuff? Not so much. It will be mentioned, but with nowhere near the same enthusiasm.

His work deserves respect, even if it makes me mutter under my breath. He prints both sides, and his work ethic is beyond reproach. He makes endless requests for more information from the CIRM, devours meeting transcripts, and writes pretty much everyday about it. Also, Jensen lives on a boat in the Sea of Cortez. I imagine he regards me as a Pollyanna-ish hyper-loyalist who finds it difficult to see anything wrong with the program — which is pretty much true. So we each know where the other stands, we sit together at the meetings sometimes, and as an ocean person he is welcome at my fire.

Sometimes the critics had good points to make. For instance, Marci Darnovsky of the Center for Genetics and Society often argued for stricter inspection of the egg donation process. When a woman is given a hormone treatment to increase her production of eggs, she may be putting her body at serious risk. But the treatment attacked was not something uniquely made up for researchers — it was part of the IVF procedure, used worldwide.

Her organization was a long term opponent of Proposition 71. But when it came to women's safety, she and the program were on the same page. Not only was a prohibition against donor compensation written into Proposition 71 (I disagreed with that; a woman should have the right to sell her eggs for research, just as she can sell them for reproductive purposes), but also one of the board's first acts was a three-day conference to study egg donation safety issues.

But if some of the criticism was positive and helpful, some was pure harassment. Dana Cody came close to admitting that. Leader of one of the unsuccessful lawsuits against CIRM, Ms. Cody was asked if the court's decision was a failure for her group. She said, "For 18 months, the California program was unable to do its research."

Research delayed is research denied. As pioneering scientist Irv Weissman put it, "Those who delay stem cell research [...] must take responsibility for lives that are lost, which might have been saved with more timely therapies."

"Do you know Jesse Reynolds?" asked Gloria, one morning before breakfast.

"Shave-headed guy, wears an ear ring, talks trash about the California stem cell program?"

"He is calling for Bob Klein to resign," said Gloria.

"Again?" I said.

Whenever I think of Jesse Reynolds and his group, the Center for Genetics and Society, I remember the time he turned out the lights on the California stem cell program.

He had been standing in the doorway of one of the CIRM meeting rooms. It was a telephone conference with people calling in from across the state. Jesse had arrived a moment late, and was leaning in the doorway before entering, waiting for a lull in the conversation.

He shifted his weight, accidentally brushing his hand across the light switch — and plunged the room into darkness. He fumbled for the switch, found it, turned the lights back on.

It was accidental, of course. But for me, it was symbolic of what he and the critics really wanted to do — to turn out the lights on the California stem cell program.

Reportedly in favor of embryonic stem cell research, Reynolds' group had nonetheless opposed Proposition 71 all though the campaign, and had never stopped.

And now he — together with John Simpson of the Foundation of Taxpayer and Consumer Rights and the *Sacramento Bee* — was asking Bob Klein to resign?

Excerpts from his organization's news release on January 29, 2007[1] outlined his accusations:

"Public interest group calls for CIRM Chair Robert Klein and Board Member John Reed to step down."

"The Center for Genetics and Society today called on California Institute for Regenerative Medicine (CIRM) Chair Robert Klein, and Board Member John Reed to step down following their admitted actions that appear to violate conflict-of-interest laws.

"This is a case of improper and potentially illegal efforts by a CIRM board member, following advice from CIRM Chair Robert Klein, to influence the allocation of public funds," said CGS policy analyst Jesse Reynolds. […]

"Reed has admitted that he lobbied CIRM staff in an attempt to reverse a decision rejecting a grant proposal by a scientist at the Burnham Institute of Medical Research, where he is president.

"CIRM Board Chair Robert Klein advised Reed to write the letter. […]

"[…] It's time for Klein to step down," [said Reynolds].

"The Center supports embryonic stem cell research but opposed the 2004 ballot initiative that established CIRM [and just one year ago] in January 2006 called on Klein to resign.[2]

"The *Sacramento Bee's* editorial page and the Foundation for Taxpayer & Consumer Rights have also called for Klein's resignation."

If you only heard the critics' side, it sounded bad. Only — I was there. If you ever have the good fortune of attending a meeting of the governing board, the ICOC, please do so. Take a seat in one of the rows and rows of folding chairs. You will probably see me, a white-haired person in a gray suit.

And at the front? There is where the ICOC sits: our 29 members, every one a champion of their field. White cloth-decked tables, microphones, water glasses, and folded cardboard name tags.

And on this special day, a nametag read: Dr. John Reed. I wished there was a little spotlight on it, because he had been absent for more than a year, and now was back.

Dr. Reed (no relation, though I would be glad to call him kin) was Chief Executive Officer of the Burnham Institute for Medical Research in San Diego.

His scientific field has the grim title of *apoptosis*, or cell death. Understanding how and why cells die is crucial in the battle for cure. If cells die too soon, you have a stroke; if they do not die soon enough and multiply out of control, that's cancer.

The Institute for Scientific Information identified Dr. Reed as the world's most frequently cited scientist in the decade from 1995 to 2005. He authored over 800 research

---

[1] Public interest group calls for CIRM Chair Robert Klein and Board Member John Reed to step down [Internet]. 2007 Nov 29 [cited 2015 Feb 4]. Available from: http://www.geneticsandsociety.org/article.php?id=3808

[2] Public interest group gives the California Institute for Regenerative Medicine poor grades for its performance [Internet]. 2006 Jan 18 [cited 2015 Feb 4]. Available from: http://www.geneticsandsociety.org/article.php?id=315

publications, wrote more than 50 book chapters, and is the named inventor for more than 70 patents.

Why was he gone for a year? Dr. Reed voluntarily absented himself from ICOC deliberations while California's Fair Political Practices Commission (FPPC) investigated something he did.

ICOC members like Dr. Reed are forbidden to participate in decisions which might bring financial grants to their home organizations.

David Smotrich, a scientist with Reed's organization had applied for a grant through the California program. The proposal was reviewed by a committee of scientists from out of state, then voted on by the ICOC. As required, Dr. Reed took no part in that decision. So far, so good.

The project was judged excellent, earning one of the highest scores given. But when the CIRM staff and legal department did their "due diligence", making sure the grant fit every rule and requirement, they found what might be a fatal flaw.

One of the requirements for this grant was that the recipient had to be a full-time employee of his or her home institution. Burnham Institute was asking for the money, therefore Dr. Smotrich had to be a full-time employee of Burnham.

Was he? Maybe yes, maybe no — it depended on how you defined full-time. According to an article published in the *San Diego Union-Tribune* in November 22, 2007[3], Smotrich had a "full faculty position and privileges at the Burnham [a small institution which] does not have a hospital, so it does not have full-time clinicians."

Suddenly, the whole project was dead — on a potentially disputable technicality.

There was no official review process for complaints. Also, the decision had already been made, so was it allowable now for Dr. Reed to voice his concerns? It was clear he could not use his power of voting or discussion to influence the project's chances when it was before the ICOC. But after the decision was made, could he speak up then, like at a football game where coaches do not interfere during the action but may argue with the referee after the play?

Was this a new situation, not yet covered by CIRM laws?

Also, there was this, from the ICOC's conflict of interest policy:

"[…] Executive officers of research institutions [who, as part of their responsibilities], oversee and advise researchers in their institution […] shall not be deemed to have a conflict of interest (under this provision). Recusal [or not voting on grants to their organization], however, is required […]."

This was official policy, and it seemed to me (a non-lawyer) to contain wiggle room. Did Dr. Reed's right to "oversee and advise" allow him to explain his institution's definition of a full-time employee? That appeared to be a gray area.

With a new organization, rules are seldom crystal clear. They must be worked out over time, as was being done. The process of making the regulations had begun with

---

[3]Somers T. Top Burnham official accused of conflict of interest. *San Diego Union-Tribune* [Internet]. 2007 Nov 22 [cited 2015 Feb 4]. Available from: http://legacy.utsandiego.com/news/business/biotech/20071122-9999-1n22stems.html

the writing and passage of Proposition 71, but it did not end there. After the voters said yes, there was an immediate major conference with the National Academy of Sciences to figure out best practices, and the ICOC (with input from the Sacramento legislature) had been working on the rules ever since.

Feeling a mistake was being made, Dr. Reed contacted Chairman Klein, asking him what to do.

Bob said he didn't have the ability to evaluate the information in the administrative review, so if Dr. Reed thought there were errors, he should make them public by writing a letter to the scientific staff.

This was done. Dr. Reed wrote a six-and-a-half page letter detailing the problem, and sent it to Dr. Arlene Chiu, Head of the Science Department. That turned out to be a mistake.

The CIRM legal department immediately caught the error and stopped it cold. General Counsel Tamara Pachter (now working with the program) pointed out that CIRM had no appeals process, (something that would be later corrected) and therefore the matter could not be considered.

The grant did not receive funding. But that was not the end of the matter.

Enter John Simpson of Consumer Watchdog, a frequent critic of our program. He filed a complaint with California's Fair Political Practices Committee (FPPC), alleging a conflict of interest, and called for the resignations of both Dr. Reed and Chairman Klein.

To me, such a penalty was like the electric chair for a parking ticket, but Simpson felt otherwise.

As Simpson was quoted as saying in the *California Stem Cell Report*, November 21, 2007: "This is not trivial […]. When you hand out millions of dollars in public money, you have to play by the rules. [Smotrich] didn't meet the eligibility rules […] and waiving them would have been unfair to everyone else."

How did CIRM feel about it?

In an interview with David Jensen, CIRM Acting President Dr. Rich Murphy said, "It is important to remember that Dr. Reed sent his letter after the ICOC had approved the grant […]. [He] mistakenly believed that conflict rules would not prevent him from providing technical information regarding the status of a faculty member […]. As soon as CIRM staff received the letter, counsel advised Dr. Reed that he must refrain from contacting the staff and board members regarding a grant to the Burnham and advised staff to disregard his letter. It therefore had no effect on CIRM's process […]."

The Fair Political Practices Commission studied the accusation for more than a year.

Meanwhile, the rules were clarified, and a complaint process was set up, whereby scientists denied grants could voice their grievances. Dr. Reed voluntarily took himself off the board until the matter was resolved. His absence was like a sports team forced to play with a top athlete sidelined. In addition to being incredibly intelligent, Dr. Reed was dedicated, clear, and caring.

Everyone waited. At last, the answer came in a public letter from Kourtney C. Vaccaro, Chief of the Enforcement Division of the Fair Political Practices Commission, on January 7, 2009:

"In our view, by submitting a 'letter of appeal' to CIRM staff, Dr. Reed intended to influence a decision that had the potential to affect his economic interests. However, [...] it appears that Dr. Reed attempted to influence a prior-made governmental decision that could not be appealed [...]. Thus, although this matter raises ethical concerns, we are closing this matter with a warning letter [...]. Dr. Reed is advised that failure to comply with provisions of the Act can result in an enforcement action [...], including monetary penalties of up to $5,000 for each violation [...]."

That is harsh language, especially when you consider it took the Commission itself, experts in the field, over a year to figure it out. There was no penalty. Dr. Reed had objected to a decision which could not be undone, therefore no crime had been committed.

None of this implies that Reynolds and Simpson were wrong to raise their points. Democracy depends on critics; like antibodies, their reactions are often painful to the body politic, yet they perform a vital function. All the fuss and furor made everyone stop and think.

Above all, the system worked. CIRM's legal department spotted the error and blocked it, while the transparent nature of our program (allowing and encouraging citizen involvement) made the blunder clear, accessible, and fixable.

The incident was done. The Fair Political Practices Commission had closed its file.

There he was, skinny and tall, shaking hands with his fellow board members. One of America's greatest scientists was back with the ICOC. His life had moved forward: the molehill of controversy small beside the mountain of accomplishment, benefiting all.

Welcome home, Dr. Reed.

And Bob Klein? He was exactly where he ought to be, doing his work as Chairman of the Board.

P.S.: Who were these 29 board members about whom the opposition was so upset?

1. David Baltimore, Nobel Prize winner at age 37, President, California Institute of Technology.
2. Robert Birgenau, Chancellor of the University of California Berkeley, former President of the University of Toronto.
3. Keith Black, Director of Neurosurgery at Cedars-Sinai, scientifically published at age 17.
4. Susan Bryant, Dean of the School of Biological Sciences at University of California Irvine, has published more than 100 scientific papers.
5. Michael Friedman, Distinguished Clinical Researcher, Chief Executive Officer of City of Hope.
6. Michael Goldberg, life sciences businessman, Inc., Magazine's Entrepreneur of the Year.
7. Brian Henderson, Dean of the Keck School of Medicine at the University of California, considered one of the world's pre-eminent authorities in cancer epidemiology.
8. Edward Holmes, Dean of the School of Medicine at University of California San Diego, continuously funded by the National Institutes of Health as a researcher from 1975–2000.
9. David Kessler, seven years Commissioner of the US Food and Drug Administration (FDA), Dean of the School of Medicine at University of California San Francisco.
10. Sherry Lansing, patient advocate for cancer, recently retired chair of the Motion Pictures Group of Paramount Pictures, founder of Stop Cancer as well as numerous philanthropies .

11. Gerald Levy, known for his research on the heart and thyroid gland, Vice Chancellor of medical sciences at University of California Los Angeles, previous-President of the Association of Professors of Medicine.
12. Ted Love, biotech expert, President and Chief Executive Officer of Nuvelo, former Vice President of product development and regulatory affairs at Genentech.
13. Dr. Richard Murphy, multiple-award-winning professor, neurotrophin researcher, President and Chief Executive Officer of the Salk Institute.
14. Tina Nova, renowned for research and earned patents in biotechnology, winner of the Athena Pinnacle Award, a Soroptimist Woman of Distinction.
15. Ed Penhoet, Vice Chairman on the ICOC, President of the Gordon and Betty Moore Foundation, immediate past Dean of the School of Public Health, University of California Berkeley.
16. Phillip Pizzo, Dean of Stanford School of Medicine, founder of the Children's Inn, a temporary home for children undergoing treatment at the National Institutes of Health.
17. Claire Pomeroy, Dean of the School of Medicine at University of California Davis, Professor of Internal Medicine, Microbiology, and Immunology.
18. Francisco Prieto, diabetes research expert, President of the Sacramento Sierra chapter for the American Diabetes Association.
19. John Reed, pioneer in cellular research, recognized by the Institute for Scientific Information as the world's most cited scientist in all areas of research from 1997–1999.
20. Joan Samuelson, founder and President of Parkinson's Action Network, patient advocate representative to the Medicare Consumer Advisory Committee.
21. David Serrano Sewell, patient advocate for ALS and Multiple Sclerosis communities, Deputy City Attorney of San Francisco.
22. Jeff Sheehy, HIV/AIDS advisor to San Francisco Mayor Gavin Newsom, Deputy Director for Communications at the AIDS Research Institute at University of California San Francisco.
23. Jonathan Shestack, founder of Cure Autism Now, the largest provider of support for autism research and resources in the country.
24. Oswald Steward, Chairman and Director of the Reeve-Irvine Research Center for Spinal Cord Injury at University of California Irvine. (In pride I must add that he also oversees the Roman Reed Spinal Cord Injury Research Act, named after my son!)
25. Leon Thal, leader of a national consortium of more than 80 centers called the Alzheimer's Disease Cooperative Study, Chairman and Professor of the Department of Neurosciences at University of California San Diego.
26. Gayle Wilson, California's First Lady from 1991–1999, member of board of directors of Gilead Sciences Inc., member of board of trustees of the California Institute of Technology, holds a bachelor's degree in biology from Stanford University.
27. Janet Wright, patient advocate for heart disease, practices invasive cardiology as a partner of Northstate Cardiology Consultants, board member of the American College of Cardiology.
28. Marci Feit, newest member of the ICOC board, appointed by Lieutenant Governor Cruz Bustamante to replace Phyllis Preciado of Fresno (Dr. Preciado had to sign off

from the board when she moved to Oregon), Chief Executive Officer of the leading non-profit hospital in the Tri Valley Area. (As Bob Klein said about Ms. Feit, "She has spent her entire 32-year career devoted to serving patients.")
29. Bob Klein, Chairman.

Champions, all.

In 2007, Sherry Lansing was awarded the Jean Hersholt Humanitarian Award by the Academy of Motion Picture Arts and Sciences. Her dedication to the battle against cancer made her a natural leader on the Independent Citizens Oversight Committee. She embodies the care and determination which is typical on the board.

ICOC board member Dr. Ann-Marie Duliege is a senior executive with over 20 years experience in the pharmaceutical industry, currently chief of strategic development and head of immune-oncology at ChemoCentryx, Inc.

15 LAST DAY IN THE WOLVERINE STATE

The king of the weasel tribe, wolverines are legitimately tough. Weighing about 40 pounds, they are small but very fast and have incredibly powerful jaws. They have been observed driving wolves and bears off their kill. Even today, the Michigan state animal may hide or destroy a human hunter's traps, before breaking into his cabin and fouling his supplies, harassing the rival until he leaves the territory.

It was 4:30 a.m., just before the 2008 Presidential elections, and my seventh (and last) day in Michigan. I scrunched closer to my desk in the office of Michigan Citizens for Stem Cell Research and Cures, and tried to think of one more stem cell message or one more group to contact.

Let's see... cancer? I typed the word into the Google search bar, added "stem cells" and "University of Michigan", and —

Okay, here we go.

One article from the *Muskegon Chronicle*[1] returned this information: "Cancer researcher Michael Clarke left his job at the University of Michigan in 2005 to take a position at the Stanford Cancer Center [in California]. Clarke oversaw laboratory studies that were the first to isolate stem cells from breast tumors [...]."

Isolating stem cells from breast tumors? Heavyweight stuff — but why would an obviously excellent researcher leave his home state?

I had never met Dr. Clarke, but I could guess. Almost certainly he wanted freedom to do his research, and money to do it with. California offered grants to embryonic stem cell researchers and protected their rights of research. In Michigan, the same work could get you thrown in jail for up to ten years, and fined up to ten million dollars.[2]

Freedom, funding, and the hope of cure versus jail, fines, and suffering for the patients — that was Michigan's choice, and America's.

In the Presidential race, Democratic candidate Barack Obama had pledged he would reverse the Bush stem cell policy. Republican candidate John McCain (formerly a supporter of embryonic stem cell research) had changed his position to please the

---

[1] Lupo L. Embryonic stem cells could be ballot issue. *Muskegon Chronicle* [Internet]. 2007 Oct 28 [cited 2015 Feb 4]. Available from: http://blog.mlive.com/chronicle/2007/10/embryonic_stem_cells_could_be.html

[2] Human cloning laws [Internet] [updated 2008 Jan; cited 2015 Feb 4]. Available from: http://www.ncsl.org/research/health/human-cloning-laws.aspx

Religious Right, saying he would sign the South Dakota Women's Health and Human Life Protection Act (HB 1215), which would criminalize the research.[3]

Both the McCain Presidential platform and the Republican national policy called for an outright ban on embryonic stem cell research, both public and private. And his choice for Vice President? Sara Palin, another known opponent of the research.

But I was here in Michigan to advocate for Proposal 2 — that federally permitted stem cell research would also be legal in the wolverine state.

Patient advocates like Laura Jackson put a human face on the controversy. Jackson, a young woman paralyzed in a cheerleading accident, spoke to the Michigan legislature, stressing that "embryonic stem research […] represents the true pro-life position because it could save human lives and eliminate human suffering."[4]

With too many obstructionists in office to allow positive legislation to pass (they blocked State Representative Andy Meisner's bill for more than a year), patient advocates organized a citizens' initiative called CureMichigan — and the Wolverine State responded.

In 20 weeks, Michiganians gathered almost 600,000 verifiable signatures and were on the ballot.

Gloria and I flew out to volunteer. (My airfare and hotel was covered by a Michigan stem cell group.) I was blissful. For seven days I could focus on the fight for Michigan stem cell research freedom, surrounded by people who felt exactly as I did. Since I like to work the early shift, starting before three in the morning, they gave me my own key to the office. Sometimes when I turned the key, the day crew was still there from the night before.

I revised my 50-stem-cell-letter-writing kit, adjusting it for Michigan. Volunteers could pick and choose from the half-hundred sample letters, modifying as they wished.

Perhaps the most useful tool was the *one-sentence political action letter*. Simplicity itself, it was just, "I_____ oppose/support _____." That and your contact information (so the reader knows you are a real person) will get the job done. You can add more of course, but that one sentence is what matters, and many political aides stop reading once they get there because that is all they really need to know.

The other side was busy too. MiCause (Michigan Citizens Against Unrestricted Science and Experimentation) led the anti-research effort, joined by Right to Life Michigan and the Michigan Catholic Conference.

As usual, the anti-science folk distorted the science. Even the name of their organization implied a falsehood — "Michigan Citizens Against *Unrestricted* Science and Experimentation"? Stem cell research came with all kinds of restrictions, including paragraphs written into Proposal 2. Glance at the actual text of Proposal 2 below. Unrestricted research?

---

[3] McCain would sign South Dakota abortion ban [Internet]. 2006 Mar 12 [cited 2015 Feb 4]. Available from: http://nomoreapples.blogspot.sg/2006/03/mccain-would-sign-south-dakota.html

[4] Lessenberry J. Stemming the tide. *Metrotimes* [Internet]. 2007 Dec 26 [cited 2015 Feb 4]. Available from: http://www.metrotimes.com/detroit/stemming-the-tide/Content?oid=2190573

A Proposal to Amend the Constitution of the State of Michigan by adding a new Article I, Section 27 as follows:

> Article I, Section 27.(1) Nothing in this section shall alter Michigan's current prohibition on human cloning .
>
> (2) To ensure that Michigan citizens have access to stem cell therapies and cures, and to ensure that physicians and researchers can conduct the most promising forms of medical research in this state, and that all such research is conducted safely and ethically, any research permitted under federal law on human embryos may be conducted in Michigan , subject to the requirements of federal law and only the following additional limitations and requirements:
>
> (a) No stem cells may be taken from a human embryo more than 14 days after cell division begins; provided, however, that time during which an embryo is frozen does not count against this 14-day limit.
>
> (b) The human embryos were created for the purpose of fertility treatment and, with voluntary and informed consent, documented in writing, the person seeking fertility treatment chose to donate the embryos for research; and
>
> > (i) The embryos were in excess of the clinical need of the person seeking the fertility treatment and would otherwise be discarded unless they are used for research; or
> >
> > (ii) The embryos were not suitable for implantation and would otherwise be discarded unless they are used for research.
>
> (c) No person may, for valuable consideration, purchase or sell human embryos for stem cell research or stem cell therapies and cures.
>
> (d) All stem cell research and all stem cell therapies and cures must be conducted and provided in accordance with state and local laws of general applicability, including but not limited to laws concerning scientific and medical practices and patient safety and privacy [...].

As for the overall accuracy of the anti-research side? Former Representative Joe Schwarz, Republican, said, "They simply are not telling the truth. [...] I cannot use the word inaccurate, this is worse than inaccurate."

Sometimes it was hard to tell if they were talking about stem cell research!

"Human life is not a commodity that can be created and killed in an effort to generate profits and patents," said Michigan Catholic Conference President and CEO Sister Monica Kostielney in a statement on October 12, 2007, to the *Kalamazoo Gazette*.

What an astonishingly dishonest sentence. Never in history has embryonic stem cell research created or killed anybody. If there is no pregnancy, no womb, and no baby, where is the human?

Their propaganda was everywhere.

On April 11, 2007, it was reported by the *Grand Rapids Press* that "more than 500,000 Catholic households statewide […] were mailed packets from the Michigan Catholic Conference signed by the state's seven bishops. A 12 minute DVD and brochure outlined the church's opposition to embryonic stem cell research […]."

That mailing, complete with DVDs, was sent out — twice — during the campaign, to *every Catholic family in the state*.

We had started out ahead in the polls: 50% for, 41% against. But the relentless propaganda had an effect, and it had leveled off to 47:43.

The advertisements were expensive falsehoods, and some were quite ridiculous.

One showed a man in a cow costume, presumably their idea of an example of what happens with stem cell research?

Another showed an actor shoveling wheelbarrows full of greenbacks, while the announcer's voice said Michigan couldn't afford to spend millions on stem cell research — even though *Proposal 2 asked for no money at all*.

The opposition pretended the bill was about cloning, exploiting the fear of a horror movie fantasy. The announcer intoned like the voice of doom, while the video showed fake companies with made-up names like "Crop Clones" on them.

If they were worried about cloning, all they had to do is to *read the first sentence of the bill*:

"Article I, Section 27. (1) **Nothing in this section shall alter Michigan's current prohibition on human cloning.**"

For me, the cruelest deception of all is the propagated falsehood that we don't need embryonic stem cell research because adult stem cell research is already bringing cures to 70 (or 58 or 100 or whatever number they were using that day) chronic diseases. Adult stem cell research is valuable; where it works, it should be used. But it is shameful to pretend adult stem cells already have a cure for Parkinson's, spinal cord injury, cancer, Alzheimer's, Lou Gehrig's disease, and more — when there is none.

And on our side?

Former President *Bill Clinton* came to Michigan. Clinton stated that embryonic stem cell research "was the pro-life position […]. Pro-life, pro-health, pro-science, and definitely pro-Michigan. There isn't a person here who doesn't know someone who could be benefited by this work […]. This is about letting Michigan do what can be done in almost every other state."

Also present and accounted for was the Californian who had done more than any other human on earth to advance stem cell research — Bob Klein. There was an old television show called the *Six Million Dollar Man*: here was the Six Billion Dollar Man! His leadership helped California invest in stem cell research: three billion for the cost and three billion in interest. He had his company to run, Klein Financial Enterprises, Inc., a real estate endeavor, plus his job as chair of the board of directors for the California stem cell program. His schedule barely left him time to breathe. But he was here to help Michigan raise funds for its fight.

Bob shook hands with A. Alfred Taubman, Michigan's hero. Taubman put his resources on the line for biomedical research in general and stem cell research in particular. Without his vision and generosity, there would have been no Proposal 2 campaign.

Both men applauded the accomplishments of Marcia Baum, Mary Smyka, and Sophie Eichner of Michigan Citizens for Stem Cell Research and Cures. In the past year, these folks made 170 visits and presentations on stem cell research to churches and social clubs — and in the last 30 days of the campaign they did an additional 100!

"Everybody knows someone with a terrible condition who cannot be cured right now," said Sophie Eichner of Michigan for Stem Cell Research and Cures, "Like my husband, who has cancer and diabetes, and my mother, who has rheumatoid arthritis."

People like Edsel Ford, great-grandson of Henry Ford, who has a son with type 1 diabetes.

Scientists like Sean Morrison, Doug Engel, and Jack Mosher were not sitting idly by, either — they raised their voices on behalf of research. Men and women from both sides of the political aisle stood up and were counted, even if there was a price to be paid.

Joe Schwarz, for example: Republican and a former Member of Congress. Joe sacrificed his political career to advance stem cell research. In Michigan, people in government were often afraid to speak out against the powerful "right to life" special interest group. If Joe had kept quiet about supporting stem cells, he would very likely have won his race. But he showed the "content of his character," as Martin Luther King said, and spoke out strong for stem cell research. The Religious Right targeted and defeated him — but Michigan and America are the richer for his integrity.

Rick Johnson, another Republican and former speaker of the house, was a Cure Michigan board member. Democratic champions definitely stood tall. People like Governor Jennifer Granholm, Senator Gretchen Whitmer, and other Wolverine State legislators were fighting in the trenches. Wheelchair warrior Senator Carl Levin made time to help.

Patient advocate Danny Heumann had been fighting to ease stem cell restrictions since 2004. Paralyzed in body but never in spirit, Danny is a motivational speaker, with so much energy the wheels of his chair almost catch fire.

We had outstanding young leaders like Mark Burton, Chair of the Committee to pass Proposal 2: vibrant, and vocal. Behind the scenes, folks made things happen, such as Amber Shinn, communications director for the effort. You might not see her in the papers, but you would see her work, bringing people together. When she and I spoke, her first thought was for others, and she reminded, "Don't forget to mention Laura DePotter, The Rossman Group, Minda Nyquist, Chris DeWitt, Erica Barrera, Kelly Danczyk" — and a bunch more.

Have you noticed that sometimes the people who do the most work get the least credit? Dedicating a piece of their lives to something great were John Simon, Rick Johnson, Jill Alper, Mark Mellman, Joe Slade White, Chris DeWitt, Mark Pischea, Kelly Rossman, Cheryl Bergman, Heather Ricketts, Carrie Jones, Kris Caswell, Karen DeMott, Laura DePotter, Mary Anne Servian, Traci Riehl, Brett DiResta, and many more outstanding individuals.

Right now, Michigan's Proposal 2 was the center of the stem cell universe.

But the Wolverine State was not alone. Across the nation, similar stem cell firestorms were being fought. Sometimes we won and sometimes we didn't.

In Missouri, 2006, priests at fairgrounds gave away plastic models of fetuses, saying scientists wanted to experiment on babies like these.

But a group called Missouricures led the battle for Amendment 2 to change anti-research laws. People like Donn Rubin, John Danforth, Liz DeLaperouse, and David Eagleton united the state's advocates. Michael J. Fox encouraged the research, despite insults from the mouth that roared, Rush Limbaugh. Senatorial candidates Clair McCaskill and Jim Talent had an epic battle, with stem cells as a point of disagreement. In the end, Talent changed his mind and came over to the pro-research side. Good on him.

Opposing Amendment 2 were actress Patricia Heaton of *Everybody Loves Raymond* and Jim Caviezel, star of Mel Gibson's movie, *The Passion of the Christ*.

You can see both their television advertisement and Michael J. Fox's at this URL: http://www.foxnews.com/story/2006/10/26/patricia-heaton-not-everyone-loves-michael-j-fox/

Missouri won, as did Connecticut, New York, Maryland, Massachusetts, and Illinois.

New Jersey both won and lost: under the leadership of Governors Jon Corzine and Richard Codey, New Jersey tried for a $450 million program. Campaign leader Russ Oster worked ceaselessly, as did Dr. Wise Young and his right hand person, Patricia Morton, mother of paralyzed Peter Morton, but it was not enough. The year before, Governor Corzine had authorized the expenditure of $150 million for stem cell laboratory construction, but research grant funding was all too limited. Unfortunately, a tactical error was made by offering a $450 million research bill in a non-Presidential year. Republicans always turn out to vote in "off-year elections" while Democrats all too often do not — we lost by a small margin.

But still the fights were being fought. Every battle won brings us closer to the day when funding regenerative medicine will be the accepted norm in every state across this land.

On our last day in Michigan, Gloria and I visited to the Detroit Zoo, one of the best in the world. Gloria's favorite moment was watching a polar bear swimming underwater in a glass-walled tank. But I wanted so much to see my favorite animal, the wolverine. I had never seen one in real life, just pictures, and now, here we were in one of the last strongholds of the amazing animal.

Gloria had injured her knee in a fall, so she was on a motorized vehicle, but I could not wait for the slow-moving tram.

Around a grassy knoll, I ran, breathing hard —

There it was, a wolverine, standing on its back legs, arms out, crouching like a martial artist. It looked at me, its brown and black fur rippled in the faint breeze, and its narrow head aimed at me. It seemed absolutely fearless and calm, embodying the spirit of the untamed wild.

One week later, Proposal 2 was approved by the voters of Michigan.
And today?

The front page of University of Michigan's stem cell research website, http://www.stemcellresearch.umich.edu/, reads: "The University of Michigan has recently emerged as a national leader in the three main types of stem cell research: embryonic, adult, and reprogrammed cells […]."

16 GETTING UP IN THE MORNING AND GOING TO — WASHINGTON?

Rep. Diana DeGette (D, CO), Dan Perry, Alliance for Aging Research and CAMR; Rep. Jim Langevin (D, RI) and Rep. Mike Castle (R, DL).

Before my son's accident, I never given much thought to people in wheelchairs. They just sat around a lot, I figured. It never occurred to me to ask how a paralyzed person gets out of bed.

5:30 a.m.; I let myself into Roman's house.

A giant shadow rose. Claws clicked on hardwood floors, as Cali (short for California golden bear), all 200 pounds of golden mastiff, came to say good morning.

"Yes, yes, I love you too," I whispered, "I will take you on your walk in a minute, soon as I find your leash."

When we returned from our round-the-block adventure, Cali stayed out back. I went into the house again, as per our arrangement: one walk in exchange for her leaving her place of comfort.

I did not want to put the lights on just yet, but I knew where the bathroom was. I sought and located the "shower chair," a wheelchair with a big round hole in the seat,

In the Gold Room: Senator Dianne Feinstein, Roman Reed, Zach Hall, former President of CIRM and Don Reed: On the day President Obama overturned Bush stem cell policy.

Senator Orrin Hatch, Don and Roman Reed. Republican champion Hatch said: Stem cell research is "the ultimate pro-life position; being pro-life is not just caring for the unborn but caring for those who are living."

and a rack underneath for a plastic bucket. I put some water in the bucket and pushed the chair into Roman's room.

"Give me ten minutes," said Roman.

"Got it," I said, as if the request was something different from what he said every day, and pulled the powerchair back from the bedside, having first ascertained Roman's feet were not in it.

President Barack Obama shares a moment with Roman Reed.

With ten minutes to kill, I went out into the garage, closed the door, turned the lights on — ahh, vision! — found wet clothes in the washer, transferred them to the drier, turned it on, tossed another hamper full of laundry into the washer, added soap, pushed buttons. With the drier vibrating and another load in the washer, I turned off the garage light and came back inside.

In the bathroom I slid my hand through the handle of the big sliding door, adapted for disability use, so it could be opened without a grip. The bathroom was specially designed: central drain in the sloping floor, no barrier for the shower, a cabinet/mirror with a roll-under cave.

Some people think "handicap accessible" just means putting a bar on the bathroom wall. That might be helpful for paraplegics, who were "only" paralyzed from the waist down, but it wouldn't be enough for a quadriplegic (more correctly "tetraplegic"), with paralysis in both upper and lower body.

When you see those inspiring videos of a paralyzed person hauling a wheelchair up a mountainside, those are paraplegics. Some of our paraplegic friends could sling themselves around like Tarzan. They had chairs with almost no backs to them, and upper bodies rippling with muscle from the constant effort of wheeling their manual chairs. (Paraplegics had their own worries, like wearing out their shoulder joints from overuse.) Quadriplegics used powerchairs or were pushed.

The public knew so little about paralysis, like why a handicapped parking space was needed. A paralyzed person needs *room on the side of the vehicle* so a sliding ramp can emerge. The wheelchair driver had to go down the ramp, and then turn right or left. If

Speaker of House Nancy Pelosi, a strong supporter of stem cell research, side by side with Zach Hall, former President of CIRM.

Gloria, Roman and Don Reed outside White House.

someone parked too close to Roman's van, he was just stuck, sometimes for hours until the person returned.

On the left side of the bathroom was the cabinet full of blankets, towels, and medical supplies like "Magic Bullet" suppositories and single-use plastic gloves.

Two men not defined by their wheelchairs, Roman Reed and, former President Franklin Delano Roosevelt.

Thankfully, Roman no longer had to be catheterized for urination. Inserting those long red tubes every four hours had been a time-consuming nightmare. Fiercely independent, Roman had dispensed with catheterization altogether, carrying a discreet plastic cup for the purpose.

But the shower-bowel program was still a team effort, which is why many quadriplegics only do it every other day or so. Terri and I shared the responsibility.

I put on one plastic glove, unwrapped the suppository, and took it into Roman's room.

When I came out a moment later, I tossed the glove, washed my hands, went into the living room and collapsed on the couch for 20 minutes until —

"Okay, Dad!"

My son was half way up on his side, using the strength of his arms and shoulders.

"Forearm," he said. I grabbed his arm with my right hand, leaned back, at the same instant using my left hand to scoot his legs off the side of the bed.

The practiced maneuver went smoothly. He was seated now, balanced on the edge of the bed. When first paralyzed, sitting up had been impossible. Therapists had to fasten him on a mechanical table and slowly rotate it to vertical, while his blood flow readjusted from horizontal. It was like a brain-freeze headache, he said, a pain so intense it had made him pass out.

Pushing the shower chair closer to Roman, I unfastened one arm rest, set the brakes, grabbed the side of the chair and got ready, one foot back, bracing myself.

Roman reached to the arm rest on the far side of the chair, planted his left hand on the near side of the chair seat, and prepared to *transfer*: shifting his weight from bed to chair.

The "transfer" was an essential part of life in disability: relocating the body from bed to chair. It sounds easy: just scoot your rear end from bed to chair or wherever. Imagine a gymnast, balancing on hands, lower body off the mat, legs up and toes pointed. Now try to imagine doing that with no stomach or leg muscles.

A lurching heave of his upper body gained him an inch. Another Herculean exertion, another inch — again, again — scoot, he was suddenly aboard the chair.

As always I felt the urge to break into applause, but restrained myself. Roman dislikes fuss.

Being able to transfer was huge. At first I had to lift him for every transfer, every time he needed to go from bed to chair, chair to car. Roman is a big man, at six feet four inches and 235 pounds; my back was constantly on fire. Worse, what if I became permanently injured myself; who would help my son?

But Roman lifted weights (using Velcro strips to compensate for having no grip) three times a week, regaining upper body strength until he could transfer himself, as he had just done.

"Armrest," he snapped. I found it on the floor, reinstalled it with a click. I backed up (easier to pull than push), tugging the chair across the room, down the short hallway.

In the bathroom, I put a towel across his shoulders and turned on the room heater. His body does not regulate temperature and he sometimes shakes with cold, even in hot weather.

"Phone," said Roman. I got it, and he went to work on his emails. Roman works constantly on the cure research effort, and there is a lot to it.

Usually I leave him alone at this time. I would go into the other half of the house, say good morning to Terri, help the boys get ready for school. I could hear preliminary noises already; someone flushed a toilet. How easy it is for most of us to use the restroom.

But today I had a question.

"Rome?" I said. He looked up. "How would you like to go to the White House?"

I had almost deleted the email. Why would I be getting a note from the White House? Probably just a fundraiser, I figured. But then I casually opened it, glanced at it — and froze.

Would I like to be present when newly elected President Barack Obama reversed the Bush stem cell restrictions? I could bring two guests.

I could not believe it, at first, but I called up Kareem Dale, the President's Disability Director, with whom I had worked a couple times, and asked him the obvious questions. Yes, it was for real. No, there was no assistance with travel costs.

Which is why on March 9, 2009, Roman, Gloria and I credit-carded our way to Washington.

"This is the most secure room in the world," said the Military Police person, "Concrete walls, guard planes on patrol overhead." It wasn't like the old days, when you could just walk into the White House and take the tour, and maybe wave at the President when he walked by. After 9–11, everything changed. When those planes crashed into the Twin Towers, something gentle and trusting went out of our lives.

But here we were, in the White House, on our way to meet the President.

After multiple security checks, we hurried down long tunnels, until we came at last to —

The Gold Room. High-ceilinged, full of scientists, doctors, politicians, and patient advocates.

Bob Klein was there, of course — I found out later he had been invited twice! And there was Zach Hall, whiskery and smiling, the President of the California stem cell program.

Senator Dianne Feinstein said hello. I remembered how kind she was, personally donating $500 toward Roman's rehabilitation expenses, as had Representative Pete Stark.

It was great to finally meet Dan Perry of the Coalition for the Advancement of Medical Research and wheelchair warrior Senator Jim Langevin. Even the Speaker of the House Nancy Pelosi came over. I remembered her wonderful pro-stem cell speech about how we are all just one diagnosis away from an incurable disease or disability.

Shaking hands with Kareem Dale, Presidential Adviser for Disability, I realized he was blind. The Obama team was very inclusive, following the disability motto, "Nothing about us, without us." There he was, one of the "deciders".

"Ladies and gentlemen, the President of the United States," the announcer said.

We were on our feet before we realized it, lifted on shared energy like an invisible elevator, clapping till our hands ached — at last, at last, a friend in the White House!

He was slender, taller than he looks on television, and a good-looking man of course, but how best to characterize President Barack Obama? I had an emotional impression which was hard to put into words, but it was a sense that he genuinely cared and would find a way to turn concern into accomplishment.

His words were important, but I was too excited. He could have read from the phone book and I would have applauded. I looked up the speech on CNN later to find out what he said:

"In recent years, when it comes to stem cell research […], our government has forced what I believe is a false choice: between sound science and moral values […]. As a person of faith, I believe we are called upon to care for each other and to ease human suffering. We have been given the capacity and will to pursue this research — and the humanity and conscience to do so responsibly."

Until that moment I had not known if I could trust him.

Our family had supported Hillary Clinton. We loved her husband as President Bill, of course, but Hillary had her own special magic, and we achingly wanted her in the White House. When she lost the nomination, we were heartbroken. But as loyal Democrats we had supported Obama, reminding the country why we so desperately needed a pro-research President.

In the campaign, Obama had pledged to reverse the Bush stem cell restrictions. But words are one thing, deeds another. Campaign promises are easy to forget. Plus he was a little busy.

The Obama Presidency had been in crisis mode since the first day. The country was on the brink of what could have become the worst Depression in American history. Economists used phrases like "edge of the abyss" and banks were about to go under. Corporations were laying off people and outsourcing American jobs to other countries.

Unemployment could have reached Depression-era levels of 15–20%. George Bush had spent an estimated one trillion dollars on the Iraq war, plus lowering taxes on the rich, turning the Clinton surplus into debt.

In that chaos, it would have been all too easy for a stem cell promise to be forgotten. But like Cesar Chavez, Obama had begun his political life as a community organizer. He knew what a promise meant to grassroots folks.

And now here he was, no fuss or self-congratulation, just doing what he had said he would do, honoring his word. He signed several copies of the bill, nodded briefly, and headed for the exit.

We had been told there would be no pictures allowed. But I wanted him to meet Roman. I hurried to the President, and said, talking very fast before the bodyguards could intervene, "My son was paralyzed in a college football accident. We passed a law named after him, the Roman Reed Spinal Cord Injury Research Act — could you please come over and say hello?"

Barrack Obama, President of the United States, turned toward my golden lion son.

As they shook hands, Roman asked, "Mr. President, could we impose on you for a picture?"

Obama smiled, a hint of mischief.

"I think they said there was no time for photographs."

"But *you* did not say that," said Roman, grinning at the President.

"You're gaming the system, aren't you?" said Barack Obama.

And our President took a picture with our son.

17 THE BOY WHO LOVED STANFORD TOO MUCH

KNOCKNOCKNOCK!

When I opened the door of the apartment in Los Angeles, a haggard-looking woman stood there, shaking. Her hair was matted and stringy, her clothes rumpled like she slept in them. Her eyes were wild and her words came in explosive bursts.

"Tomorrow morning at ten o'clock — you will hear a little boy — screaming. Do not call the police — he has autism, he just does that — okay?"

I nodded warily, wondering if she was the one who had the mental condition. She hurried away. That, I thought, was the end of that.

But at 10:00 a.m. the next morning, I heard a scream. From our second-story window I could look down into the neighbor's backyard.

In a sandbox in the middle of the yard, a little boy was sitting, all alone. He was playing with toy trucks — and screaming. It went on for about five minutes. His face showed no recognizable emotion. Then he stopped, arranged his trucks in a neat little row, and went inside.

Autism. One child out of 88 is affected, which is almost one percent of the population. For a lifetime of autism, the price is high in financial terms alone, an estimated three million dollars per person.[1]

I had almost no knowledge of it, aside from the movie, *Rainman*, in which Dustin Hoffman played an autistic man. He seemed a sort of genius with problems: capable of instantly memorizing phone books and counting cards at Las Vegas, but terrified of physical contact, shrieking if hugged.

Autism, it seemed, was not one condition, but many. The term Autism Spectrum Disorder sums it up: a multiplicity — Asperger syndrome, Rett syndrome, Angelman syndrome, Timothy syndrome, Fragile X syndrome, and more.

Some autistic folks require care all their lives. Others live independently, and can be valuable employees if their interest coincides with a company's need. Some computer companies regard autistic employees as highly desirable, as they are capable of staying focused for long periods of time.[2]

---

[1] Autism has high costs to society [Internet]. 2006 Apr 25 [cited 2015 Feb 5]. Available from: http://archive.sph.harvard.edu/press-releases/2006-releases/press04252006.html

[2] Ten Wolde H. SAP looks to recruit people with autism as programmers. Reuters [Internet]. 2013 May 22 [cited 2015 Feb 5]. Available from: http://www.reuters.com/article/2013/05/22/us-sap-

They often obsess on a subject, like the boy in a red sweater in my eighth grade English class. He was a cheerful kid, full of energy — but he would only talk about Stanford University.

When I stood outside my classroom to say hello to my kids at the beginning of each period (everyone deserves a smile and hello), he would always say something new about Stanford University.

And when it came time to write the essay of the week, no matter what the subject was, he would somehow bring it around to Stanford University: sports, science, whatever — Stanford University did everything best. If the subject was sharks, he would sit at his desk, face contorted — and then smile, and I would know he had figured out a Stanford University connection. It might begin like, "If sharks were studied at Stanford..."

Counselors, parents, and I met and talked. He was on a limited school day — I could keep him in the class, or not; grade him, or not. Of course I kept him; every child deserves school, and he was no trouble. I think I gave him a provisional "A" for all the good writing he did about Stanford University. He finished out the year, and his smiling face is still clear in my memory.

I always wondered what happened to him when he grew up. Was he one of the lucky ones who found a job and became independent, or did he end up institutionalized? I never knew.

Fast-forward to 2006 and a Los Angeles meeting of the California stem cell board, the ICOC, and a "spotlight" presentation — on autism.

The spotlights focused on a chronic disease, so we could learn the nature of the enemy and how it might be defeated.

This one featured Jonathan Shestack of Cure Autism Now. Shestack is slender and intense, as if something inside will not let him rest.

I arrived late and missed half the presentation. But what I did see shook me.

It was a home movie. At 18 months, Jonathan Shestack's little boy Dov looked healthy and normal. He kissed Daddy with that wide open mouth the way babies do, like they are going to bite, but don't — usually. Everyone smiled, remembering our own kids at that age.

The movie was a series of brief clips of the boy growing, with Shestack's terse narration. It seems there are milestones a child should attain — which did not happen for Dov. He had not learned how to point a finger at an object, or make eye contact with someone, or even to speak fluently: at 13 years, he had a working vocabulary of about 20 words.

The movie ended, and CIRM's Melissa King took the microphone. Now, Melissa is always calm in the face of adversity. If Martians landed, she would announce that lunch might be postponed.

But as she read a public letter to us, she kept losing her place. She would apologize and go on, blinking rapidly, a track of shining under her eyes.

The letter was from a mother, Nicki P., talking about her child, Christopher. At two, Christopher was normal; a cheerful intelligent boy.

---

autism-idUSBRE94L0ZN20130522

But on March 21, 1999, he said, "I'm finished, all finished," repeating the phrase for days, before "going into a freefall, to a world only accessible to him," said his Mom.

"At age 9, he cannot read, write, or even hold a pencil appropriately. His limited speech is unclear, sometimes even to me. He cannot access many of the skills he had as a toddler, and needs help with almost all daily living skills. [My fear is that] he will end up in a home that will not understand his two-to-three-word requests, and he will not have me to advocate for him […]."

I made contact with her later on. Her son at age 17 was functioning at the level of a two-year-old.

"Sometimes, when I am asleep," she said, "he wakes up, comes quietly into my bedroom, and dives through the air to land on top of me — like he used to do, when he was two."

Ms. P. tried to help others in her situation. Working from her home, she set up a phone center for the parents of autistic children so they could at least have someone to talk to when the crises came: when the marriages broke up, when there was no money, when the child would eat only cornflakes and nothing else, or the caregiver had been too long without sleep.

And she was quoted in a *San Jose Mercury News* article about a mother for whom the strain was too much.

Elizabeth H. shot and killed her 22-year-old autistic son, and then herself. She had been trying to get help for him, but state budget cuts got in the way. Apparently, when he was of school age, assistance was available, but once he turned 21, the aid stopped. The condition, however, did not.

Ms. P., full-time caregiver to her own son, spoke about what the woman faced.

"Elizabeth […] was not just a mother. She was her child's nurse, his advocate, his playmate, his cook, his personal hygiene assistant, and his communicator.

"Our system has crumbled, not simply because of budgetary constraints, but because of the way funds are allocated," she said, "The disabled, elderly, and chronically ill are being left behind […].

"Parents and autism experts are worried the problem will only grow worse, as hundreds of thousands of autistic children nationwide become adults over the next decade even as states cut more social and medical services […]."

But the California stem cell program is taking on the challenge.

Meet Dr. Alysson R. Muotri, of the University of California at San Diego. A slender and energetic man, originally from Brazil, Dr. Muotri had been trying to gather tissue samples from autistic children.

"Taking blood samples from autistic children is […] not nice," he said when I interviewed him, "They scream and fight. […] So, we came up with the Tooth Fairy program."

When an autistic child loses a baby tooth, they can donate it to Dr. Muotri's program.

"It was so successful we got more teeth than we needed, and had to stop the outreach," he said.

Dr. Muotri had three main goals for his research. First, there had to be a model of the disease so he could test drugs on it and see what helped and what did not. Second,

Alssoun R. Muotri fought autism by what might be called the Tooth Fairy Technique.

he wanted to find a "molecular footprint" for the disease. That way newborns could be tested; the earlier the diagnosis, the better the chance for a positive outcome. And lastly, he hoped to reverse the disease.

The California stem cell program offered a chance. As his weapon of choice, he picked induced pluripotent stem cells, similar to embryonic stem cells, but made from patient body tissues. Why not human embryonic stem cells? He said he wanted stem cells directly from an autistic person, and so they had taken tissue samples — the baby teeth.

Dr. Muotri changed those tissue cells into embryonic-like cells. Then, in a process taking about three months, he converted the cells into brain cells, like those of the patient.

Now he had diseased brain cells to work with in the lab, almost like a living autistic brain.

He found that neurons (nerve cells) derived from autistic patients make less connections than do those from healthy individuals. Diseased astrocytes — a star-shaped cell type in the brain — contributed to these connection shortfalls. If healthy astrocytes were put in contact with autistic neurons, they could help recovery.

Dr. Muotri found a drug — Insulin Growth Factor One — that the malfunctioning neurons seemed to like, *and which reversed the effects of autism*. The experiment is new as this is written, but there is reason to hope. Most of his work has been done with human cells in a dish. Naturally he wants to go ahead, possibly trying for one of CIRM's disease team grants, which provides enough funding to go all the way into human trials.

How does Dr. Muotri feel about this?

"The possibility of helping people with autism is so exciting. When I talk to parents now, I see hope in their eyes, that there is a light at the end of the tunnel.

"And when I get up in the morning every day, I *run* to my lab."

18 STEM CELL SUMMIT, STEM CELL WORLD

Jose Cibelli: Michigan stem cell research expert, adviser to Andalucia.

Since the World Stem Cell Summit began in 1994, I have attended nine times. I missed this year's (2014) event, only because I had to finish this book before the deadline. Organized by Bernie Siegel's Genetics Policy Institute (GPI), the summits are jam-packed with cutting edge science, old friends to enjoy, and new ones to meet. Every summit had its own unique magic, but for me the ultimate was the one in 2009, in Baltimore, Maryland.

To enter, you had to walk *through the sky*: a glass-walled bridge, arching from the Hilton Hotel across to the Baltimore Convention Center — blue sky and clouds all around.

Was it big? There were three connected auditoriums large as football fields and ceilings so high you could lose an airplane. There were side rooms for lectures and presentations from companies, colleges, states, and nations — and forests of scientific posters, with researchers standing by to explain.

Day one, we were welcomed by the Governor of Maryland, Martin O'Malley: tanned, charismatic, a superb speaker — no wonder some mention him as Presidential material.

Jeannie Fontana, Grant Albrecht: Grant is receiving a Genetics Policy Institute Advocacy Award.

"Maryland and California work together on stem cells," he said, "Healing people is our American pride! Stem cells are weapons of mass *salvation* [...]. Maryland has spent $56 million on stem cell research in the last three years [...]. With only 2% of the nation's population, Maryland controls 8% of America's biotech [...]. We always fought above our weight class!"

The only problem was an embarrassment of riches: we had too much quality! Mike West was speaking. Owner and founder of Biotime Asterias, the scientist/businessman essentially began biomedicine, bringing together ace scientists Jamie Thomson and John Gearhart to form Geron, Inc.

But I missed his speech, because I had the chance to be on a National Public Radio interview show. With a joke, Dan Rodricks of Maryland Public Radio introduced us, "The doctors are in the house!" I appreciated the promotion, but there was only one physician present: Dr. John McDonald, Christopher Reeve's neurologist. Dr. McD was the Director of the International Center for Spinal Cord Injury at Kennedy Krieger, and a champion of focused rehabilitation for paralytics. He wanted paralyzed people to do vigorous exercise and fought to restore their lost functions.

Dan Rodricks asked me how I felt about the future of stem cell science.

I replied, "It was like [...] being in the ocean, and on the far horizon, you see the tops of trees. It's just a patch of green, but there's an island attached, and it's real and new and waiting to be explored."

Meanwhile, back at the summit...

The international aspects alone were worthy of a piece of your life.

Dr. Fanyi Zeng spoke. The Secretary-General of the Chinese Society for Stem Cell Research is a tiny vigorous exclamation point of a person, resembling the actress Zhang Ziyi, star of the epic film *Crouching Tiger, Hidden Dragon*. Dr. Zeng's English is flawless, but she went through a lot of science very fast; my pen and brain had trouble keeping up.

Bernie Siegel makes it look easy: Sharing joy with an audience of research supporters.

Dapper and elegant, like he just stepped off the cover of GQ magazine, was Dr. Jose Cibelli of Michigan and Spain. Working at Andalusia, Dr. Cibelli helped develop Spain's Program for Cell Therapy and Regenerative Medicine. A founding member of the California stem cell program, Jose was expert in all three of the most advanced forms of stem cell research — human embryonic stem cells (hESC), somatic cell nuclear transfer (SCNT), and induced pluripotent stem cells (iPSCs) — a lot of big words and a lot of science!

The Executive Director of Andalusia's program, Dr. Natividad Cuenda, spoke about her agency's work on cellular reprogramming. Her goals were to translate into useful forms the results obtained in three research programs: regenerative medicine, genomic medicine, and nanomedicine. Andalusia has multiple tissue banks for their scientists, such as spinal fluid banks, stem cell banks, tumor banks, and a DNA banks. Dr. Cuenda directs the agency to plan governmental resources, share knowledge, facilitate cooperative research, and sponsor non-commercial clinical trials. I was glad Dr. Cuenda is on our side — but I am not sure she sleeps very much!

Speaking of international aspects, Canada had spent $80 million on a Stem Cell Network to develop standardized methods for stem cell researchers.

"Investment in research is an answer to our moral obligations: an act of faith as well as reason," said redheaded Canadian spokesperson Drew Ryall.

Canada set up a symbolic linking of arms for research. They wanted people to visit their website — http://stemcellcharter.org/ — and put their names in support of stem cell research. It was so easy, I did it myself, and I am computer-challenged. In fact, I got so excited about this idea, I emailed the Canadians after the event, asking for a paragraph of explanation — and here it is:

"The Stem Cell Charter […] is a one-page document that outlines a framework to move stem cell science forward responsibly. The Foundation aims to create an international grassroots community of scientists, business people, policy makers, patient advocates and members of the public […]. We're asking everyone to visit http://stemcellcharter.

org/, read and sign the Charter and send it to friends, family and colleagues. While you're on the site, check out 'Rock Star Scientists' and some mini-videos about different areas of stem cell science."

Michael Werner spoke about bringing the stem cell world together through the new **Alliance for Regenerative Medicine**. This has now become the premiere alliance for patient advocates, formally taking the place of the Coalition for Advancement of Medical Research, CAMR.

Former Congressman Jim Greenwood spoke. Greenwood was now leader of the Biotech Industrial Organization, or BIO, representing more than 1,200 biomed companies.

"We need a policy environment conducive to growth," said Mr. Greenwood. Even now, when the Bush limitations were largely repealed, there was reason to fear "mischievous amendments".

"Also, we need proactive leadership to drive the research, not merely permit it," he concluded.

Dean Tozer, representing Advanced Biohealing, spoke about Dermagraft,[1] a *healing "skin" for diabetic skin lesions*, which costs $1,425 each.

That seemed a tad expensive — but then I thought about the pressure sores that paralyzed people get: ulcers on heels or buttocks which can rot flesh to the bone, and keep a person bedridden for months. Complications from a pressure sore led to the death of Christopher Reeve.[2] Biospherix[3] offered a new kind of "clean room", the X-vivo Workstation, perfectly sterile environments — and it appeared to come in separable units for ease of relocation.

Alain Vertes, an executive of Roche, the giant pharmaceutical company, spoke about his company's recent purchase ($46.8 billion!) of Genentech.[4] Roche asserted his "prudent optimism" about regenerative medicine. I had a chance to speak with Vertes later. His son's name Romain was similar to my son's, Roman. They shared a name — and we shared a dream.

Devyn Smith of Pfizer gave a talk titled "Creating Partnerships with Large Pharma?" Dr. Smith stated that Big Pharma (and you don't get much bigger than Pfizer) needs to cooperate with the regenerative medicine movement for a very simple reason: self-interest.

For the first time in decades, Big Pharma was losing money — a loss of one to three percent this year, if I understood correctly. Why? Their patent exclusivity was expiring.[5]

---

[1] http://www.dermagraft.com/
[2] Szabo L. Spinal cord injury patients often succumb to bed sores. *USA Today* [Internet]. 2004 Oct 24 [cited 2015 Feb 4]. Available from: http://usatoday30.usatoday.com/news/health/2004-10-24-spinal-usat_x.htm
[3] http://www.biospherix.com/
[4] Pollack A. Roche agrees to buy Genentech for $46.8 billion. *The New York Times* [Internet]. 2009 Mar 12 [cited 2015 Feb 4]. Available from: http://www.nytimes.com/2009/03/13/business/worldbusiness/13drugs.html?_r=1&
[5] Anderson K. This patent cliff 2014 chart shows how much revenue Big Pharma will lose. *Money Morning* [Internet]. 2014 Feb 18 [cited 2015 Feb 4]. Available from: http://moneymorning.com/2014/02/18/patent-cliff-2014-chart-shows-much-revenue-big-pharma-will-lose/

If they did not hold the patents on their medicines, other companies might dip into what had been an exclusive source of money. They might need to be open to new methods of healing and sources of revenue.

Congressman Mike Castle (R-Delaware) was there! With Senator Diana DeGette, Congressman Castle had co-authored the Stem Cell Research Enhancement Act, twice passed by both houses of Congress — and twice vetoed by President Bush.

How great to have the chance to greet him personally, to shake his hand and say thank you.

Bob Klein, Chairman of ICOC, was up next. Due to the many public meetings that California's stem cell program has, I sometimes hear Bob speak a couple times a week, but I never miss the chance; he always itches the brain.

For instance, one press article had criticized CIRM, claiming that perhaps 5% of research buildings we funded might not be able to meet their construction schedules. "The article had it backward," said Bob, "In today's economic climate, a 95% success rate is phenomenal!"

New York had not one but two substantial stem cell programs.

One, the New York Stem Cell Foundation , was privately funded, and led with elegance by Susan Solomon as Chief Executive Officer, and Kevin Eggan as Chief Science Officer. They have some terrifically ambitious and worthwhile projects, such as building a bank of "2,500 stem cell lines, representing the genetic diversity of the United States and the world". They are also one of the few research centers in the world to develop SCNT. There was even a bone regeneration program going gangbusters.

The second New York program, NYSTEM, had a spectacular $600 million funding commitment over 11 years.[6] (As of this writing, one of their top scientists, Lorenz Studer, has been working for more than a decade to find a way to restore the dopamine cells taken away by Parkinson's disease. My impression is that he will succeed, and in the very near future.)

New York scientists developed what might be the world's first patient-specific ALS cell line, a huge breakthrough in Lou Gehrig's disease, so the progress of the disease can be followed in a Petri dish instead of a terminally ill person.

Cure, of course, is for everyone, but no one deserves it more than those whose bodies were broken in service to our country: the soldiers.

A panel discussion was held on Regenerative Medicine for Wounded Soldiers and Civilians — what happens to our brave young men and women when they are carried home wounded from the battlefield? What if there was a way to make them whole again, to regrow their arms or legs? The Defense Department has a $250 million grant to try to regenerate limbs — but they are only allowed to use adult stem cells! This was a mistaken policy begun under the Bush Administration, which should be reviewed and adjusted.

Linda Powers of Toucan Enterprises took us on a whirlwind tour of World Biomed. Ms. Powers is a venture capitalist, a source of money (and direction) for biomed startup companies.

---

[6] NYSTEM. Empire State Stem Cell Board strategic plan. 2008 May. 44 p. Available from: http://stemcell.ny.gov/sites/default/files/documents/files/NYSTEM_Strategic_Plan.pdf

Another speaker pointed out that Germany gets stem cell assistance from Israel. This symbolized that former enemies can work together for the good of their people! Germany also announced the signing of its memo of cooperation with the California stem cell program. (Israel also signed one, later, in 2014.)[7] Cancer researcher Dr. Curt Civin of Maryland was best known for isolating stem cells from blood (for which he won the 1999 National Inventor of the Year Award), but he also pointed out the vital necessity of keeping track of state efforts: quoting Supreme Court Justice Louis Brandeis, "The states are laboratories of democracy".

The Juvenile Diabetes Research Fund received a richly deserved advocacy award. This endlessly laboring group not only works to raise awareness of diabetes, but funds hundreds of millions in research, and is always where the fighting is done in defense of scientific freedom.

For grassroots advocacy, Danny Heumann accepted an award on behalf of Cure Michigan and Michigan Citizens for Stem Cell Research and Cures. Michigan has worked so hard for its research freedoms, and cheerful Danny was the perfect person to share the joy.

Stanford University's Dr. Irv Weissman, bearded and burly, gave off a friendly warmth, like a woodstove in Winter. Often called the "Father" of adult stem cell research, the man everybody calls Irv is a tremendous advocate for full stem cell research.

Wise Young pointed out serious shortcomings in the current stem cell research funding policy — such as SCNT and parthenogenesis being not eligible for funding. "We must raise our voices, tell Congress and the President. There is no law which prevents our funding [these methods of research], but the NIH is *restricting itself* [...]."

There were so many good people everywhere you looked, like Amy Comstock-Rick, then President of CAMR, and a dedicated leader for Parkinson's Action Network.

Dr. Jane Lebkowski of Geron had the unenviable task of speaking on a day when we were all dying to know what was going to happen with her company and the FDA and the spinal cord injury human trials of embryonic stem cells. She could not, of course, tell us what was going on; nothing must be allowed to jeopardize this great leap forward.

Peter Kiernan of the Christopher and Dana Reeve Foundation (CDRF) spoke about the CDRF's important survey on the number of paralyzed people in America:

"According to a study initiated by the CDRF, nearly one in 50 people live with paralysis — approximately 6 million people. That's the same number of people as the combined populations of Los Angeles, Philadelphia, and Washington DC."

Kiernan spoke about the opposition, which he dubbed the "armies of the night" and were better organized than us, and that advocates should learn to speak with a united voice.

Josh Basile is a soft-spoken new power in the advocate arena. Josh said an ocean accident paralyzed him and took his voice — but he got it back through advocacy. (Four years later, Mr. Basile became a law attorney.)

---

[7] Germany and California announce collaboration to advance stem cell research toward cures [Internet]. 2009 Sep 17 [cited 2015 Feb 4]. Available from: http://www.cirm.ca.gov/about-cirm/newsroom/press-releases/09172009/germany-and-california-announce-collaboration-advance

Maryland patient advocate John Kellerman, from the Maryland Stem Cell Research Commission, was fighting both Parkinson's and cancer. His beautiful daughter stood beside him as he expressed his dream that he might one day *dance* at her wedding.

Organized lunch talks with experts let the conversations continue up close and personal — sharing thoughts, making friends, and swapping business cards.

We even had a world television premiere: the global premier of BioBusiness.TV, a ten-part series titled *Stem Cell Review* starring top names in our field.

The NIH's Ann Hardy spoke on the importance of health surveying, so we can know the population affected, as well as the nature of the diseases we are up against.

A joyous shout-out should also go to Dr. Elizabeth Blackburn, who had just won the Nobel Prize for her work with telomeres, which are important parts of chromosomes. A former member of the Bush Presidential Bioethics Commission, she had been kicked off the board for supporting hESC research.

I had the privilege of speaking on an advocates' panel with great co-workers.

Co-workers like Mary Wooley, President of Research! America, who stressed that we needed to raise the awareness of science in America. Ask a citizen to name athletes or movie stars, and he or she can rattle off dozens of names. But when asked to name scientists, most can only think of Einstein. But these are folks whose work saves lives?

"And can you pass the Starbucks test?" she asked. If you saw your Congress person at a Starbucks, drinking coffee — would he or she know you? As advocates, we need to be in such close touch with our representatives that they would recognize us instantly.

One amazing individual was Howard Zucker, Assistant Director-General of the World Health Organization. As a college student he designed neurological experiments performed during astronaut trips to outer space. Dr. Zucker wrote a paper for the Federal Working Group on Regenerative Medicine; Bernie Siegel regards it as a blueprint for the biomed industry.

When it was my turn to speak, I had fun, and hoped the audience did not suffer too much. I speechified on the need to support biomed, and to encourage the new industry with governmental support and tax shelters, as we did with the computer industry. Biomed should not be so bashful — it should think of itself as a new Defense Department, for it saves lives and is the foundation of a new and permanent economy.

During the question and answer period, Dr. Mark Noble stood up from the audience. White-haired and muscular, he looks more like a retired football player than the pioneer of stem cell research that he is. In 1983, he co-discovered the first precursor cell isolated from the central nervous system. In the 26 years since then, he has not only lived in the lab more than two average scientists put together, but has also never feared to take a stand politically, defending our research.

When he said he had a "difficult question," I got a little nervous. After 15 years listening to scientists, I can usually keep up (if they talk slow) — but a difficult question?

But here it came: "What can *scientists* do better, to assist advocates?"

"Above all, keep doing what you are doing," I said, "You are doing the impossible with the invisible, using microscopes to find ways to fight incurable disease. If you win, we win."

There are also chores no scientist wants, but which must be done if the research is to survive.

First, scientists have to be involved politically: don't let opponents of research make the funding decisions.

Secondly, scientists should use small words. People process new information at an 8th grade level. If I talk to a garage mechanic, he has to talk slow, or I won't have a clue. Scientists have the equivalent of a 20th grade education. If they talk to the public like they talk to each other, we will not "get it" — and we can't support what we don't understand.

Talk to a teenager. If their eyes glaze over, rethink your verbal strategies.

And finally, let us help you. Hook up with a patient advocate group in your area of expertise. Speak at our meetings, tell us what you are doing, and share your difficulties. Are you having funding problems? Let us know. Maybe there is something we can do. Is there a legislative bill which may threaten your research? Work with us, we are millions.

We were at the awards ceremony dinner, leaning back, bellies full of great food. I was just looking around to see if I could wangle an extra dessert, when Bernie's wife Sheryl Siegel walked up to the microphone and thumped it to be sure it was working.

"Several months ago," she said, "I wanted to have a birthday party for Bernie — and he said no, how could he have a birthday party? His friends lived all over the world, who would he not invite?"

Which was perfectly true. International Bernie lives on a plane more and more these days, speaking, networking, and making friends for the cause. So Bernie said no — and Sheryl cheerfully ignored him.

"*This* is Bernie's birthday party," she said, holding up a cake with candles, "And you are all his friends."

The ruckus we raised let her know that she was right.

19 HOW NOT TO HAVE SEX IN A PERSONHOOD STATE

Here is a riddle, but it is not a joke. A married couple has just made love. They live in a "personhood" state. Could one or both be in violation of the law?

Answer: If the wife used birth control pills, personhood laws might charge her with murder — and the husband might be accessory to the crime!

Ridiculous? Yes. But it is not a joke — in a personhood state.

"Personhood" is a challenge to Roe v. Wade, the Supreme Court decision which allows a woman to terminate her pregnancy. There would be numerous other consequences as well, including an attack on stem cell research, but criminalizing abortion is the overriding purpose of personhood.

The idea of personhood springs from a sentence in the Roe v. Wade decision[1]:

"If this suggestion of *personhood* is established [...], the fetus's right to life would then be guaranteed specifically by the 14th Amendment" (italics mine).

Presently, a woman may legally terminate her pregnancy up to the point of viability. After the fetus can live on its own (viability), abortions are against the law. But if the meaning of "person" was changed to a fertilized egg, abortions at any stage could become illegal.

Here is a key quote from a recent personhood law, Amendment 62, from Colorado:

"Section 32. **Person defined**. [...] the term 'person' shall apply to every human being from *the beginning of the biological development* of that human being" (italics mine).[2]

The beginning of biological development. We are talking about the fertilized egg — such as that which a married woman often sheds as part of her monthly cycle. It passes from the body in a liquid state (often called a "heavy flow day"), and is disposed of; but according to the "Personhood" folks, that shed embryo have full legal rights under law!

First, look at Colorado's constitution, the normal way as it is right now.

"Section 25: Due Process of Law. No person shall be deprived of life, liberty or property without due process of law."

---

[1] *Roe vs. Wade* (1973) 410 U.S. 113, 93 S.Ct. 705, 35 L.Ed.2d 147.
[2] Colorado Fetal Personhood, Initiative 62 (2010). *Ballotpedia* [Internet] [updated 2014 Oct 16; cited 2015 Feb 5]. Available from: http://ballotpedia.org/wiki/index.php/Colorado_Fetal_Personhood_Amendment_62_(2010)

But under personhood? Add in that new phrase "from the beginning of the biological development" and the meaning changes to: "No person, from *the age of a fertilized egg,* shall be deprived of life, liberty or property without due process of law."

Under personhood, every fertilized egg has full rights in a court of law. What would this mean? How would life change in a personhood state?

For birth control, only abstinence or "barrier methods" (such as condoms) would be legal. This would criminalize the actions of millions of women. An estimated 99% of all marriage-age American women have used contraceptives, typically the easy-to-use birth control pills.[3]

Kristy Burton-Brown, founder of the Colorado Personhood Initiative, said, "Does the birth control pill cause abortions? In a word, yes. The birth control pill really does cause abortions."[4]

So if birth control pills cause abortions, and abortions are illegal at any stage, and 99% of all American women use or have used birth control pills — virtually an entire sex just became criminals. But wait, there is more nonsense; much more.

Although billed as a pro-life measure, personhood could prevent childless couples from having a baby — if it bans the IVF procedure.

IVF has brought children to an estimated five million families around the world. It is a technical procedure: mixing sperm and egg in a saltwater-filled Petri dish. About 20 blastocysts are made; the healthiest one or two fertilized eggs are implanted in the mother's womb. And the others? These are frozen and stored, or thrown away.

Personhood supporters describe the IVF procedure as "mass extermination of human beings."

Here are their words: "The Colorado Personhood Amendment [...] would force medical scientists to come up with ethical alternatives to mass production and mass extermination of human beings at [...] fertility clinics."

If IVF clinics are truly guilty of "mass extermination of human beings," then doctors, nurses, and parents around the world would be complicit in the "crime".

And what about those forced "ethical alternatives"? No one is sure what they would be.

Would every blastocyst have to be implanted? That was what Nadya Suleman, dubbed the "Octomom", insisted upon — whereby she ended up with eight children at one time.[5]

Embryonic stem cell research would be illegal, because it involves blastocysts, the fertilized eggs that would otherwise be thrown away — but which now would be protected

---

[3] Guttmacher Institute. Contraceptive use in the United States. 2014 Jun. 4 p. Available from: http://www.guttmacher.org/pubs/fb_contr_use.html

[4] Burton-Brown K. Does the birth control pill cause abortions? In a word, yes [Internet]. 2011 Apr 25 [cited 2015 Feb 5]. Available from: https://thelostgenerations.wordpress.com/2011/04/25/does-the-birth-control-pill-cause-abortions-in-a-word-yes/

[5] Nadya Suleman. *Wikipedia* [Internet] [updated 2015 Feb 1; cited 2015 Feb 5]. Available from: http://en.wikipedia.org/wiki/Nadya_Suleman

under law. In the name of protecting microscopic "people", real-life individuals like my paralyzed son would be denied the hope of embryonic stem cell research.

The backers of personhood themselves admit this, saying: "It will ban human embryonic stem cell research. Human embryonic stem cell research is nothing more than human harvesting [...]; the creation and destruction of human beings for medical experimentation."[6]

This is, of course, nonsense.

Stem cell research, as the name implies, is cells, cells, nothing but cells. It involves living tissue, not a life. A fertilized egg in a petri dish (or microscope slide) is not a child and can never be one. Without the nurturing shelter of the womb, it is biologically impossible to make a baby. Where there is no pregnancy, there can be no babies — this is not rocket science.

Personhood believers are working hard to put their laws on the ballot.

But surely national politicians steer clear of such nonsense? To the best of my knowledge, no Democrat in Washington supports personhood. The same, unfortunately, cannot be said of Republicans, including several Presidential candidates.

Candidates like Paul Ryan, Mitt Romney's pick for Vice President. He co-sponsored the Sanctity of Human Life Act (HR 212), which gives "all the legal and constitutional attributes of personhood" to fertilized eggs. Ryan stayed quiet on the personhood issue during the Presidential campaign, but after the election, he co-sponsored another personhood bill.

And where did Republican Presidential candidate Mitt Romney stand on this issue? He pledged to "absolutely" support a Constitutional amendment to impose personhood on the nation.[7]

Wherever personhood legislation has been brought to the ballot, it had been crushingly defeated.

The citizens of Mississippi, perhaps the most conservative in the country, rejected personhood as too extreme.[8] Coloradans voted it down three times.[9] The Oklahoma Supreme Court found it unconstitutional.[10]

But it has not gone away.

---

[6] Scare tactic alert [Internet] [cited 2015 Feb 5]. Available from: http://www.personhoodusa.com/blog/scare-tactic-alert/

[7] Fact sheet: Mitt Romney supports dangerous "personhood" amendments like the one a court just struck down in Oklahoma [Internet]. 2012 May 1 [cited 2015 Feb 5]. Available from: http://www.plannedparenthoodaction.org/elections-politics/newsroom/press-releases/fact-sheet-mitt-romney-supports-dangerous-personhood-amendme/

[8] Pettus EW. Mississippi "personhood" amendment vote fails. *Huffington Post* [Internet]. 2011 Nov 8 [updated 2012 Jan 8; cited 2015 Feb 5]. Available from: http://www.huffingtonpost.com/2011/11/08/mississippi-personhood-amendment_n_1082546.html

[9] Colorado Fetal Personhood, Initiative 62 (2010).

[10] Bassett L. Oklahoma personhood measure struck down by Supreme Court. *Huffington Post* [Internet]. 2012 Apr 30 [cited 2015 Feb 5]. Available from: http://www.huffingtonpost.com/2012/04/30/oklahoma-personhood-fetal-personhood-bill_n_1465657.html

In Colorado, despite losing almost three to one on the previous attempts, supporters are trying again, for the fourth time.[11]

Other states have similar bills to offer: North Carolina, Georgia, Michigan, Montana, Iowa, and Louisiana.[12]

Republican Presidential front-runner Rand Paul even tried to insert personhood into a flood insurance bill! He literally wanted to hold up flood insurance for "5.6 million flood-prone properties" until "the Senate votes to recognize that life begins at conception."[13]

Unbelievable.

Mississippi's repeat effort is particularly irritating, because its backers show so little respect for the voters which previously defeated it, saying in a *Huffington Post* article dated March 6, 2013:

"The voters [...] were confused. They didn't understand the last amendment."

Confused? The voters heard the personhood arguments, understood them, and then rejected them.

But the threat remains: to women's reproductive rights, to the IVF method of producing a family, and to embryonic stem cell research.

Personhood: a word to remember, a threat to guard against.

NOTE: As this book goes to press, the current GOP front-runner candidate for President, Scott Walker, has just revealed to right-to-life groups that he supports personhood....

http://www.nytimes.com/2015/02/23/us/politics/scott-walker-hardens-tone-on-social-issues-to-woo-christian-conservatives.html?_r=0

---

[11] Draper E. Personhood Colorado, supporters will also back Brady amendment *The Denver Post* [Internet]. 2013 Oct 1 [cited 2015 Feb 5]. Available from: http://www.denverpost.com/breakingnews/ci_24215280/personhood-colorado-submits-petitions-fourth-try-at-abortion

[12] Sargent G. Coming this fall in the Senate races: big fights over personhood. *The Washington Post* [Internet]. 2014 Apr 15 [cited 2015 Feb 5]. Available from: http://www.washingtonpost.com/blogs/plum-line/wp/2014/04/15/coming-this-fall-in-the-senate-races-a-fight-over-personhood/

[13] Sheppard K. Rand Paul demands fetal personhood in flood insurance bill. *Mother Jones* [Internet]. 2012 Jun 27 [cited 2015 Feb 5]. Available from: http://www.motherjones.com/mojo/2012/06/rand-paul-demands-fetal-personhood-flood-insurance-bill

20 FIGHTING THE KILLERS: LEUKEMIA AND CANCER

Dr. Catriona Jamieson: Who says glamor and science can't go hand in hand?

I was helpless when my sister Patty died, three thousand miles away. The only connection between us was a faintly buzzing phone line.

"She's in a coma," my mother said, "She can't hear you."

"Put the phone by her ear," I said, and spoke loudly.

"Patty, it's me, it's Donny." She called me Donny, a name I hated from anyone else.

Things had been going badly for me that Pennsylvania Winter. I had lost my job and could not find another. I worked day labor for minimum wage, when there was any work to be had. One week I made $8 and the Young Men's Christian Association rent was $12, so I had to keep dodging the man behind the desk.

Suddenly, things got better. I would be working at the York Barbell Company, where all the greatest weightlifters in America trained. I had borrowed some money to make the pay call.

"I am okay now," I said, shouting, desperate to get through, "I have a job, I am fine, and — "

Silence for a moment.

Then, very softly, her voice. "That's nice, Donny... I love you," she said.

Irv Weissman: Pioneering stem cell researcher, now working on what maybe a cure for cancer. (Photo by Chris Vaughn.)

And then she died.

"It was like she was waiting to hear from you," said Patty's best friend, Gloria Jean Aceves, later to become my wife, who had been there in the hospital room. "And then she just slipped away."

We had not really understood, at first, when Patty came home with the unexplained bruises on her arms. The doctor frowned and ordered some tests. When they came back, he shook his head, and ordered them repeated.

"Leukemia," he said at last, "There is nothing we can do."

Patty told her boyfriend Nye Morton that she wanted to break off their relationship.

"I am going to die," she said, "And they say it is going to be bad. I don't want you to go through that, to see me like that — I want you to remember me like I was."

"Oh, no," said Nye, "I feel just the opposite. If we are going to be parted, I want us to spend every moment we can together first — so let's get married."

I hitchhiked from Pennsylvania to be at the wedding in California. She was so beautiful.

One night while she was still in the house, I knocked on the door of her room, and she said, "The wig is at the cleaners."

The disease had taken her hair. I told her I did not care about that.

"I just didn't want to freak you out," she said. I made my face stay normal when I came in.

She was 23 years old. She died in a hospital surrounded by doctors and nurses, and there was nothing anybody could do.

Leukemia. Even now, half a century later, that miserable cancer of the blood which took my sister is still incurable. Despite the most aggressive treatment, patients with

Maple Leaf scientist Dr. John Dick believes the concept of the cancer stem cell is crucial in the battle against leukemia.

Acute Myeloid Leukemia (AML) have only about a 30% survival rate. If it comes back after remission, there is only a ten percent chance of survival.[1]

But the California stem cell program is fighting leukemia now. We have learned the nature of the enemy, and, perhaps, a way to defeat it.

The most powerful weapon of the California stem cell program is the Disease Team Grant.[2] For these awards (as much as $20 million each), California scientists set up a dream team, develop a plan to challenge an "incurable" condition — and a schedule to bring it to human trials within four years.

As this is written, two disease teams are taking on leukemia: one headquartered at Stanford University, the other at the University of California at San Diego.

On the San Diego team, Drs. Dennis Carson, Catriona Jamieson, and Tom Kipps are cooperating with Canadian scientist John Dick: each country paying its own way.

Stanford's team is led by Irv Weissman, and includes Ravindra Majeti, Branimir Sikic, and Bruno Medeiros, in cooperation with U.K. scientists Paresh Vyas and Denis Talbot, both at the University of Oxford.

What do the two international teams have in common? A new way of looking at cancer:

"Some leukemia cells possess stem cell properties [and are] resistant to treatment. [Existing drugs fail to] eliminate the cells [which persist], continuing to grow, spread, invade and kill […]."[3]

---

[1] Leukemia fact sheet [Internet] [cited 2015 Feb 5]. Available from: http://www.cirm.ca.gov/our-progress/disease-information/leukemia-fact-sheet
[2] RFA 09-01:CIRM Disease Team Research Awards [Internet] [cited 2015 Feb 5]. Available from: http://www.cirm.ca.gov/our-funding/research-rfas/disease-team-research-i
[3] Development of Highly Active Anti-Leukemia Stem Cell Therapy (HALT) [Internet] [cited 2015 Feb 5]. Available from: http://www.cirm.ca.gov/node/5487/review

Stem cells, it seems, have an "evil twin": the leukemia stem cell (LSC) which is not wiped out by chemotherapy or other radiation treatments. The LSC starts the leukemia, developing a bunch of bad cells.

The chemotherapy or radiation treatments kill most of these secondary cells, but not the LSCs. This is remission, when patients get better for a short time. Most of the leukemia is gone, but the cause of it remains, waits, and then returns.

Leukemia, of course, is a form of cancer. Think of the leukemia cancer cells as little monsters with labels on their backs. These "surface markers" help the immune system identify cancer cells as threats, giving out what are called "EAT ME" signals. If the body's immune system detects them, it will send fighting cells (macrophages) to kill and eat them.

Unfortunately, LSCs know how to hide. They grow a little cloak, a protein coating known as CD47, to cover the surface markers, hiding the "EAT ME" signal. The body's immune system no longer recognizes the LSCs as threats, and does not attack them.

The San Diego team has identified and named one of these LSCs, calling it ROR1. But to recognize an enemy is not enough; it must also be fought and killed, *and in a way that does not destroy healthy cells.*

Tom Kipps of the San Diego team invented a "heat-seeking missile" to go after ROR1. This is an antibody, called UC961.

Everything depends on beating the CD47 cloaking device, so that it can no longer effectively shelter the cancer.

The Stanford disease team is led by stem cell pioneer Irv Weissman, who has also been working to defeat CD47 for more than a decade, ever since he found the cloaking device on mouse leukemia stem cells.[4]

What is the difference between the two team's approaches? Dennis Carson said of Weissman's group, "We are developing a poison to attack the leukemia; they are changing the body's response so that it can fight the blood cancer." If the cloaking device could be removed, the body's own defense system could attack the monster cells. The Stanford team has developed an antibody, named HU5F9, which is designed to peel back the cloak so the microscopic monsters can be killed.

From Dr. Weissman's official report[5]:

"Cancer stem cells [...] produce a cell surface 'invisibility cloak', [CD47, which is] a 'don't eat me' signal. [Our] antibody counters the cloak, allowing the patient's natural immune system [...] to eliminate the cancer stem cells."

This may turn out to be a way to fight not only leukemia, but all forms of cancer.

Read the next sentence carefully; it is wonderful.

---

[4]Sifferlin A. A single antibody to treat multiple cancers? *Time* [Internet]. 2012 Mar 28 [cited 2015 Feb 5]. Available from: http://healthland.time.com/2012/03/28/a-single-antibody-to-treat-multiple-cancers/

[5]Development of therapeutic antibodies targeting human acute myeloid leukemia stem cells [Internet] [cited 2015 Feb 5]. Available from: http://www.cirm.ca.gov/our-progress/awards/development-therapeutic-antibodies-targeting-human-acute-myeloid-leukemia-stem

"We have now demonstrated that HU5F9 is effective at inhibiting the growth of [...] tumors, including [cancers of the] breast, bladder, colon, ovaries, [brain, head and neck], and multiple myelomas."

How important is the California stem cell program in this process?

"Without CIRM, this program would not have left the starting gate. It was turned down for funding by the National Cancer Institute. [It has been] taken to the current stage mainly by CIRM funding [...]."

What comes next? Human clinical trials, as soon as the FDA says yes.

Is one of these approaches (or both) the answer to cancer? No one can say for sure. We cannot predict the pace of science any more than we can know which of us will get the deadly diagnosis.

But one thing is clear: without research funding, cure *cannot* happen.

21 THE THIEF OF MEMORY

Larry Goldstein: Scientist, explainer, advocate.

In the classic comedy *Splash*, an old woman comes to work oddly dressed: wearing her bra on the outside of her sweater. It was just a funny moment, never explained. But if the movie had followed her life from then on, it would have ceased to be a comedy.

Forgetting the order of putting on clothes is a sign of Alzheimer's disease (AD).

The woman might have been a trusted executive in the company. But now, her career was over. She could no longer be allowed to write checks, because she might pay bills two or three times each, or not at all. At home she would soon forget who her husband was, and her children, and her own name. At night, she might need to be locked in her bedroom, so she would not wander away. She would need to be watched and cared for 24/7, and the only escape was death.

This dreaded condition is sweeping America. Five million Americans have AD, roughly one in 60. Among those of us aged 65 and over, one in ten have AD today.[1]

And now, jump ahead to January 11, 2011: President Barack Obama has just signed the National Alzheimer's Project Act (NAPA) into law. It was a bipartisan effort, passed by both houses of Congress after a dedicated effort by many patient advocate groups, including a petition signed by more than 110,000 citizens, gathered by the Alzheimer's Association. It was spearheaded by champions like Maria Shriver, Sandra Day O'Connor, and Newt Gingrich.

(So far so good, right? But wait…)

According to a 2011 report in *USA Today*, "NAPA's aim is to create a coordinated national strategy that deals with Alzheimer's, a brain-wasting condition [with an annual medical cost of] *$172 billion* […]" (italics mine).

It is useless when a program tries to fight disease — with NO MONEY.

As also pointed out by Allan Vann[2], a retired public school teacher and advocate for patients and families coping with AD, "Without power to authorize new federal spending or redirect current federal spending by NIH, NAPA cannot realistically hope to accomplish its goals."

Alzheimer's folks are our loved ones, your family and mine. How are we going to take care of them? As live-in attendants cost five to seven thousand dollars a month, most families try to provide care for folks with AD at home, dividing up the chores. Everything you do for a baby may need to be done for an AD sufferer — and some may even be physically aggressive toward those who are trying to provide for their needs, people they no longer recognize.

A friend of ours is facing this right now. The husband had been a man of great responsibility, with a near genius IQ. Now he literally does not know the difference between a book and a television "clicker". If you asked him for a clicker, he would hand you item after item, guessing blindly.

One night he let himself out of the house and walked away — until he came to a bridge over a freeway. He was halfway across the overpass when he became confused and just stood there. A female passerby, in an incredible act of kindness, recognized his symptoms and physically detained him, stopping her car and getting out, and then corralling him with her car door while calling 911.

Marriages break up in exhaustion. Some caregivers become so stressed they die before the person they are caring for; others lose their jobs for taking too much time off for caregiving.

National cost estimates of AD run from a "low" of $100 billion a year, to the middle-of-the-road estimate of $172 billion, to a high of $300 billion. (If that last cost estimate

---

[1] Alzheimer's disease [Internet] [updated 2014 Jul 25; cited 2015 Feb 6]. Available from: http://www.cdc.gov/aging/aginginfo/alzheimers.htm

[2] Vann AS (2014). The National Alzheimer's Project Act… missed opportunities from a caregiver's perspective. *J Am Geriatr Soc* **62** (5): 966–967.

was divided equally among our population of three hundred million, it would mean a thousand dollars from every man, woman, and child.)

And now we have a national Alzheimer's program — with no funding? That is unacceptable.

California is fighting AD. To find out more, let's visit with Leeza Gibbons. If you watch television, you probably know Leeza, for ten years the host of television's *Entertainment Tonight*. But you may not know that she has dedicated her life to fighting Alzheimer's.

In an interview with Camille Peri, *Caring.com*'s features editor[3], Leeza said, "I remember being in my mother's house in South Carolina, helping her make the bed, and she was watching my every movement closely, trying to mirror what I was doing. Then she stopped and looked at me, and I said, 'What's the matter, Mom?'

"She said, 'You're a very nice lady. How do I know you?' And I said, 'You know me because today I am your daughter, yesterday I was your daughter, and I'll always be your daughter.'

"And she said, 'Oh'."

Leeza is the driving force behind a beautiful charity, Leeza's Place, which seeks to aid and comfort Alzheimer's caregivers. Leeza also volunteered several years as a board member of the California stem cell program.

If you hunt around a little bit at the California stem cell program website, www.cirm.ca.gov, you will find a list of diseases and "incurable" conditions that California is attempting to research and cure. Click on "Alzheimer's disease" and you will find scientists working to solve the puzzle, one piece at a time.

One such champion is Larry Goldstein at the University of California, San Diego. If you ever have the chance to hear Dr. Goldstein speak, don't miss it. He makes the research understandable while never sugar-coating the difficulties involved. He favors a variety of approaches to curing this terrible disease, beginning with a micro-model of Alzheimer's itself, contained in a Petri dish.

When an airplane crashes, investigators search the wreckage for the "black box" to find out what went wrong. That box records pilot information, which serves as clues to the causes of the crash.

But what if there was a black box to point out danger *before the plane crashed*? That is what the Goldstein model of Alzheimer's might be like, so we could study the chain of molecular events leading to the outbreak of the disease, recognize warning signs, and then stop or delay it.

Right now scientists can only learn from patients in the advanced stages of the disease, which may have begun decades earlier, and not to mention that it is unlikely for anyone to donate a piece of their living brain. But if the disease could be followed and studied cell by cell at every stage — in a dish of salt water — we might find the weak spot and then prevent, cure, or delay the onset of AD. Delaying the onset of AD would be hugely

---

[3] Peri C. Talking with Leeza Gibbons: providing support for Alzheimer's caregivers [Internet]. [cited 2015 Feb 6]. Available from: https://www.caring.com/interviews/interview-with-leeza-gibbons-about-her-mom-s-alzheimer-s

helpful: both to affected individuals and their families, and also to the economy, saving potentially billions of dollars.

But to save billions of dollars (and lives!), we must be prepared to spend millions. Dollars to our scientists are like guns to our soldiers; they cannot fight without them. Some say we cannot afford cure research, but that's wrong — we cannot afford *not* to fund it!

As Larry Goldstein put it, "We are woefully underinvested in the fight against disease, given its cost and burden on families. [...] I am hopeful that we can make progress with new stem cell-based tools, given the complexity of the Alzheimer's problem. But we need many more approaches to solve this terrible and growing problem."

Matthew Blurton of University of California, Irvine also added, "The NIH budget is not able to fund enough AD research. The current outlook is that *only about 3% of grants reviewed* by the National Institute of Aging [the primary funding source for AD research] *will be funded*"[4] (italics mine).

In a statement of intent, Leeza Gibbons said, "Regarding the lack of a budget for the Alzheimer's program, [...] without dollars behind the awareness, our national program is at risk of being just hype and very little help.

"In the fight against Alzheimer's, we need three things: awareness, care and a path to cure [...]. The California Institute for Regenerative Medicine (CIRM) is systematically using stem cells as a sword to slay the dragons of disease and life-limiting illnesses, including the thief of memory called Alzheimer's.

"It is part of the California commitment to attacking neurological disease, so one day we will be able to say: this assault on millions is just a memory."

Leeza Gibbons, champion for Alzheimer's cure research.

---

[4] Fagan T. San Diego: what — three percent? Money woes trump science at SfN [Internet]. 2010 Nov 19 [cited 2015 Feb 6]. Available from: http://www.alzforum.org/new/detail.asp?id=2616

22 THE ANTI-SCIENCE SOCIETY

As a country, America has long since decided to support human embryonic stem cell research. We had the great debate for more than a decade, and both sides were heard at length.

The result? A recent national poll by Harris Interactive[1], under the Nielsen company, shows 73% support for human embryonic stem cell research. According to Humphrey Taylor, Chairman of the Harris poll, "There is now overwhelming public support for using embryonic stem cells in biomedical research. Even among Catholics and born-again Christians, relatively few people believe that stem cell research should be forbidden because it is unethical or immoral."

There are very good reasons for such support, number one being to protect our families.

But there are also economic reasons to support full stem cell research.

Consider the threat of going bankrupt, or becoming homeless, through medical debt.

In a study by researchers at Harvard Medical School, it was found that half of all **bankruptcies** in the U.S. occurred in the aftermath of a serious medical problem.[2] Steffie Woolhandler, Professor of Medicine and one of the authors of the study, asserted that "unless you're (a billionaire like) Bill Gates, you're **one illness away from financial ruin**."

As for **home foreclosures**? Associate Professor Christopher Robertson at the University of Arizona examined homeowners going through foreclosure in four states and found that medical crises contribute to half of all home foreclosure filings, and that medical causes may put as many as 1.5 million Americans in jeopardy of losing their homes each year.[3]

---

[1] Harris Interactive. Embryonic stem cell research receives widespread support from Americans. 2010 Oct 7. 5 p. Available from: http://www.harrisinteractive.com/NewsRoom/HarrisPolls/tabid/447/mid/1508/articleId/579/ctl/ReadCustom%20Default/Default.aspx

[2] Himmelstein DU, Thorne D, Warren E, Woolhandler S (2009) Medical Bankruptcy in the United States, 2007: Results of a National Study. *Am J Med* **122** (8): 741–746.

[3] Robertson CT, Egelhof R, Hoke M (2008) Get sick, get out: the medical causes of home mortgage foreclosures. *Health Matrix* **18** (65).

There is also the extraordinary (and growing) size of the disability population. The World Report on Disability estimates that 15% of the world's population are disabled.[4] That's around **1,050,000,000 people** with disabilities!

This staggering number (one billion people with disabilities!) is increased by an aging population and a global increase in chronic health conditions, such as diabetes, cardiovascular diseases, and mental illness.

And in America?

According to a report from the U.S. Census Bureau[5], about 56.7 million people — 19 percent of the population — had a disability in 2010, with more than half of them describing the disability as severe.

We must provide CARE for these our loved ones, until there is a CURE. That care/cure conundrum is a major problem of our day, but how often do politicians talk about it?

What politicians do talk about — a lot — is the national debt. But they do not seem to realize that chronic disease to the deficit is like gasoline to a campfire.

Add up the annual costs of just four chronic conditions: stroke and heart disease ($432 billion), diabetes ($174 billion), lung disease ($154 billion), and Alzheimer's disease ($148 billion)[6] — that adds up to **$908 billion**, close to a trillion dollars.

And the national debt? According to the conservative Heritage Foundation, "In 2013, […] the deficit dropped to 'only' $642 billion."[7]

If the costs of disease grow so high that people go bankrupt, who gets stuck with the bills? You know this. Either the taxes go up to pay for needed social programs, or the government borrows and the debt goes up.

You would think those who worry about the national debt would be leading the charge for medical research. Unfortunately, those who shout the loudest about the national debt are often those who deny sufficient funding for scientists trying to bring cures — which could bring the deficit down!

Let's look at two of what I consider anti-science politicians. One is a Republican; the other, I regret to admit, is a Democrat.

Former Governor Mike Huckabee is a man who would be President. In 2012, he was one of the Republicans' top candidates for President, leading the nomination charge at 24% compared to 14% for Sarah Palin and Mitt Romney, and is so again today.[8]

---

[4] World report on disability [Internet]. 2011 [cited 2015 Feb 6]. Available from: http://www.who.int/disabilities/world_report/2011/report/en/

[5] Nearly 1 in 5 people have a disability in the U.S., Census Bureau reports [Internet]. 2012 Jul 25 [cited 2015 Feb 6]. Available from: https://www.census.gov/newsroom/releases/archives/miscellaneous/cb12-134.html

[6] The impact of chronic diseases on healthcare [Internet] [cited 2015 Feb 6]. Available from: http://www.forahealthieramerica.com/ds/impact-of-chronic-disease.html

[7] Boccia R, Fraser AA, Goff E. Federal spending by the numbers, 2013: Government spending trends in graphics, tables, and key points [Internet]. 2013 Aug 20 [cited 2015 Feb 6]. Available from: http://www.heritage.org/research/reports/2013/08/federal-spending-by-the-numbers-2013

[8] Jensen T. Huckabee ahead nationally [Internet]. 2011 Jan 21 [cited 2015 Feb 6]. Available from: http://publicpolicypolling.blogspot.sg/2011/01/huckabee-ahead-nationally.html

Getting ready for the 2016 elections, he is again leading in the important Iowa polls.[9]

Huckabee would be a disaster for medical research. He supports personhood, granting full legal rights to every fertilized human egg. He has endorsed personhood bills in Georgia, Colorado, and even one for the entire country, the Human Life Amendment.

According to a report by the *Denver Post* on February 25, 2008, "Republican Presidential candidate Mike Huckabee […] endorsed a proposed Colorado Human Life Amendment that would define personhood as a fertilized egg. The former Arkansas governor and Baptist minister also supports a human-life amendment to the U.S. Constitution."[10]

The Democrat is **Daniel Lipinski** of Illinois. Now, it is rare for a Democrat to oppose embryonic stem cell research. Of 535 Representatives and Senators, only a total of 18 Democrats voted against the Stem Cell Research Enhancement Act, and one of these was Daniel Lipinski.

Mr. Lipinski also authored what I consider a deceptive bill, the Patients First Act (HR 2807). His legislation was developed with the Family Research Council, arguably the most powerful Religious Right lobbying organization in the country, and was approved by the Bush White House.

About as "pro-patient" as a cobra in a hospital bed, the Lipinski bill would have stacked the deck against embryonic stem cell research, prioritizing National Institutes of Health grants in favor of adult stem cell research.

Speaking in response to the Lipinski bill, Sean Tipton, then-President of the Coalition for the Advancement of Medical Research, had this to say:

"[…] What the sponsors of this legislation […] are trying to do is to appear that they are for stem cell research when they oppose [its] most promising form — embryonic stem cell research."[11]

Anti-science folks like Lipinski and Huckabee should really organize a club. But what could they call it? They need a proper name which sums up their attitude. Flat Earth Society? No, that is already taken. Apparently there are folks who (still) appear to genuinely believe that if your boat sails out far enough, you will fall off the edge of the world.

I suggest a clearer title that is simple, direct, and easy to remember: the Anti-Science Society.

Or A.S.S. for short.

---

[9] CNN Political Unit. Huckabee narrowly tops 3-straight 2016 Iowa GOP polls. *CNN* [Internet]. 2014 Apr 16 [cited 2015 Feb 6]. Available from: http://politicalticker.blogs.cnn.com/2014/04/16/huckabee-narrowly-tops-3-straight-2016-iowa-gop-polls/

[10] Draper E. Huckabee endorses "personhood" amendment. *The Denver Post* [Internet]. 2008 Feb 25 [cited 2015 Feb 6]. Available from: http://www.denverpost.com/breakingnews/ci_8360651

[11] Hall R. Adult stem cell research puts patients first, proponents say. *CNSNews.com* [Internet]. 2008 Jul 7 [cited 2015 Feb 6]. Available from: http://cnsnews.com/news/article/adult-stem-cell-research-puts-patients-first-proponents-say

23 STEM CELL THANKSGIVING

Patricia Olson: CIRM's scientific guardian of the gates.

November 25, 2010.

Across the street from Monster Park, San Francisco, California, is a place where nightmares are fought. We'll go there in a moment, but first…

Today is Thanksgiving. I am thankful for so much. My beloved Gloria, wife of 41 years, did NOT have breast cancer. My father, Dr. Charles Reed, aged 88, survived a heart attack and is back to playing tennis three times a week. My son, Roman, continues to challenge paralysis, working every day to advance the cause of cure. My daughter, Desiree, is having an exciting career as a sports attorney. Of my grandchildren, Roman Junior is hitting homeruns, Jason's Tae Quan Do is a thing of beauty, as is Jackson's playing of the violin, and our youngest, Katie Sadie Desiree Reed, at 18-month-old should already be called Katherine the Great.

And after that?

My Thanksgiving joy is a statement from Bob Klein stating that he may ask voters to approve another bond measure on the 2016 ballot to keep the stem cell program going.

http://www.utsandiego.com/news/2014/feb/20/robert-klein-cirm-stem-cell-billion-2016/

Bob's term in office as Chair of the governing committee of the California stem cell program is coming to an end next month. You might think he would be glad to wave goodbye and retire to a well-deserved rest. But no, he may try to make another miracle.

In 2003, when he announced his plans to pass a $3 billion stem cell program, the general opinion was — no chance. The state was mired in a financial downturn and the opposition was spreading their anti-science poison; how could one small corner of medical knowledge so excite the world?

But California said yes! And the world won.

Let me tell you why — better yet, come to a meeting and see for yourself. First, go to the website, https://www.cirm.ca.gov/. Click on "Board Meetings", get the date of the next one, and find a site that's closest to you.

The official location now is 210 King Street, San Francisco, California, across the street from Monster Park. It iappears to be a small place, its entrance wedged between two massive buildings. It is easy to miss the modest sign with letters in raised iron: CALIFORNIA INSTITUTE FOR REGENERATIVE MEDICINE. But just push open the double glass doors, turn left, ride the elevator to the third floor, and there you are.

So many great people have worked there, some we know and others we should: from the first office manager Jennifer Rosaia to scientist Arlene Chiu, the spinal cord injury expert.

Here, smiling people fight the most dreaded conditions known to man. There are no hospital beds or laboratories and test-tubes, but it is a life-and-death struggle nonetheless.

This is where California's millions go out to stem cell research — and to what purpose?

Think of eternal pain, or loss of body control, or the slow subtraction of the mental self: conditions so awful that if done to prisoners it would violate the Geneva Convention against torture. But the California stem cell program, if it hurts, we fight it.

These are real-life problems, as down to earth as you can get. For example, think about something undignified, which we all have to do several times a day. No matter how elevated our position, we all need the restroom; how miserable life becomes when we cannot.

A friend who has prostate problems got caught short once while driving. He stopped at a gas station and tried to deal with the difficulty, but the pathway was blocked. In agony, he finally called Highway Patrol, and they sent an officer with a catheterization kit. My friend described it as the most humiliating moment of his life, for a stranger to put on a plastic glove, grasp his equipment, and shove a plastic tube inside — but the urine *had* to come out.

For many paralyzed people, catheterization happens 4–6 times a day. Not only is this time-consuming (and may require an attendant), but urinary tract infections are likely.

But what if somebody could take away that particular paralysis problem?

I was sitting in the conference room at the CIRM, with Roman beside me plus maybe ten to 12 other people — and a lot of folks connected to the meeting via conference phone.

The meeting was to reconsider a rejected grant proposal for a man who hoped to return bladder control to paralyzed people.

This was Dr. Leif Havton of University of California Los, Angeles and University of California, Irvine. He was an expert on the "spinal tail" injury. About 20 percent of all spinal injuries involve the tip of the spine, the part you hit if your chair is yanked out and you sit down hard. A breakage there can bring paralysis, neurological pain, and loss of bladder, bowel, and sexual function.

Roman and I knew Leif Havton's work. Over the past nine years, the small California law named after my son had funded Havton seven times: small grants, adding up to half a million dollars.

Havton applied for a CIRM grant of $1.6 million over three years, to use embryonic stem cells to patch up the spine and restore bladder control to paralyzed rats.

And he wanted to work with "chronics". A newly paralyzed person has an "acute" injury; after three days, the injury is considered chronic, which is far more difficult to deal with. Most attempts to heal paralysis are done close to the time when the injury occurs. There may be some natural recovery which the therapy could enhance; also, early intervention might lessen the wave of secondary damage, which happens when the body releases chemicals which chew up the spine, increasing the severity of the original wound. The Havton grant was important!

His grant request had been denied. The panel of reviewers were split; some thought it was great, others felt it was too ambitious. But the CIRM at that time had a policy called "extraordinary petition" for scientists who feel that mistakes or misunderstandings on the part of the reviewers may have caused the rejection of their grants.

(The policy has since been removed. It was an awkward process and was time-consuming for an overworked staff, but still it gave one more chance to the scientists, and its absence is a loss.)

Dr. Havton was allowed to call in from Southern California and make his case. He had a lot to say, and this was his field of expertise.

But he would not go unchallenged. Sitting in the office with us was Dr. Patricia Olson. If I was a scientist trying for a grant, I would absolutely want her on my side — not against me. Patricia Olson can be ferocious. She does not yell, but her intensity level? Wow. She fights to protect CIRM money as if it was her own.

Olson came at Havton's proposal like a lawyer in a trial, and he came back just as strong. Their words were like swords of arguments and counter-arguments, scientific words which whizzed over my head.

When it was time for public comment, Roman and I both spoke. We were in favor of the grant, not only because it seemed important, but also because we knew the man. Leif Havton is integrity personified; he will not tell you something cheerful just to make you happy, but will always stick to the facts he knows and the hypothesis he is exploring, and he never confuses the two.

Almost everybody on the ICOC board had something to say, but Dr. Oswald Steward could not speak on the subject because he worked at the University of California at Irvine where the research might take place.

When everyside had been heard, Melissa King called the roll. One by one their votes were taken. It was close. I counted on my fingers, yes, yes, no, yes, no — was it enough — Yes!

Leif Havton.

Dr. Leif Havton would be given the first instalment of his three-year grant. The rest will come if he met the agreed-upon milestones — and Patricia Olson would be looking over his shoulder.

Afterwards she came over and said she hoped we understood why she presented the negative side. Roman said, of course! That was her job; California would be ill-served if she did anything less. I mumbled something similar. But it was more than that — she was magnificent. She knew every dollar spent was one dollar less. By grudging every penny, she fought to make the program work at peak efficiency and to fund only the very best.

That is the spirit of the CIRM: patient advocates and government officials working together to "turn stem cells into cures" — as Roman Reed said, in what became the CIRM's official motto.

I never leave that building without a smile on my face and a renewed respect for the men and women who make it work. They are what I'm thankful for on this Thanksgiving day.

As this is written, our ten-year lease on the San Francisco address has run out. CIRM's new headquarters will be at 1999 Harrison Street, Oakland, three blocks from the beautiful Oakland museum. We appreciate the warm welcome of Oakland Mayor, Libby Schaaf.

24 SWIMMING FROM ALCATRAZ

Desiree Reed-Francois, shown here with brother Roman Reed, swam to shore from Alcatraz Island.

Sunday, June 20, 2010. I was waiting for my daughter, Desiree Reed-Francois, to swim in from Alcatraz, the old prison island in San Francisco Bay. She was taking part in the 30th annual "Escape from the Rock" swim.

There was no reason to worry, I told myself; Desiree was in top shape. As Deputy Athletic Director for the University of Tennessee, she worked out every day and ran full marathons routinely.

But there had been that article about great white sharks being spotted in the Bay… Desiree reminded me that her chances of being attacked by sharks were less than being struck by lightning — but logic was no help.

"See you in a bit," I said, careful not to say goodbye. She smiled, tugged on the official yellow swimming cap. Then she waved and ran barefoot into the mob of participants, heading for the boats that would take them to Alcatraz. Their voices merged into the hollow sound of footfalls on the dock, and then were gone.

Well, I told myself, get something done, work on your notes.

The 8th annual meeting of the International Society of Stem Cell Researchers (ISSCR) had just come to town for the period between June 16–19. Writing an article about it was like summarizing an encyclopedia.

The place where it was held, Moscone West, was luxurious but practical, like a palace designed for hard work. High-ceilinged and immaculate, it had an army of helpful people.

As I entered one jam-packed auditorium, five huge screens showed Alan Trounson, President of the CIRM, greeting everyone.

Jamie Thomson, inventor of human embryonic stem cell research, was there, as was Shinya Yamanaka, inventor of induced pluripotent stem cell (iPSC) research — like Babe Ruth and Hank Aaron in the same room!

Although English is not Dr. Yamanaka's first language, his presentation was clear and followable. His slides were brief, lessening information overload. Also, he understood the need for *pauses* when he spoke — small silences around complicated parts — so the brain could catch up. He described refinements to the iPSC process, changing skin cells to embryonic-like stem cells.

Yesterday, there had been a chance to meet him personally at the Japanese consulate. I had gotten there early, taking a trolley car to an older portion of San Francisco. Unfortunately, the building was old, and not wheelchair accessible. There was no ramp, and the stairs were steep. Since Roman would be joining us, I scurried around trying clear a wheelchair path. Rod McLeod, Japan's Consulate Coordinator for Public Relations, and other kind folks volunteered to lift and carry Roman and his power chair (separately) up the flights of stairs.

When Roman and Gloria arrived, I explained the situation. Roman immediately said no — had I forgotten how much his power chair weighed? He was about to turn the van around and go home. Wait a minute, I said, and ran inside.

Professor Yamanaka graciously came outside and visited with Roman for a while, leaning casually into the window of the van. He had a special interest in spinal cord injury, he explained, having once been a neurological surgeon.

Meanwhile, back at the ISSCR, half of one floor was reserved for the biomedical exhibitors.

Cellartis, the Swedish company, was present, and I enjoyed meeting Petter Björquist, who talked about a liver regeneration project his company was working on. In my country alone, an estimated 60,000 people are on borrowed time, hoping for a transplanted liver, without which they will die. But what if they could regrow their own?

The Node (http://thenode.biologists.com/) was a new online portal for people to share news about developmental biology. Cross-fertilization of ideas is a recipe for progress, and open platforms such as these are useful.

Or, The Company of Biologists (http://www.biologists.com/) where scientific magazines, such as the *Journal of Cell Science* and the *Journal of Experimental Biology*, are sold — and the profits go toward funding research.

There was a place where you could order tumor tissues — CELLnTEC advanced cell systems (http://www.cellntec.com/). Somewhat ironic — most people wanted to get rid of tumors; here you could buy one!

Scientists' posters, with brief descriptions of their projects and ideas, are the heart and soul of every convention. Sometimes ten feet long and four feet high, posters are crowded with small print and pictures, each one a piece of the puzzle: a possibility of cure. There were so many! On Thursday's poster session, I spent an hour just counting those which involved embryonic stem cell research — one hundred and one! And the next day they would put new posters up.

Katrin Grasme of Germany wrote on a question that has puzzled science for centuries: how do salamanders regrow their own spinal cords? The little amphibian grows a tube of protective flesh (a neurotube) around the wound. Inside this tube, the cells at the injury site reprogram themselves to the embryonic stem cell state — in a manner not unlike the iPSC process?

And there was something new: a poster "teaser session" where each scientist was allowed two minutes to talk about his or her project. I did not get a chance to see this in action but the idea seemed great. The scientists, under time constraints, must organize their thoughts and be clear.

Speaking of clarity, that was one thing San Francisco Bay did not have. The Bay's life-rich gray-green water is jammed with plankton that forms a shroud and shuts off vision at just a few feet, so you never know what might pop out in front of you.

There are sharks, of course, most terrifyingly the great whites, predators of sea lions, several of which were barking from the rocks. If the prey was there, could the predators be far?

"Swim in the middle of the pack," I had advised Desiree, "don't be the one that gets picked off!"

"Yes, Dad," she had said politely, just as if she had not heard my shark tips ever since she learned to swim at nine months.

Between 1972–1987, I had swum with sharks almost every day, as a professional diver for Marine World Africa USA. I should be out there with her, I thought, but reality tugged at my sleeve. Almost two decades had passed since then. A mile and a half swim for an out-of-shape oldster? I would probably need to be rescued myself.

I couldn't see any of the swimmers yet; just a ship, an enormous black freighter, cruising by. Get out of the way, I muttered, don't you know my daughter is trying to swim home?

The ISSCR is the United Nations of stem cell research. Countries represented here included this year's host, the United States of America, plus Germany, Sweden, Russia, Spain, Finland, Austria, Australia, Argentina, Italy, France, Japan, Belgium, Britain, Brazil, Mexico, Singapore, China, and more.

What a delight to see again Dr. Fanyi Zeng of Shanghai, China. The inventor of "Tiny" (the world's first iPSC-made mouse), Dr. Zeng is not only a terrific scientist, but an ideal unofficial ambassador — multilingual and personable — a friend to bring together nations.

International cooperation is a must: already more than a dozen nations had partnered with CIRM, and closer to home, several states as well. ICOC Chairman Bob Klein and Governor Jim Doyle of Wisconsin had just signed one of the all-important documents of cooperation.

Keynote speaker Fred "Rusty" Gage spoke about the hippocampus, a part of the brain which may break the rules and regenerate itself. If hippocampus nerve cells could regrow, perhaps the rest of the brain and spine might be influenced to do the same?

On Dr. Irv Weissman's panel was the young and promising Marius Wernig, who once showed up at a formal event wearing a black tuxedo and white tennis shoes — but tonight he was sartorially conservative. Too bad, I missed the shoes. Like Fanyi Zeng, Wernig had begun his career as a world-class musician — a violinist and composer of concerts.

Owen Witte, a student of Weissman's, took over the lab when Irv was overseas. In his absence, Witte hired a woman scientist named Ann Tsukamoto — who later became Dr. Weissman's wife! Witte was working on a way to fight prostate cancer. I squirmed a little, remembering the undignified part of the physical exam. But cancer steals lives as well as dignity.

Emanuelle Passeque addressed the process of aging as affected by DNA damage. As I become more and more ancient, such matters become interesting.

The sea before me now seemed so empty, and vast. The Bay is famous for its currents.

On the bench next to me was a swimmer who had tried the event last year.

"You turn your head to take a breath, and a wave smashes you in the face," she said, explaining why she had to accept a rescue then, and was not attempting it this time.

Swimming all the way in from Alcatraz seemed increasingly impossible.

There was a problem with the ISSCR's convention: there was no panel on patient advocacy.

Advocates are the in-betweens — not scientists, but supporters of those who are. Every convention on medical research needs our involvement.

Why? Because without public funding, the jobs will not be there for the scientists. We are the public, and if we are not involved, the politicians won't care.

Too many researchers have no clue of the tremendous battles going on; they just fill out the grant request forms, and blindly hope there will be some money for them. But what if there is not enough money to provide the jobs for the scientists?[1]

What if there are laws (with jail terms) prohibiting the research? In some states you can be put in jail for what the advocates made legal in California.

---

[1] Vastag B. U.S. pushes for more scientists, but the jobs aren't there. The Washington Post [Internet]. 2012 Jul 7 [cited 2015 Feb 7]. Available from: http://www.washingtonpost.com/national/health-science/us-pushes-for-more-scientists-but-the-jobs-arent-there/2012/07/07/gJQAZJpQUW_story.html

A patient advocate panel at the ISSCR convention would have discussed what scientists and patients can do to protect and expand research funding — essentially tips on legislative judo.

There are only about four thousand scientist members in the ISSCR. But in America alone, an estimated one hundred million people suffer chronic disease or disability — and they have families. What politician could ignore a constituency like that?

Four thousand scientists and one hundred million patients, plus families — should we not work together? Naturally, I found a microphone and spoke up about my concerns, and the new ISSCR President, Dr. Elaine Fuchs of Rockefeller University, was kind enough to suggest I write up my request and email it to her, which of course I did.

Some of my conference notes were incomprehensible, my handwriting being so bad. Someone (my scribbled notes did not identify him or her) said, "The regenerative industry is developing before your eyes. What we do now becomes the standard, so we have to be careful; our watchword must be safety, safety, safety!"

Safety. In the distance I could see kayaks and policemen on jet-skis, herding the swimmers, guiding them towards the shore.

One body was brought in on a liferaft — what, what? — but he sat up. He was fine, just exhausted.

Athletes were coming in now, graceful men and women, the professionals. The first swimmer bounded ashore after only 28 minutes in the water. But there were empty waves behind him.

Come on, Babe, get home, I found myself muttering, wringing my fingers.

Another wave of swimmers came ashore; Desiree was not among them. I began to play little games with my mind — that person was tall and slender, could that be her? But it was not. Not her, not her, not that one either, and then —

There she was, tall and strong, like a goddess of the sea, huge brown eyes alight with energy.

I heard myself shouting:

"DESIREE, THAT'S HER, THAT'S MY DAUGHTER!"

And then we were driving home, chattering like magpies, about the sea lion which had popped up beside her, and the tremendous force of currents like underwater waves —

And the scientists returned to their struggle.

25 BRIDGE TO A NEW LIFE

To be a stem cell researcher or biomedical lab worker — is that not a noble career? Imagine if your daily chores meant the opportunity to save lives and ease suffering, while taking on the greatest financial challenge to humanity today — the budget-crippling costs of chronic disease.

But how does a college student get that actual first biotech job?

Imagine the following job interview:

Boss: "What can you do?"

Applicant: "I can do anything!"

Boss: "When someone says they can do anything, usually that means they have not done anything yet. What experience do you have?"

Applicant: (painful pause) "Well, how can I get any experience, if nobody will give me a job?"

What if, instead, the applicant said: "I had a year's experience at a world-renowned institute"?

The California stem cell program is trying to help deserving young candidates get that edge — that all-important foot-in-the-door first job — as part of their college experience![1] Here is how the "Bridges to Stem Cell Research" program puts deserving students into a world-class laboratory.

First, the colleges themselves apply for a Bridges grant. Michael Yaffe, CIRM's Associate Director for research activities, currently oversees the Bridges program. As he puts it, "The Bridges grants [are] available to all colleges (public and private) except for research-intensive universities […]. Students at the big research universities already have opportunities to do projects in stem cell labs […] while those at smaller schools may not […]."

Selected colleges are provided with funding for the program, including stipends for the students. Students take basic stem cell training at their home college for the first year, followed by a second year in a lab position at one of the associated research institutions, during which they receive up to $30,000 as a stipend, as well as experience working at the lab.

---

[1] Funding opportunities [Internet] [cited 2015 Feb 7]. Available from: http://www.cirm.ca.gov/our-funding/research-rfas/bridges

At the end of the two-year program, they will have produced a Master's degree thesis,[2] the research for which was done in a world-class lab. This is priceless practical experience.

"Every college has a slightly different program," says Yaffe, "Three Junior Colleges, for example, have a certificate program designed for those who might want to go directly to work in a lab job at a biotech company or at a research institute."

They now have a degree qualifying them to go on to further education to become a research scientist, or go to work immediately as a fully qualified and experienced lab technician.

Up and down the state, from Humboldt to San Diego, colleges (16 in all) are participating in the Bridges program. I visited one of these at the San Francisco State University (SFSU) to interview Dr. Carmen Domingo, who directs the program there. (You can meet her and her students in a three-minute video at http://www.youtube.com/watch?v=GoY0n56wYkY.)

She was bursting with pride and enthusiasm, and her classroom was the same. These were students going through their first year of the stem cell program. After a base of knowledge in stem cell science had been established by the end of the first year, they would spend the second year at one of University of California San Francisco's partner institutes (while receiving a stipend of up to $30,000), including: the University of California at Berkeley, the University of California at San Francisco, Stanford University, the California Hospital at Oakland Institute for Research, and the Buck Institute for Aging Research.

As a teacher years ago, I organized a multi-cultural history club to point out contributions of various ethnic groups. It was a joy to see students from various cultural backgrounds in the SFSU program — African American, Hispanic, Asian, as well as Caucasian. According to Dr. Domingo, nearly 70% of the young scientists were either first in their family to achieve college, or a member of an under-represented minority.

Does the program work? The answer lies in the students themselves: look at the graduates of the first three years of the program; all of whom are working in the field, or continuing on to a higher degree.[3]

---

[2] Senate Bill 471, The California Stem Cell and Biotechnology Education and Workforce Development Act of 2009, requires the California Department of Education — in consultation with the CIRM and representatives of the biotechnology industry — to promote stem cell and biotechnology education and workforce development in existing programs, such as the California Partnership Academies, the California Career Resource Network, multiple pathways, and other existing programs.

It also requires that stem cell biology be included in the Science Framework that specifies what is taught in the science classroom. The bill requests the University of California, San Diego in consultation with CIRM, to include stem cell and biotechnology information in the California State Summer School for Mathematics and Science, as specified.

[3] CIRM Bridges: building California's stem cell research workforce [Internet] [cited 2015 Feb 7]. Available from: http://www.cirm.ca.gov/our-progress/video/cirm-bridges-building-californias-stem-cell-research-workforce

For instance, one such student, Myra G., said, "I'm doing my research [...] on cystic fibrosis, a genetic disorder that causes sticky mucus to line the lungs and airways. Generally, people [with this condition] die before the age of 40 [...]. The Bridges program introduced me to literally hundreds of uses for stem cells to treat both disease and injury. In addition, the program allowed me to combine my interests and perform research in a lab that was using stem cells to model and correct genetic disease. The Bridges program provided me with hands on experience [...]. Regenerative medicine is a relatively new field in respect to disease modeling and I hope to be part of it for many years to come, something I wouldn't have considered if not for this program [...]."

Another student, Ian B., "I found out about the CIRM Bridges program when I started looking for states and universities that conducted stem cell research. I had just graduated from Arizona State University with a degree in biology but didn't know what to do with my life. I thought of my best friend who suffered a spinal cord injury when he was 16 and was told he would never walk again [...], and of my grandparents who died of Alzheimer's, [and] my father-in-law with early onset Parkinson's. [...] Stem cell research was the one field that [might] have helped them all. It is because of the CIRM Bridges program that I will be continuing my education and research on [...] stem cells at a world class institution. The CIRM Bridges program gave me the foot-in-the-door help I needed to get into the field of my dreams."

The California stem cell program is advancing the field of regenerative medicine for the good of all humanity. Maybe you should be part of that tremendous effort too.

To find out more, visit http://www.cirm.ca.gov/

Carmen Domingo, San Francisco State University's Director of CIRM's Bridges Program.

26 SKIDDING ON ICE

Jane Lebkowski: Chief Scientist throughout the Geron/Asterias human trials.
Dr. Lebkowski faced every challenge, never gave up, and triumphed.

Before his van crashed, Roman Reed was having a great day. It was near midnight and snow was falling steadily when my paralyzed son reached the Los Angeles grapevine, halfway over the mountains.

With his favorite road music from Metallica booming over the sound system, he was driving home after giving a speech at the University of California at Irvine. He was there to rally support for the new paralysis research bill (Assembly Bill 190, Wieckowski, D-Fremont). AB 190 was intended to restore funding to the Roman Reed Spinal Cord Injury Research Act.

The original bill, "Roman's Law", was small: just $1.5 million a year, but it brought in much more money than it cost. For a total investment of $14.6 million over its first ten years, the successful research had attracted $63.8 million in additional funding from the National Institutes of Health and other sources, new money for California. (The final figures were $17 million and $84 million.)

But the funding had been yanked. The program itself had been renewed, unanimously — but without funding. Because money was so short in Sacramento, the new bill would be self-funding: paid for by traffic ticket add-ons: three dollars extra, the money to go to paralysis cure research —

The van hit a patch of black ice. It spun around twice, ramming into the guardrail with thunderous impact and the screech of metal tearing.

Roman was in the driver's seat, frantically working the hand controls. The 300 pound wheelchair behind him flipped like a toy, breaking in half — as the van went through the guardrail and stopped, one wheel dangling over the abyss.

Silence, but for the tick… tick… tick… of the cooling engine block.

And there Roman stayed, unable to reach his cell phone, until a pair of Good Samaritan truckers stopped to help him in the snow. A tow-truck got the van to a nearby Wendy's, where a kindly manager let it park in the lot overnight, after the truck drivers lifted Roman into a booth.

"Oh, hi, Dad," he said on a borrowed cell phone, "I had a little accident, no big deal…"

By an amazing coincidence, Gloria's sister Leah had been driving North not far from him. She and husband Steve immediately turned their car around and drove back onto the snowy grapevine. They got Roman out, just before snow closed the road.

When Roman got home, his 20-month-old daughter, Katherine, looked at the dried blood on his face and exclaimed, "Daddy hurt!" She ran for the box of baby wipes and went to work on her father's face, scrubbing the blood spots vigorously on his bruised lips. He kept smiling.

So Roman would be out of action for a while, right? Well, not exactly. With the help of Craig's List, Roman located a 12-year-old van, a couple new batteries, a second-hand wheelchair, and was, in the words of Willie Nelson, "On the road again."

A greater blow awaited. The Geron trials were canceled. The world's first embryonic stem cell human trials — on spinal cord injury! — would go no further, aside from monitoring the people who had already received the cells. This was research for which "Roman's Law" had provided initial funding. And now, not only was Geron cancelling its human trials, but was leaving the stem cell field altogether.

Roman and I attended the farewell party of the Geron Stem Cell Research Department. We sat beside Jane Lebkowski, Chief Scientist, Joe Gold, sixteen years at Geron, and Kate Spink, organizer of the world's first human trials with embryonic stem cells.

We were proud to be there. As Geron's former CEO Dr. Thomas Okarma said in a letter supporting the renewal of Roman's Law:

"[…] Geron Corporation succeeded in achieving FDA (Food and Drug Administration) concurrence to begin the world's first human trial of our embryonic stem cell-based therapy for spinal cord injury. This would not have been possible without support from the Roman Reed Program […]."

For eight years, Geron had met every objection of the FDA piling up 20,000 pages of correspondence. They were winning. And then, money problems got in the way.

Here was a roomful of giants: 66 of the world's best stem cell research scientists and technicians. They had succeeded in the initial stage of the world's first human embryonic stem cell trials. The safety trials for GRNOPC1, a stem cell product intended to ease paralysis, had been successful.

They should have been lauded as heroes; instead they were losing their jobs. They had done everything right, played by the rules all the way, and they had triumphed. By FDA requirements, the first trial for any new drug or therapy must be related to safety. Could the cells be injected into a newly paralyzed person without doing them harm? Opponents of the research said the stem cells would cause cancer: it didn't happen. There was no harm done to any of the patients. The second step would have meant larger quantities of cells into other newly injured patients, perhaps five times as much. But that step, as well as the final one to try still larger quantities of stem cells on a much larger group of patients, would not take place with Geron.

Why were the trials stopped? The immediate answer is financial.[1] To take a new therapy or medicine to market is incredibly costly. The price was just too high for Geron, a small company.

Political harassment was also part of the picture: endless hate-filled propaganda from the Religious Right and its allies in the Republican party, plus actual political attacks, laws proposed to fine scientists, and even jail them for research. There was also favoritism shown to the "less controversial" adult stem cell research, funded lavishly compared to embryonic ones. In 2008, human adult stem cell research received $297 million in federal funds — embryonic stem cell research received only $88 million. Even now, under the far more progressive Obama Administration, adult stem cell research received nearly triple the funding that embryonic stem cell research received: $341 million compared to $125 million.[2]

But the people of Geron had changed the equation. Never again should a newly paralyzed person be told what my son and those before him were told — that there was no hope.

Katy Spink spoke to her colleagues, "Time after time over the past eight years, I have seen you deliver on goals that the rest of the world considered impossible [...]. Together, we have pushed back the bounds of medicine forever."

Joe Gold stood up and added his voice, "The methods we've discovered to grow human embryonic stem cells [are now] tools being used by researchers all over the world [...]. We made a path that not only educated the FDA but also provided a route for others to bring [...] life-changing therapies to people who desperately need them [...]. We have all been part of something that has changed the world."

And Jane Lebkowski?

---

[1] Brown E. Economics, not science, thwarts embryonic stem cell therapy. *Los Angeles Times* [Internet]. 2011 Nov 21 [cited 2015 Feb 8]. Available from: http://articles.latimes.com/2011/nov/21/health/la-he-geron-stem-cell-20111121

[2] Estimates of funding for various research, condition, and disease categories (RCDC). National Institutes of Health (U.S.); 2015 Feb 5. Available from: http://report.nih.gov/categorical_spending.aspx

"You faced social, religious, scientific, financial, regulatory, clinical, and media challenges, yet you overcame all these hurdles with rigor, integrity, and grace […]. The technology will continue to be developed because you have shown that it is possible […]."

"And finally," she said, "let us remember the words of Senator Edward Kennedy:

"The work goes on, the cause endures, the hope still lives — and the dream shall never die."

27 HELEN KELLER AND STEM CELL RESEARCH

Though blind and deaf, Helen Keller, shown her with her great teacher Anne Sullivan, inspired the world with her courage. (Wikipedia photo.)

When I was ten I fell on a bamboo stick, which injured my right eye. I was fortunate, recovering partial vision of 20/400, meaning I see at 20 feet what others see at 400. But it was a terrifying experience. I had only one good eye left; what if I lost that one too? I would practice being blind, blindfolding myself with gym socks and stumbling around the room.

Not to read, or watch expressions change on a face, or see the colors of the sky??

Helen Keller was born healthy on June 27, 1880, in Tuscumbia, Alabama. In what was almost a preview of her life, the toddler learned to communicate by sign language; little Martha Washington, six-year-old daughter of the family cook, understood her, and could interpret. In time Helen developed over 60 signs, by which she let her wants be known.

Dennis and Catherine Bowes Rickman: Two champion scientists, fighting blindness with stem cells.

At nineteen months, Helen Keller contracted meningitis, and became both blind and deaf.

Her parents spoiled her and she became wild, flying into a fury if denied anything. The home was filled with silent rage, and people were afraid of the ferocious little girl.

A doctor, J. Julian Chisholm, referred the family to Alexander Graham Bell, inventor of the telephone, who then recommended Helen to the Perkins Institute for the Blind — from which came the partially-sighted Anne Sullivan.

When Sullivan arrived, she gave a doll to seven-year-old Helen, then grabbed her hand and spelled "d-o-l-l" into her hand. It did not work. For more than a month, the connection was not made; Helen did not understand the constant grabbing and tickling of her hands. Frustration mounted, until one day she flew into a rage and smashed the doll.

But then the teacher put Helen's left hand under running water from a pump while spelling "w-a-t-e-r" into the palm of her right hand. Helen froze, then grabbed Anne Sullivan's hand and spelled the word back — then dragged her around the room, grabbing one object after another, desperate to know everything's name, to re-connect with the universe.

At 12, Helen became friends with Mark Twain, who coined the phrase "Miracle worker," referring to Ms. Sullivan. Twain brought the young Miss Keller into contact with Standard Oil magnate Henry Rogers and his wife Abbie; they paid for Helen's education.

Ms. Keller learned to speak and to "read lips," putting her fingers on the speaker's mouth. In time she would cover the mouths of Presidents, including Dwight D. Eisenhower. With the Braille system of raised dots on a page, she learned to read and write — and eventually authored 12 books.

She cofounded the American Civil Liberties Union, fought for women's right to vote, workers' need for safe conditions, socialism, religious tolerance — and is even credited with introducing the Akita dog to America after being presented with one by the Emperor of Japan!

She wrote words that, as Shakespeare said, "will echo down the corridors of time," including the immortal line: "Life is either a great adventure, or it is nothing."

Her philosophy and ethic fits with the subject of patient advocacy: "I long to accomplish a great and noble task, but it is my chief duty to *accomplish small tasks as if they were great and noble. The world is moved along not only by the mighty shoves of its heroes, but also by the tiny pushes of each honest worker.*" — emphasis added.

In 1964, Helen Keller received the Presidential Medal of Freedom.

But there is another Helen Keller story, which I read long ago, but cannot verify.

An aggressive reporter was interviewing Ms. Keller. He kept asking the same question: what was her life like, being both blind and deaf? Again and again, he asked; finally she snapped back:

"You want to know what my life is like? *My life is a black hole.*"

It is not enough to simply admire Helen Keller and draw inspiration from her courage. For those who truly honor her memory, we need to know what she focused on in the last years of her life — and carry on from her example.

Helen Keller fought for cure, challenging the conditions which afflicted her, raising money for medical research. If she were alive today, there is no question in my mind that she would be active in the stem cell struggle, using research to fight blindness and deafness.

How she would have loved the California stem cell program!

Deafness cure is being systematically advanced at the lab of Dr. Stefan Heller at Stanford University.[1] With the aid of research grants from CIRM, Heller is working to use stem cells (both embryonic and induced pluripotent) to regenerate tiny hair cells within the ear — the vibrations of which translate to sound.

"Chickens," said Dr. Heller in a recent interview, "Chickens can regenerate hair cells in their inner ear. If humans could do the same, we might not go deaf!"

Ms. Keller would have enjoyed meeting the Saturday Night Live comedian Will Forte. Forte was friends with another stand-up comedian, Dennis Rickman, PhD, whose day job was in medical research, trying to use stem cells to defeat blindness. Forte helped

---

[1] Heller S. 2014 Research Report [Internet]. 2014 Dec 10 [cited 2015 Feb 7]. Available from: https://hearinglosscure.stanford.edu/2014/12/2014-research-report/

Rickman raise $10,000 to start a program called the Stem Cell Initiative for Eyes (SCIfeyes) at Duke University in Raleigh, North Carolina.[2]

I spoke to Dr. Rickman and his wife, Dr. Catherine Bowes Rickman.

He advocated the ethical use of both adult and embryonic stem cells for scientific research, and had a very good reason for doing so — in 1995 he had been diagnosed with leukemia.

After a two-year search, a young woman in Germany was found with bone marrow like his. She then shared her stem cells with him.

This gave Dr. Rickman more than ten years of additional life. He used those years to fight for sight for others, even though he was losing his own.

Several forms of eye disease attack the retina — the inner lining at the back of the eyeball — which turns light into vision. Stem cells might restore the damaged retina.

"I'm not naïve enough to think that it will be done in the next few years," he said, "[But] when I'm not able to do this work anymore, there will be someone else carrying it on."[3]

He passed away February 21, 2010.

But the work he did continues today. Dr. Rickman's wife, Catherine Bowes Rickman, PhD, is a recognized authority on Age-related Macular Degeneration (AMD), the most common form of adult onset blindness.

"He had the vision, but not the funding," she said of her husband in a personal communication.

Two friends of the Rickman research family are Mark Humayun and Dennis Clegg — currently collaborating on a disease team grant from the California stem cell program.[4] Are they the ones who will have the great breakthrough? The international project also involves David Hinton, also from the University of Southern California, and Pete Coffey of University College, London, U.K.

The problem they face is immense. The National Eye Institute estimates one in five people over 65 suffer AMD. In their public write-up for the CIRM grant, they state:

"Retinal degeneration represents a group of blinding diseases [...]. By 2020, over 450,000 Californians will suffer [...] vision loss or blindness due to AMD, the most common cause of (eye disease) in the elderly. Part of the retina, called the macula, [lets]

---

[2] SNL comedian to help launch stem cell research fellowship at Duke Medical Center [Internet]. 2005 Nov 17 [updated 2005 Nov 18; cited 2015 Feb 7]. Available from: http://corporate.dukemedicine.org/news_and_publications/news_office/news/9398

[3] Padget A. SNL comedian promotes local stem cell research. The Chronicle [Internet]. 2005 Nov 22 [cited 2015 Feb 7]. Available from: http://www.dukechronicle.com/articles/2005/11/22/snl-comedian-promotes-local-stem-cell-research#.VOoBc_mUdqU

[4] Stem cell based treatment strategy for age-related macular degeneration (AMD) [Internet] [cited 2015 Feb 7]. Available from: https://www.cirm.ca.gov/our-progress/awards/stem-cell-based-treatment-strategy-age-related-macular-degeneration-amd

people read, visualize faces, and drive [...]. The disease initially causes damage in central vision, [leading to] legal blindness."

With AMD, the center part of the visual field is a black spot, with only a fringe of vision around the edges.

"Put a thumb in front of your eye," said Dr. Humayun when I interviewed him, "Then imagine trying to deal with that all day long."[5]

What keeps a person's retina in good shape?

"A layer of cells at the back of the eye called the retinal pigment epithelium (RPE) provides support, protection, and nutrition to [...] the retina."

If the support cells go bad, the retina deteriorates. But there maybe a way to prevent or even reverse that negative process. Using embryonic stem cells, the team hopes to nurture the cells which benefit the retina.

Dr. Humayun continued, "Recent advances in knowledge and technology of human embryonic stem cells bring new hope for the development of cell replacement treatment [...]. RPE cells derived from human embryonic stem cells are a potentially unlimited and robust source for regenerating RPE [...]."

Will it work? This time, thanks to the California stem cell research program, the scientists will have funding — almost $19 million — to find out.

Helen Keller's battle — and Dennis Rickman's — will go on.

---

[5] Facts about age-related macular degeneration [Internet] [updated 2013 Jul; cited 2015 Feb 7]. Available from: https://www.nei.nih.gov/health/maculardegen/armd_facts

28 HOW TO MEND A BROKEN HEART

> "Hearts will never be practical, until they can be made unbreakable"
> — L. Frank Baum, *The Wonderful Wizard of Oz*

It was 9:20 on a recent Sunday morning; time to pick up Gloria. On Sundays she walks to Church, and after the service we hold hands and take a walk together.

So I turned off the computer, trotted downstairs, out the door, up the street, turned right — and there was Gloria. She was too early. I usually had to wait outside the church ten or 15 minutes, shifting foot to foot, reading my notes or — was something wrong?

The color was gone from her face. Midway through the service, she said, pain had struck in the middle of her back and chest. One arm felt heavy. She broke out into a cold sweat.

Heart attack, must be. All those symptoms, plus that is how her family usually dies. Gloria is high-strung, close to my age of 70, and a little on the heavy side…

So of course we rushed to emergency, right? Well, not exactly.

Gloria's brother Marty (just beginning his cancer radiation treatments) and sister Brenda (recovering from hers) were coming to the Bay Area, and Gloria had promised to take them to an Oakland A's baseball game. Not until next day could I could persuade her to go to the hospital.

But there at last she was, snug and safe in the emergency room. There was a needle in the back of her hand, wires taped to her chest, an oxygen clip in her nose, not to mention monitors beeping, the squeak of nurses' rubber-soled shoes —

"I'm bored," said Gloria.

Heart attack. The number one cause of death in America — claiming approximately one million lives annually.[1]

Some have heart conditions that are not immediately fatal, but which put them at risk.

The American Heart Association has estimated that 5.7 million Americans currently suffer from heart failure. Cardiovascular disease costs an estimated $286 billion in direct (out of pocket) and indirect (time lost from work) expenses.

---

[1] Heart disease: scope and impact [Internet] [cited 2015 Feb 8]. Available from: http://www.theheartfoundation.org/heart-disease-facts/heart-disease-statistics/

Stanford's Joseph Wu is fighting endstage heart failure, using embryonic stem cells to take on threats to people like the author's wife Gloria.

All those lives, all that money. But what can we do, if the heart goes bad? We could try and make a mechanical replacement, like a robotic pump.

Here is a riddle. What is the connection between the invention of the mechanical heart and the cartoon character "Tigger" in Walt Disney's *Winnie the Pooh*?

Answer: Paul Winchell.

Paul Winchell was a world-famous puppeteer (his classic TV shows with Jerry Mahoney and Knucklehead Smiff ran for years), and was also the voice for cartoons including Tigger and Gargamel from the *Smurfs*. He was medically trained and had a tremendously inventive mind, holding more than 20 patents — including the very first patent for a mechanical heart (U.S. Patent #3097366).

The first robotic heart put into a person was the Jarvik-7, invented by Robert Jarvik. Jarvik himself "denies that any of Winchell's design elements were incorporated into the device he fabricated for humans, [which was] successfully implanted into Barney Clark in 1982."[2]

But when it was time to patent the Jarvik heart, which was developed at the University of Utah, they were unable to do so because Winchell held the patent. The college asked him to donate his invention to them, which he did.

The controversy continues to this day. Dr. Henry Heimlich, inventor of the Heimlich Maneuver, said, "I saw the heart, I saw the patent, and I saw the letters. The basic principle used in Winchell's heart and Jarvik's heart is exactly the same."

---

[2] Paul Winchell [Internet] [cited 2015 Feb 8]. Available from: http://www.jarvikheart.com/basic.asp?id=72

In any event, the mechanical heart was a huge step forward. It was especially valuable as a temporary assist between operations, to keep the patient alive while waiting for a substitute heart.

Real heart transplants are practical. The problem is there aren't enough hearts to go around.

Former Vice President Dick Cheney had to wait 20 months for a transplanted heart. At last a heart was found that matched, and his life continues today.

Might there be a stem cell answer? Joseph Wu is trying to find out. Working with Deepak Srivastava of the Gladstone Institute and Wolfram Zimmerman of the University Medical Center Goettingen in Germany, Dr. Wu is battling heart attack damage.

Their weapon is tissue engineering, which involves the development of biological substitutes to restore, maintain, or improve the functioning of a body part.

Dr. Wu said, "Our proposal seeks to use [...] tissue patches seeded with human embryonic stem cell-derived cardiomyocytes [heart cells for treatment of] heart disease in small and large animals [...]."

Have you ever fixed a flat tire on a bicycle and had to glue on a rubber patch?

Tissue engineering may work similarly, except the patch becomes part of the patient.

As quoted from his fund request statement, Dr. Wu said, "[...] Novel therapies with stem cells in combination with supportive scaffolds [are emerging as a promising] avenue. Engineered tissues have now been used to make new bladders for patients [...], and more recently new trachea (in the throat — dr) for patients with late stage tracheal cancer.

"Our team intends to push the envelope by developing human tissue-engineered myocardium — heart muscle for treatment of] heart failure [...].

"At the end of three years, we are confident we will be able to derive a lead candidate that can move into [...] preclinical development. These discoveries will benefit millions of patients with heart failure in California and globally."[3]

And Gloria?

"I found no evidence of lasting damage," the doctor said.

We had dodged a bullet.

With exercise and proper diet, my angel will be free to harass her husband for years to come.

---

[3] Heart repair with human tissue engineered myocardium [Internet] [cited 2015 Feb 8]. Available from: https://www.cirm.ca.gov/node/22431/review

Gloria Reed, heart attack survivor: long may she reign in the Reed Household! Shown here with spectacular son Roman and happy grinning Grampa, Don C. Reed.

29 WHY WE CAN'T AFFORD *NOT* TO CURE PARALYSIS

Hans Keirstead: Called stem cell's Elvis Presley, a rock star for research.

Californians Hans Keirstead and Mark Tuszynski are world leaders in paralysis cure research.

What they have in common is that both received small grants from the Roman Reed Spinal Cord Injury Research Act before they became the superstars they are today. Success with these little grants led to larger ones from institutions like the NIH, the Christopher and Dana Reeve Foundation (CDRF), and the California stem cell research program.

Mark Tuszynski is the quieter of the two. Interviewing him, one focuses on the science.

He was a skeptic initially about using stem cells to replace nerves lost after spinal cord injury.

"I did not think it would have much of an impact," he said, "Because of the difficulty in growing axons through the injured spinal cord, and the complexity of neural brain circuits. How would stem cells navigate through the maze of the spinal cord?"

But Tuszynski's opinion changed after his group started working on neural (nerve) stem cells at the urging of a paraplegic scientist, Paul Lu, in his center for neural repair at the University of California, San Diego.

Quiet but determined, Mark Tuszynski is like rain falling softly, bringing an end to the drought. When paralysis becomes a condition that can be healed, and may that day come soon, people standing up from their wheelchairs will owe him a round of applause.

"We tagged the nervous system-derived stem cells with a green fluorescent marker, which allowed us to track the cells after putting them into the spine," he said.

"To keep the cells from washing out of the spinal cord, we embedded them in a gel. Finally, we added proteins to stimulate the cells' survival.

"To our astonishment, the grafted nerve cells extended new connections (axons) over long distances, some nearly the entire length of the spinal cord."

Usually, when a non-scientist like myself looks at a photograph of a damaged spine, it is hard to tell what is going on. One can see a V-shaped gap in the spinal column, or a little tentative fuzz which might or might not be new growth — it's hard to get excited.

This was different. At a recent meeting of the California stem cell program, the audience gasped at the micro-photograph. Green-marked nerves leaped across the gap in the injured spinal cord, bridging to the other side. The new nerve cells were bio-marked with green, so you could follow the growth — it was like the whole spine was slathered with lime green paint.

And what did it mean, in practical terms? Was there any recovered motion?

"Injured rats with completely severed spinal cords recovered significant motion, including the ability to move every joint of their limbs," said Dr. Tuszynski, "But they could not support their weight on their legs. So there is still room for improvement."

Tuszynski and his colleagues are moving forward to test the procedure on non-human primates at the University of California at Davis. Success or failure there will determine if the approach is brought to human trials.

And after that?

"Chronic," he said. That is a huge word to a paralyzed person. It refers to an older injury. Anyone who has been paralyzed more than a couple of weeks has chronic paralysis and under current conditions is likely to remain as they are.

Most experiments are done with new injuries, called "acute". That is the right time to try to help, the sooner the better, but it also makes it hard to tell if it was the treatment which helped, or something else. For instance, newly paralyzed Roman received injections of Sygen, as well as heavy exercise — which one helped him? Both? We cannot tell. But if improvement is gained for a chronic patient who has been paralyzed for some time, you can know for sure that was a successful treatment.

And the second scientist? Dr. Hans Keirstead is a cheerfully dramatic and outgoing individual who loves to fly his own helicopter — and is a master in the world of business.

Professor of Anatomy and Neurobiology at the University of California at Irvine, Hans founded the Sue and Bill Gross Stem Cell Research Center, and holds a leadership position at California Stem Cell, Inc.

Working with his father, Hans built hospitals in Africa, distributing 600 wheelchairs and several tons of multi-vitamins.

His stem cell expertise involves him in many endeavors, including human trials of a way to fight cancer. ("Not my invention," he hastens to add, 'The approach was invented by Dr. Robert Dillman of Hoag Hospital. I led the manufacturing scale-up, and the commercialization.")

Keirstead is attacking paralysis four ways: (1) to reduce the damage from the body's reaction to new injuries; (2) to eliminate the spinal cord scar and allow nerve communication between brain and body; (3) to reinsulate damaged nerves in the spinal cord; and (4) to develop new motor neurons, great long nerves which stretch from spinal cord to muscle.

Why did spinal cord injury call to him?

"I wanted to help with brain injuries and conditions," he said, "This was a great place to start. The spinal cord is anatomically simpler than the brain, and controls functions that are easier to measure than the brain."

Treatments which help spinal cord injury may also help traumatic brain injury, ALS, stroke, cerebral palsy, spina bifida, Parkinson's, muscular dystrophy, multiple sclerosis, Spinal Muscular Atrophy (SMA; that cruel condition which paralyzes and kills children, often before the age of two), and more.

And if the research does not get funded so that paralysis continues, uncured?

According to a major survey initiated by the CDRF, 5.6 million Americans are paralyzed — 1.9 percent of the population in 2008 when the survey was taken — which is almost one person in 50.[1]

Check this out. First, round off the 1.9% to two percent for ease of calculation. If you divide that into the population of your home state, you can tell roughly how many people are paralyzed there.

For instance, my home state of California has a population of 38 million. Two percent of 38 million is 760,000 paralyzed people — and they have extraordinary medical expenses to pay.

---

[1] Paralysis facts and figures [Internet] [cited 2015 Feb 8]. Available from: http://www.christopherreeve.org/site/c.mtKZKgMWKwG/b.5184189/k.5587/Paralysis_Facts__Figures.htm

According to CDRF's estimates, the first year of paralysis alone will cost a high quadriplegic (paralyzed in both upper and lower body like Christopher Reeve) $1,023,924 — and subsequently $171,808 every year for as long as he or she lives.

The cost depends on where the injury is located in the spine. The higher up, the worse it is, and the more expensive. Christopher Reeve had a C-1 (first cervical vertebra, right before the brain) injury. That is the worst.

Roman's injury was lower down, injuring the fifth and sixth cervical vertebrae. Someone in his condition faces medical bills of roughly $739,874 in the first year and $109,077 annually thereafter.

A paraplegic (with injuries below the neck) will incur $499,023 first year expenses, followed by $66,106 every year after that.

Since few people are rich enough (or have enough insurance) to handle such costs, they turn to the government for help — programs like MediCal and MediCare which are paid for by taxpayers; i.e., you and me.

Everybody who pays taxes is paying to take care of people with paralysis.

Isn't cure a better way?

As Aileen Anderson, Director of the CDRF Spinal Cord Injury Core, said, "Even a modest treatment […] could save $770,000 over the lifetime of one [spinal cord injured patient] not only to impact quality-of-life and independence of [the] patients, but also reduce the shared costs of healthcare and loss of productive employment."

Aileen Anderson: determination and excellence.

30 THE WAR WE MUST NOT LOSE

Logically, the war on stem cell research should be over. Not only do we have a President who supports it, but our country is essentially united on the issue.

Recent polls show a huge 72% support for embryonic stem cell research, including a majority of Republican voters (58%), Democrats (82%) and Independents (73%). Only 12% of the public oppose using stem cells for biomedical research.[1]

How many issues are there where at least three out of four Americans agree?

And yet the opposition keeps piling on the anti-science legislation.

Below are five federal attacks on the research, followed by a far larger list of state assaults. It was written in 2011, and needs updating. Look at it only as an idea of the attacks we face.

As you glance through it, look out for two things.

First, notice the letter R after each of the sponsors' names? Each of these anti-research bills was sponsored by a Republican.

Second, look at the titles of the bills where they are available. None are labeled anti-stem cell research. Politicians can read polls too. They do not want the blame for obstructing cure.

1. HR 110: Human Cloning Prohibition Act of 2009 (Fortenberry, R-Nebraska) — would make an advanced form of stem cell research (nuclear transfer) a federal crime.
2. S 99: Ethical Stem Cell Research Tax Credit Act of 2009 (Vitter, R-Louisiana) — stacks the deck by establishing tax credits for adult stem cell research, while prohibiting it for embryonic stem cell research.
3. HR 227: Sanctity of Human Life Act of 2009 (Brown, R-Georgia) — a personhood bill, giving full rights of citizenship to every microscopic blastocyst, including those about to be thrown away.
4. HR 877: Patients First Act of 2009 (Forbes, R-Virginia) — prioritizes stem cell research funding permanently in favor of adult stem cells and against embryonic stem cell research.

---

[1] Gardner A. Most Americans back embryonic stem cell research: poll. *U.S. News* [Internet]. 2010 Oct 7 [cited 2015 Feb 8]. Available from: http://health.usnews.com/health-news/managing-your-healthcare/research/articles/2010/10/07/most-americans-back-embryonic-stem-cell-research-poll

5. HR 1654: Cures Can Be Found Act of 2009 (Paul, R-Texas) — amends the Internal Revenue Code itself, providing tax breaks for adult stem cell research but not embryonic stem cell research.

(For a look at *current state of stem cell policies*, visit the Empire State stem cell program's list at http://nyscf.org/scmapus/.)

For an idea of the kind of threats that stem cell research still faces, read on.

Arizona: SB 1307 and HB 265 (2010) — "A person shall not [engage in nontherapeutic research that] results in injury, death, or destruction of an in vitro human embryo." Since in vitro means in the Petri dish or test tube, that means no human embryonic stem cell research.

Alabama: SB 335 (2010) — Definition of person for code construction includes sex cells from moment of fertilization.

California: Human Life Amendment (2010) — This personhood initiative did not gather enough signatures to get on the ballot, but was backed by the California Republican Assembly.

Colorado: Personhood law (2010) — This was a copy of the original personhood bill, with slightly varying language, such as "beginning of biological development" instead of "fertilization". Fortunately, the more people studied it, the more they shook their heads in disbelief; the initiative was crushingly defeated at 70.5% to 29.5%.

Florida: Personhood Amendment (2009) — An attempt to ban state funding for embryonic stem cell research was inserted into the budget; happily, it was blocked (with a line-item-veto) by then-Republican Governor Charlie Crist.

Georgia: Ethical Treatment of Human Embryos Act (2009) — The ultimate anti-woman and anti-science bill, this would subject women who have miscarriages to criminal investigation on the basis of prenatal murder. (In February 2011, Georgia's Senate filed a new state constitutional amendment, the Georgia Personhood Constitutional Amendment, or HR 5, which emphasized the paramount right to life — from the fertilized egg onward.)

Iowa: In 2011, Iowa's new Governor Terry Branstad wanted to reinstate a 2002 state law that limited the use of embryonic stem cell research. The Iowa House subcommittee also passed a bill that year pressing for personhood from the moment of conception.

Maryland: HB 925 (2009) — An amendment pushing for personhood.

Michigan: SB 647-652 (2013) — A six-pack of hyper-regulation laws, SB 647 through to SB 652 were apparently designed to make embryonic stem cell research unbelievably complicated, and with jail time for errors. Fortunately, while the bills passed in the Senate, they failed in the House. (There was also an attempt to revoke Proposal 2, the hotly debated and successful stem cell freedom initiative.)

Mississippi: HR 1091 (2013) — Senator Roger Wicker supported this personhood bill, Life at Conception Act. This is the same Roger Wicker who co-sponsored the archaic Dickey-Wicker Amendment which put personhood into law before embryonic stem cell research even began. (His co-author on Dickey-Wicker, James (Jay) Dickey famously stated that there were no homosexuals in his district.) The Mississippi House passed a law forbidding The University of Mississippi to use state funds for research that would destroy human embryos as well.

Missouri: There was an unsuccessful attempt to revoke Amendment 2, which allowed Federally-approved stem cell research, and also an attempt to pass a personhood bill in 2010. For Missouri, this is the tip of the iceberg. Despite the passage of a Constitutional Amendment to protect the research, Missouri has been continually assaulted by the Religious Right. There were 33 such attacks, most of which were organized by the Missouri Roundtable for Life whose then-leader, Ed Martin, admitted that a main purpose of their many legal actions, "to be candid, is nuisance."[2]

Montana: Montana Personhood Amendment (2010; previously 2008) — While this failed to make the ballot, Montana attempted to constitutionally enact personhood laws.

Nebraska: LB 606 (2008) — Researchers were compelled to accept these restrictions in exchange for the pledge of no new anti-research legislation by right-to-life groups, who then immediately tried to put pressure on University of Nebraska, the only place in the state where human embryonic stem cell research could go on. Fortunately, and despite tremendous pressure, an 8–8 regents vote prevented the ban.

New Jersey: A 170 and S 335 — Bills were introduced to prohibit public funds for stem cell research. Republican Assemblyman Michael Patrick Carroll's prospective Assembly Bill 1383 would insert state government into IVF family planning, indicating that a human embryo which has not been implanted two years from the date of its creation *shall become a ward of the state*.

North Dakota: HB 1450 (2010) — The Defense of Human Life Act, created by State Representative Dan Ruby against stem cell research, was passed by the North Dakota House. HB 1572, by the same author, was defeated in the Senate in 2009.

Oklahoma: HB 1442 (2011) — The Destructive Human Embryo Research Act has passed first committee and now goes to the floor of the House.

South Carolina: S 450 (2009) — The Life Beginning at Conception Act sought to establish that the right to life for each born and preborn human being is bestowed at fertilization.

Texas: HB 1109 (2011) — These amendments to the Health and Safety Code declare that the life of an individual human organism begins at the moment that the initial splitting of a human cell occurs during fertilization. According to this bill, an unborn human organism is alive and is entitled to the rights, protections, and privileges accorded to any other person. Texas has been attacked by a multitude of GOP anti-science bills, including SB 1600, HB 58, HB 1358, SB 1, SB 98, SB 174, SB 629, SB 641, SB 655, and SB 1033.

Virginia: HB 1639 (2010) — The Right to Life Bill is a personhood law which states that human beings, even preborn ones, have constitutional right to enjoyment of life

---

[2] MO Roundtable's frivolous legal complaints dismissed... again [Internet] [cited 2015 Feb 8]. Available from: http://www.firedupmissouri.com/content/mo-roundtables-frivolous-legal-complaints-dismissed-again

from moment of fertilization. Republican Governor Bob McDonnell also inserted an anti-embryonic stem cell research amendment into the budget.[3]

Wisconsin: In the state where embryonic stem cell research began, Republican Governor Scott Walker is opposed to it, and has made the statement that "scientists agree adult stem cell research holds more promise than embryonic" stem cell research, a statement which is blatantly false.[4]

In light of all this, Bernie Siegel urges, "Bills have been filed in almost every state [...] seeking restrictions on human embryonic stem cell research funding in some way, shape, or form. The advocacy community [has] work to do."

---

[3]Whitley T. McDonnell proposes budget amendment to ban stem-cell research. *Richmond Times-Dispatch* [Internet]. 2011 Feb 3 [cited 2015 Feb 8]. Available from: http://www.richmond.com/news/mcdonnell-proposes-budget-amendment-to-ban-stem-cell-research/article_c5bacbc6-d798-5755-b8d9-acb24ca463fd.html

[4]Kertscher T. Scott Walker says scientists agree that adult stem cell research holds more promise than embryonic. *PolitiFact* [Internet]. 2010 Oct 21 [cited 2015 Feb 8]. Available from: http://www.politifact.com/wisconsin/statements/2010/oct/21/scott-walker/scott-walker-says-scientists-agree-adult-stem-cell/

31 IN WHICH I GET CANCER

DON C. REED, BARBARA C. REED, DAVE F. REED, CHARLES H. REED

No one is immune from disease or accident: Don Reed, cancer; sister Barbara, cancer; brother David Reed, leg nearly amputated from motorcycle accident; and father Dr. Charles Reed, cancer survivor.

"Scoot your bottom over," ordered the nurse, as I lay on my side in the backless hospital gown.

I felt a sudden inclination to say no. If I just got up and put my clothes on, I could walk on out of there. Maybe life would go back to normal, and I could skip what lay ahead.

But the moment passed. I did as requested, and the mechanical sample-taker invaded, going deep inside, taking 12 little bites out of me. If that sounds gross and terrifying (and painful, despite the anesthetic), that is only because it was.

The test results came back, and an appointment was set up for the next day.

"You have cancer," said the doctor, by way of greeting.

Since 90% of all males get prostate cancer if they live long enough, I did not panic immediately. But then he mentioned some numbers.

Dr. Owen Witte reports directly to the President of the United States on progress for cancer cure.

This year, roughly 240,000 male Americans will be diagnosed with prostate cancer; one in eight will die of it.[1]

Because prostate cancer is (usually) slow-growing, many men never know. They die with the cancer, but not because of it. It was just not big enough to give them problems.

But for some, the cancer must be dealt with immediately. A measurement called the Gleason score identifies the cancer from 1 (no problem) to 10 (write the will).

With my score of 7, action was probably required. Options included surgery to remove the prostate gland, radiation therapy, hormone shots, or some combination of the above.

I went with the surgery first, the "radical prostatectomy."

It was a nightmare with moments of comedy. My wedding ring had to come off before the surgery, and that seemed out of the question. The only time that ring had been off was when I was a diver for Marine World 30 years ago. I was 30 pounds thinner back then, and the ring slid off my finger in the shark tank. I found it later by sheer luck, gleaming right beside the filtration grid, trembling on the edge.

"We might have to cut the ring off," they said. Apparently electricity could short circuit — and arc to the ring like a lightning bolt — and amputate my finger!

Fortunately, with the help of soap and a strong-armed nurse, the ring was removed.

While I slept, a spider-like little robot crawled across my belly and then cut deep, guided by the doctor. There were five incisions. Using the robot enabled the cuts to be more precise and smaller.

---

[1] Prostate cancer [Internet] [cited 2015 Feb 8]. Available from: http://www.cancer.gov/cancertopics/types/prostate

When I woke up, my first words came out in a squeaky voice like Elmer Fudd's in the *Bugs Bunny* cartoons, the kind that went "you wascally wabbit." "Where is my wing, my wedding wing?" I enquired.

What a joy to wake and see my family in the room. Roman and Gloria and Terri and Little Man and Jason and Katherine my little angel… I was drifting in and out of consciousness, and visitors were not supposed to stay, but I did not want Roman to leave. "Roman, stay there," I said in my Elmer Fudd voice. He understood, and would not let the nurses chase him off.

It was three days before the bleeding from the wounds stopped. I had to go to emergency once, having lost too much blood.

When the prostate was dissected, the news was not encouraging. The cancer was aggressive and moving: the Gleason score had climbed a notch higher — up to eight now, from seven.

Best-case scenario? Hopefully the cancer was gone, removed with the prostate gland. But if the cancer had spread beyond the prostate gland…

Should I get radiation therapy — six weeks of daily sessions — with such possible (though unlikely) side effects as rectal cancer?

Cancer is tough to kill. It may go into remission and come back later.

But if cancer is tough, so is the California stem cell program. It was funding several projects that were suddenly very important to me.

One project was pioneering Stanford University scientist Dr. Irv Weissman's work, which I wrote about previously.[2]

Another, led by Dr. Owen Witte of the University of California, Los Angeles,[3] involved body signals that turn cancer on — and, hopefully, off.

To say that Owen Witte is a qualified researcher is like saying Mount Everest is a hill. The author or co-author of 363 published papers, Dr. Witte is the Director of the Eli and Edythe Broad Center of Regenerative Medicine and Stem Cell Research. If I were to list all his awards, accomplishments, and responsibilities, it would take another chapter.

With a CIRM Basic Biology IV grant (#RB4-06209) of $1,254,960, Dr. Witte was studying the *pathways* of cancer.

Do you know that hippopotamuses fight each other on land? The gigantic water mammals come up onto the shore and trot to the grass. They charge at each other on special pathways worn down by their bodies, crashing into each other. The winner gets to mate, so they fight enthusiastically — but they do not kill each other. When one loses, it just leaves the pathway.

Dr. Witte's work identifies pathways that regulate the growth of normal prostate stem cells. Human cancers utilize these same pathways to promote tumor progression.

---

[2]Reed DC. Fighting the killers: California stem cell program takes on leukemia and cancer. *Huffington Post* [Internet]. 2013 Sep 20 [cited 2015 Feb 8]. Available from: http://www.huffingtonpost.com/don-c-reed/stem-cell-cancer-research_b_3956160.html

[3]Owen Witte, M.D. [Internet] [cited 2015 Feb 8]. Available from: http://people.healthsciences.ucla.edu/institution/personnel?personnel_id=45695

Cancer "dysregulates" these tiny pathways.

Was there a way to let the good stem cells control the pathways and stop the prostate cancer?

Dr. Witte says, "Our recent studies have uncovered an important [molecule called Trop 2] that is expressed on prostate cancer cells, [which] predicts poor prognosis for many tumors including prostate, ovarian, pancreatic, breast, gastric, and colorectal cancer."

Trop 2 activates cancer in what is to me a very strange way.

It is like a video game my granddaughter Katherine explained to me, called "Fruits and Bombs". On the screen is a bunch of watermelons, pineapples, and grapes — and cartoon bombs. You run your fingers across the screen, slashing the fruits, but if you touch a bomb, it explodes, and you "die".

For Trop 2 to activate the cancer, it has to be slashed (or cleaved) by enzymes called proteases. If we can block the slashing, we might prevent the cancer.

Dr. Witte explains further, "Our goal is to use [...] molecules to block the first cleavage on Trop 2. [We predict that] will be an effective strategy to prevent disease progression in prostate and other cancers [...].

"[Currently], advanced stages of prostate cancer are treated with hormonal therapy, which causes significant changes in mood, bodyweight and composition, impotence and gynecomastia (men growing breasts) in addition to the pain and suffering from the disease."

His goal? "To [...] extend life and minimize suffering of men with [...] prostate cancer."

And, he adds, "Many of the molecules that we are investigating are implicated in a range of tumors [...]. Our findings may [also] benefit patients suffering from [other numerous kinds of] cancers."

With determination, intelligence, and passion, CIRM scientists like Owen Witte fight to solve diseases like prostate cancer — which threatens every male on the planet.

Lupron. I had an injection of the female hormone replacement. It felt like a harpoon flung across the room. Apparently, if a male takes female hormones, it discourages the cancer — at least for a while. Funny; I had argued many times with opponents of stem cell research about Lupron, which was used in the IVF procedure and which (while generally safe for women) had definite side effects for a male.

I experienced all manner of gross stuff, like needing to wear adult diapers. My muscles went suddenly soft, previously light barbells became unliftably heavy, and running around the lake was impossible, though I still continued to walk. And my nipples extended like an excited woman. Oh, that's just temporary gynecomastia, reassured the doctor; some men grow actual breasts.

And the agony of indecision. Should I get radiation and, if so, what kind? I visualized it to be like a ray gun. I would get zapped once a day, five days a week, for seven weeks.

But how big should the focus of the rays be — narrow focused, or broad beamed?

Think of a circle with mouse ears: that is the shape of the prostate gland, about which every man, if he lives long enough, will become aware of. (My best friend from childhood, Ken, says that if we were cars, all males would be recalled for defective prostates.)

My cancer had spread to one of the mouse ears.

I asked for two opinions.

One doctor was convinced I should get the radiation; another was not so sure — and what kind of beam, narrow or broad?

If I got the narrow beam radiation, it would zap only the prostate — or rather, the hole where the prostate had been. The narrow beam had less side effects, but could also miss more cancer cells. The broad beam took on the whole neighborhood (lymph nodes) as well. Broad beams kill more cancer cells (nothing gets them all), but side effects could be serious — like rectal cancer.

Too much radiation, and I could end up with a colostomy bag and have to go to the bathroom out of a hole in my side. Too little radiation, and I could die before my goals were achieved. I needed at least ten years more of life, to achieve "the plan" my personal set of life goals.

First and foremost, I wanted part two of Proposition 71, which Bob had hinted at.

"At the end of 2015, we will take some major state-wide polls," he said, "And if we have earned the voters' support, we will do it again — for five billion dollars."

There were other goals as well — to watch my grandkids grow, to write five more books, and to see at least one of them made into a movie.

Above all, I wanted to see my son fulfill Christopher Reeve's prophesy: to "stand up from [his] wheelchair, and walk away from it forever."

Paraphrasing what Moses implored in the *Book of Deuteronomy*, "Behold, I place before you the means to life and the means to death; therefore, choose life."

Life with major inconveniences — or probable early extinction.

I thought about my brother David, choosing to keep his leg. His motorcycle had gone off a cliff, landing on top of his leg when he lit in a tree, shattering the bones terribly. The choice was amputation or a series of painful operations, where the pieces of bone were stacked around metal rods. I had encouraged David to keep his leg, and he had done it.

"Give me everything," I said.

It was the fourth day of the third week of radiation, and as I walked toward the cancer treatment center, I was feeling a little sorry for myself. Well… okay, I was wallowing in self-pity!

Why couldn't the cancer have waited a few years, till we have reliable stem cell cures?

The doctors had predicted fatigue and nausea, and were right on both counts.

But they had not mentioned the nightmares. I had dreams of death: waking up from a preview of my own funeral, watching the coffin closing over myself, feeling myself lowered down into the grave, hearing the thumps of shovelfuls of dirt tossed in, as the earth closed over me.

Was my body shutting down, getting ready to die?

No, no, no, the cancer social worker said, dreams like that were usual, just the body and mind brooding about the worst possibilities, as could happen with any worry.

And this was just the "honeymoon period"! Everyone agreed the first couple of weeks were the best part of the treatment. It would be worse later on, they said almost gleefully, as the radiation *accumulated in my body*. Four weeks of radiation yet to come — I imagined myself glowing in the dark — and the worst would be the two weeks afterward, when all the treatment was done.

Not to mention whatever gruesome side effects might remain. Almost anything was possible — stuff like the penis growing together, as the hole in the middle closed up…

The cancer center was busy, professional, and stocked with cheerful people.

One of the nurses had no hair on her head, undergoing cancer treatment herself.

In the dressing room, I put on the backless robe, tied it awkwardly, wondering if the word "embarrass" had its roots in "bare ass" — and went out into the men's waiting room.

One huge Samoan guy and I talked about stem cells and our grandkids — but most of the patients were in no mood for chatter. On their faces I tried to read my own emotions. Angry? Cheated? Fearful? Mourning a future that might have been?

"Mr. Reed!" said the nurse with a smile, like I was a friend dropping by everyday, and who just happened to want cancer treatment. I was glad when they used my last name. It lent the proceedings a shred of dignity, of which there was precious little, with questions or comments like: "How's your urination?" or "We'll have to shave that off," or "You'll probably have a sunburnt penis," and other such conversation.

The radiation machine was so big it had its own room. It cost $2.2 million and looked like the mechanical monster in the *Robocop* movie; a metallic beast with malicious intent.

I kicked off my shoes, lowered the adult diaper so the radiation could get at the target area, and climbed aboard the table, lying on my back. There was a sheet underneath, for ease of dragging.

I waited for my steering wheel, a foot-wide rubber circle connected to nothing. I would hold it on my chest during the procedure to keep my upper body still — ahh, there it was.

"Soup three," said the male nurse, and the female nurse tugged the sheet towards herself, just a little.

"Ant one," she said, and he tugged his side a half-inch down.

"Superior and anterior," she explained, whatever direction they needed to drag me.

Laser rays aimed at the hospital tattoos on the sides of my hips. Like crosshairs on a gun, they would line up exactly, locking the X-rays in position, so the radiation went where it should.

"See you in a while," said the nurse. The foot-thick door shut, leaving me alone with the poison.

"Grrrrrrrrr…" went the X-ray robot. I felt the nearness of monstrous metal arms closing in around me, nearer, nearer — "Zeeeeeeet!" came the radiation.

I wanted to see what was going on, but did not dare. It wouldn't hurt, they said, if I opened my eyes during the procedure. But what if they were wrong and I got my eyeballs fried? I did not want to end up like the factory workers who painted radium on the hands of watches, and who ended up getting cancer from their jobs. You'll be fine, their employers had said.

So I hid behind my eyelids, listening to the machine's buzz and growl. Sometimes it moved, sometimes it moved me.

"Grrrrrr — zeeeeet. Grrrrrrrr. Grrrrrrrr. Zeeeeeeeeeeeeeeet. Grrrrrrrrrrrrrrrrrrrr."

I remembered my sister Barb in the hospital, fighting the cancer, lying on her side, unconscious from the chemotherapy. I was not allowed to go near her, even with a mask on, because her immune system was shut down — they had taken her to the edge of death, and one breath of an outsider's germs might kill her. Barbara was still alive today, after 17 years of battling cancer, leukemia, scleroderma, and other conditions, constantly in and out of the hospital. She had endured. She had taken everything, and was still alive.

I opened my eyes, and caught a glimpse of painted cherry blossoms and pale blue sky. Was I hallucinating? I shut my eyes again, and the painted ceiling went away.

"Grrrrrrrr. Zeeeeeeeeeet." Silence. More silence.

I did not move; just waited.

"Okay, see you tomorrow," the nurse said, helping me sit up. Twenty-one treatments to go...

As I walked out into the parking lot, it occurred to me how lucky I was. I was getting a treatment for cancer. But what if I had not had the insurance? I would just rot and die. Across America and the world, how many people got no treatment at all, and just died at the will of the cancer?

It is December 30, as I write these words. Completion of this book is scheduled for January 15, 2015, two-and-a-half weeks from today. Then I give it to the publishers, World Scientific Publishing, Inc., and they will work their magic.

My cancer results? Still unknown. As long as the Lupron is in my system, it drives the numbers down, and the test will not be reliable. I had two injections (more would probably have no effect; it quits working after a while) and must wait till the effects are completely gone.

32 SICKLE-CELL ANEMIA AND THE POLITICS OF PAIN

Ted Love: A life dedicated to defeating sickle-cell anemia.

Once at the gym I was trying to explain to an African–American woman why stem cell research was so important. We had known each other almost 30 seconds, so naturally it was time for the subject to come up — and I used the example of sickle-cell disease (SCD) which disproportionately threatens black Americans.

Was she familiar with sickle-cell disease?

"Only that my Auntie passed from it," she said, "And plus I work for a biomed company..."

Oh well, open mouth, insert foot.

What is SCD? Imagine a round red ball, a blood cell, soft as if full of corn oil. Now imagine it suddenly curving into a "C"-shape, resembling a sickle, and hardening.

When the blood cells "sickle," they clog the arteries in what doctors call a sickle-cell crisis. A sickle-cell crisis may last an hour — or a day — of excruciating pain.

Like broken glass in the veins, the sickle-shaped blood cells can do devastating damage.

Donald Kohn, a researcher at CIRM, explains, "By 20 years of age, about 15% of children with SCD suffer major strokes […]. By 40 years of age, almost half of the patients have had central nervous system damage leading to significant cognitive dysfunction […], recurring damage to lungs and kidneys […], frequent hospitalizations, [and] early death […]."

Why does this condition single out African–Americans? It is apparently a defense against malaria, common to sub-Saharan Africa, but the defense is worse than the disease.

Poverty makes things much worse, increasing the damage when proper treatment is not available. The populations most affected by sickle-cell disease suffer from significant healthcare disparities, which lower the quality of care they receive for their disease.[1]

The U.S. Census declared in 2010 that 15.1% of the population lived in poverty. Of these, 9.9% are white persons while 28.4% are black persons.[2] The median family income for a white family in 2012 was $57,009 — for a black family, $33,321.[3] Because blacks are far more likely than whites to be poor, they often lack decent medical insurance.

Lacking proper medical advice, SCD sufferers often do not know the most crucial care.

For instance, according to CIRM's sickle-cell anemia fact sheet, the most common recommendation for people with sickle-cell disease is to stay hydrated. *The more water a person drinks*, the less likely it is that their abnormal blood cells will clog. Another effective treatment is a medication called hydroxyurea, which reduces crises by 50% and death by 40%, *but most adults are not treated*.[4]

As a result, the problem is worsening. More than 90,000 Americans have SCD and despite decades of research, the average life expectancy has gone down from 42 years in 1995 to death at 39 today.

President Obama's Affordable Care Act (ACA) brings reasonably priced medical care to millions of families who never had it before. This is a wonderful accomplishment, and history will thank him.

But the ACA could have been so much better if the Republican-dominated Roberts Supreme Court had not interfered with it. Initially, the program would have covered just about everyone who did not have good insurance.

For example, in 2011, the ACA *required* states to expand Medicaid eligibility to families with incomes below 138% of the federal poverty level; i.e., $31,000 a year for

---

[1] Sickle-cell anemia fact sheet [Internet] [cited 2015 Feb 11]. Available from: https://www.cirm.ca.gov/our-progress/disease-information/sickle-cell-anemia-fact-sheet
[2] Poverty in the United States. *Wikipedia* [Internet] [updated 2015 Feb 5; cited 2015 Feb 11]. Available from: http://en.wikipedia.org/wiki/Poverty_in_the_United_States#Poverty_and_race.2Fethnicity.
[3] DeNavas-Walt C, Proctor BD, Smith JC. Income, poverty, and health insurance coverage in the United States: 2012. U.S. Census Bureau; 2013 Sep. 88 p. Report No.: P60-245.
[4] Sickle-cell anemia fact sheet [Internet] [cited 2015 Feb 11]. Available from: http://www.cirm.ca.gov/our-progress/sickle-cell-anemia-fact-sheet.

a family of four. However, a 2012 Supreme Court ruling made it *optional* for states to expand Medicaid eligibility.

As a result, a rich state like Texas can ignore the suffering of the poor — because it chose to "opt out" of ACA.

As of this writing, 21 state governors (all Republicans) have opted out of ACA, shrugging their shoulders about their constituents whom they just cut off from medical care.[5]

But while poverty increases the suffering, even a billionaire is not safe. Right now, the best you can hope for is a relative with a blood match.

Which brings us to Bert Lubin, MD.

If you live in Oakland, California, you probably know Dr. Lubin, who for more than 36 years has been working to save children's lives from SCD.

According to CIRM's sickle-cell resource webpage, "At the Children's Hospital of Oakland, Dr. Lubin began the Sibling Donor Cord Blood Program, [which is] offered to families across the United States who currently have a child with a blood disorder such as sickle-cell anemia [or some other disease], *and who are expecting another child*. Following the birth of a healthy child, [cord blood] is harvested. Because cord blood is enriched with blood-forming stem cells, it is cryopreserved (frozen) and can be later used for transplantation. *A number of lives have been saved* following transplantation with cord blood units collected in this program [...]" (italics mine).

A transplant of healthy bone marrow may also overpower SCD — unless the body's immune system rejects it. If a proper "match" is found (between brother and sister, for instance), that bone marrow transplant might eliminate SCD. But not everyone has a brother or sister, let alone a healthy one — and it is still dangerous to take bone marrow from one person and give it to another. If it is not a good match, the donor's bone marrow may attack the recipient's body.

So what is to be done? If you visit the SCD information page at the CIRM website (https://www.cirm.ca.gov/our-progress/disease-information/sickle-cell-anemia-fact-sheet), you will see the approach California is taking. First, there are a bunch of "blood study" projects, to gain a better understanding of the red stuff that keeps us alive. Then come the more specific methods, building on the basic science, to try and bring an actual cure.

What if there was a way to take the patient's own bone marrow out, fix it, and put it back?

That is the approach taken by three Southern California scientists: Donald Kohn and Victor Marder, both of the University of California, Los Angeles, and Thomas Coates of the Children's Hospital of Los Angeles. They are working together on an eight million dollar ($8,834,129) project — to try and stop blood cells from sickling.

---

[5]Where the states stand on Medicaid expansion [Internet]. 2015 Feb 11 [cited 2015 Feb 11]. Available from: http://www.advisory.com/daily-briefing/resources/primers/medicaidmap.

The scientists want to "treat patients with SCD by transplanting them with their own bone marrow, [with stem cells] genetically corrected by adding a hemoglobin (blood) gene that blocks sickling of the red blood cells.

"This approach has the potential to permanently cure [SCD]. A clinical trial using stem cell gene therapy for patients with SCD will be developed by this team [which will treat SCD] patients from across the state [...]. All scientific findings and biomedical materials produced from [these] studies will be publicly available to non-profit and academic organizations in California [...]."

Will it work? California intends to find out.

Another approach is being tried by one of CIRM's board members, Dr. Ted Love, who has dedicated his life to finding a cure for sickle cell disease. His company, Global Blood Therapeutics, has just begun clinical trials with a one-a-day pill to prevent the agony of sickling.

http://www.bizjournals.com/sanfrancisco/blog/biotech/2015/01/global-blood-therapeutics-sickle-cell-disease.html?s=print

And one more thing about Bert Lubin, whom I know from many meetings with the Board of Directors for the California stem cell program.

Bert and I have a long-standing friendly disagreement about which stem cell approach is better: adult () or embryonic. He says adult, I say embryonic. But it is like the old song that goes, "potatoes, potahtoes, tomatoes, tomahtoes" — we have differing opinions, but are united in the quest for cure.

If adult stem cells turn out to be the best treatment for this particular disease, I will shake Bert Lubin's hand and say, "Point for your side!" and I imagine he will smile back and say the scientific equivalent of "told you so!"

But the winners will be the children: released from the pain of SCD.

Bert Lubin, President, Children's Hospital, Oakland Research Institute.

33 MOWGLI AND THE MATRIX: A YEAR IN THE LIFE OF THE CALIFORNIA STEM CELL PROGRAM

Pretend for a moment that you are Mowgli the jungle boy, sleeping on a branch of a tree, in the heart of the Indian forest. To you, that forest is irreplaceable: each tree is wildlife shelter — food for the body, beauty to delight the soul.

Suddenly your nose twitches, and you wake to the acrid whiff of distant smoke… a fire is coming — you have a little time — what do you do? Run, hide, pray for rain?

Or divert a river, soak the ground and save the forest?

The meetings of the California stem cell program are long, full of talk, and sometimes even a little snooze-inducing, if truth be told.

Once, a certain aging advocate (who shall be nameless to protect the guilty) fell asleep in the back of the room, and vice Chair Art Torres led three cheers — hip, hip, hooray! — to rouse the snoring sleeper.

But it is my forest nonetheless, and I am seldom happier than at these gathering of friends and warriors.

But like the smell of distant smoke, we all know danger is approaching.

Time is ticking away, toward a day when the money is gone, and there will be no California stem cell program anymore.

What do we do, when the money runs out? For me, the answer is one word: **continue**.

Proposition 71 was like a farmer's early work in the fields. The ground had to be cleared, boulders had to be removed, the soil had to be plowed, and the seeds had to be planted. Now there was something to protect. Under the ground, the seeds have broken open; pale tendrils struggle up toward the sun. Green shoots are breaking through, where only dirt and rocks had been before.

Continuing Proposition 71 will bring us to the harvest of cure. This is when the work pays off, as theories becomes therapies, and people thought to be incurable get well — not just better, but *well*.

Run with me now, through the forest of the CIRM: highlights of one year, 2010, in the life of California's stem cell program. (The *underlined dates* mark the meetings. Most quotes are from CIRM news releases, generally authored by our indefatigable media representative Don Gibbons. For more detailed information on these meetings, you may go to https://www.cirm.ca.gov/ and click on Public Meetings under Board & Meetings, followed by transcripts for the date of interest.

January 27, 2010: "[...] State Controller John Chiang introduced a resolution recommending that we should 'begin planning the CIRM's future' [...]."

February 4, 2010: "The Board [...] today authorized $40 million to fund [the] Tools and Technologies II Awards, [which are] for new stem cell therapeutic tools [...] to break through *technical roadblocks* and accelerate the path of stem cell therapies to reach patients suffering [...] (italics mine).

For example, in a statement on her grant, "The Stem Cell Matrix: a map of the molecular pathways that define pluripotent cells", Jean Loring of Scripps Research Institute says:

"A major roadblock in the development of stem cell therapies is the lack of tools for quality control [...]. We have developed a new method (PluriTest) for [identifying] stem cell populations [...]. At a cost of about $150 to $300, the data are uploaded to the PluriTest website. Ten minutes later, the user gets [...] a 'pluripotency score' [...].

"PluriTest has caught on in the stem cell research community: in the three years since the paper was published [in *Nature Methods*], almost 8,500 gene expression profiles have been [...] analyzed by PluriTest. There are more than 500 registered users from 29 countries [...].

"We are now developing new tools based on the PluriTest concept [...] for cell therapy for Parkinson's disease and multiple sclerosis, cardiac (heart) cells, and hepatocytes (liver cells) for use in transplantation and drug screening [...] to assure the quality of their cells."[1]

February 18, 2010: Responding to yet another legislative "do-over" political bill, Senate Bill 1064, leaders of our program said: "[...] We wish to express our [...] concerns regarding [SB 1064]'s potential economic impact on the state's new tax revenues and new jobs created by CIRM, [and] on finding treatments and cures for diseases and traumas [...].

"CIRM has already adopted extensive regulations to balance the opportunity for Californians to benefit from their investment in stem cell research while ensuring that the research is not impeded. These [...] include revenue sharing, access plans, and pricing preferences for public entities. CIRM's governing board adopted the regulations after extensive public hearings, a process John Simpson of Consumer Watchdog has referred to as a model for making policy [...]."[2]

February 25, 2010: "The CIRM [...] has launched an on-line stem cell education portal. The extensive set of course materials and activity resources will help [...] educators prepare the youth of California to join the fast-growing biotech economy [...]."

March 11, 2010: "[To speed therapies into doctors' offices], the Targeted Clinical Development Awards provide [funding for clinical trials] and activities that support [them]."

---

[1] The stem cell matrix: a map of the molecular pathways that define pluripotent cells [Internet] [cited 2015 Feb 11]. Available from: https://www.cirm.ca.gov/our-progress/awards/stem-cell-matrix-map-molecular-pathways-define-pluripotent-cells
[2] Klein R, Torres A, Roth D. Letter to: Alquist, Elaine K. 2010 Feb 18. Available from: http://www.cirm.ca.gov/sites/default/files/files/board_meetings/Letter_Regarding_SB%201064.pdf

April 29, 2010: Vice President and former Senator Art Torres said, "[$28 million has been] approved for Basic Biology II [...], answering fundamental questions about stem cell biology [...]." For instance, Yong Kim from the University of California at Los Angeles is studying what makes an embryonic stem cell able to become any cell of the body. The board felt his work was so important that they voted to give him four years' worth of funding, which summed up to $1,259,371.

"The board endorsed AB 1931 [...] to extend the Roman Reed Spinal Cord Injury Research Act of 1999 through January of 2016. [...] The Act, which continuously appropriates funds of spinal cord injury research to the University of California, San Diego was scheduled to expire in January of 2011. (Note: this was CIRM's recommendation, but not a guarantee that the bill would pass.)

"[The] governing board [then] voted unanimously to endorse SB 1064 with [amendments that had] been negotiated with the bill's sponsor, Senator Elaine Alquist." (This was a quiet triumph of diplomacy, with much back and forth between Senator Alquist and us.)

May 21, 2010: "The Human Biomolecular Research Institute presented disease advocacy awards to board members Joan Samuelson (founder of Parkinson's Action Network) and David Serrano Sewell at a Celebration of Stem Cell Research [...]."

June 15, 2010: "[CIRM] and the New York Stem Cell Foundation [signed an agreement to] foster collaboration between researchers funded by the two organizations and accelerate the pace of advancement of regenerative medicine."

June 17, 2010: "Wisconsin Governor Jim Doyle [and CIRM] will sign a Declaration of Cooperation in stem cell research [...]."

June 22, 2010: "[The board] approved $25 million to fund 19 projects intended to overcome immune rejection of transplanted stem cells."

August 19, 2010: The board approved the concept of major grants to top-notch teams, as much as $20 million each. Each team must bring a new therapy or medicine to human trials within four years, a very ambitious target. The board also voted to support the California Umbilical Cord Blood Collection Program (AB 52, Portantino).

August 23, 2010: "The leadership of CIRM [...] deplores the decision of U.S. District Judge Royce Lambert to freeze federal funding of all human embryonic stem cell research [as a result of the lawsuit, Sherley v. Sebelius]."

October 5, 2010: "Tomorrow, October 6, researchers, patients, advocates and students around the globe will celebrate Stem Cell Awareness Day, with special activities planned in five countries, six states, seven research institutions and 25 high school classrooms [...]."

October 7, 2010: CIRM President Alan Trounson said, "CIRM and the Andalusian Initiative share the mission to accelerate the field of stem cell research [...] through international collaborations that involve the best scientific endeavors, regardless of geography."

In approval of the program, Natividad Cuendes, Executive Director of the Andalusian Initiative for Advanced Therapies, said, "This kind of agreement and collaboration allows the sharing of scientific knowledge and promotes translational research that brings therapies to patients."

October 13–15, 2010: A group of eight distinguished scientists, advocates, ethicists, and biomedical officials will be reviewing CIRM's performance. They are: Alan Bernstein, Executive Director of the Global HIV Vaccine Enterprise; George Daley, Director of stem cell transplantation at the Dana Farber Cancer Institute; Sir Martin Evans, Nobel Laureate; Igor Gonda, Chief Executive Officer of Aradigm Corporation; Judy Illes, Professor of Neurology at the University of British Columbia; Richard A. Insel, Chief Scientific Officer for the Juvenile Diabetes Research Foundation; Richard Clausner, formerly Executive Director of the Bill and Melinda Gates Foundation's Global Health program; and Nancy Wexler, Professor of Neuropsychology at Columbia University.

October 21, 2010: "[Funding was] approved for 19 awards worth [$67 million in total], designed to move good ideas out of the lab and into [clinics and hospitals]."

For instance, Mark Zern, from the University of California, Davis, said, "In California, as in all parts of the U.S., there are not enough livers available for transplantation for all the people who need them. The result is that many more people die of liver failure than is necessary. One way to improve this situation is the transplantation of liver cells rather than whole organ transplantation [...]."

October 26, 2010: The finance sub-committee discussed how biomedical businesses could borrow research money from CIRM — and pay it back.

At a time when construction unemployment in California reached Depression-era levels, in some counties as high as 30 percent (!), CIRM and donors funded and opened *three laboratories in seven days* — and provided 25,000 job-years of work.

These laboratories comprise the Eli and Edith Broad CIRM Center at the University of Southern California, *another* CIRM-Broad cooperative effort at the University of California Los Angeles called, the Broad Stem Cell Research Center at the California Institute for Regenerative Medicine, and the world's largest stem cell laboratory, Stanford University's Lorry I. Lokey Stem Cell Research Building.

December 3, 2010: Bob Klein accepted the nomination for a second term as Chair — but only for as many months as it would take to find a "replacement" for him.

December 8, 2010: Members of the eight-person External Advisory Panel (EAP), who in October investigated CIRM's progress to date and trajectory moving forward, reported: "[...] CIRM is at a flash point in its history, moving from stage 1 to stage 2." Alan Bernstein, Executive Director of the Global HIV Vaccine Enterprise, said, "We were very impressed with the speed with which stage 1 got going and the number of programs that are underway [...]."

Suggested improvements include these three key recommendations:

1. "Pave a path from fundamental to translational research, medicine, product development, and healthcare delivery [...]."
2. "[...] Significantly increase breadth of community outreach and education programs [...]."
3. Work more closely with the biomedical industry.

December 15, 2010: Bob Klein was officially re-elected (for 6–12 months), assisted by biomedical expert Duane Roth and former Senator Art Torres.

"While there are those who ask for me to serve a year, I believe that in six months we can accomplish the essential contribution [to] the transition to a new Chair," said Klein, "I will remain highly committed to the stem cell initiative, but after eight years, my family needs my focus."

Was this happening? Could Bob Klein actually be retiring?

34 TO WHOM GOES THE KINGDOM?

Legend has it that when Alexander the Great lay dying, his generals gathered around him. Tension mounted. Each warrior glanced at the others, and then back at the man who had conquered the world's greatest empire — and finally someone dared to ask the question:

"To whom — goes the kingdom?"

Alexander of Macedon smiled faintly, and said, "To the strongest."

After his passing, the generals fought amongst each other, and the empire crumbled into dust.

Compare his passing to that of one of America's greatest Presidents.

When Franklin Delano Roosevelt was approaching his fourth term as President, he requested that Harry Truman be his running mate as Vice President.

"This time, it matters," said Roosevelt.

When he died, many worried his achievements would be swept away.

But Roosevelt had chosen his successor well. Truman was a tough fighter, a savvy insider, and a man of unbending honor. Because of his tenacity, America kept what it had won, and Social Security and unemployment insurance still provide a safety net for those in need.

Bob Klein was stepping down, retiring from the Chairmanship of the board.

Which example of a leader's passing would California face? Alexander, called the Great, but whose kingdom vanished? Or Roosevelt's careful departure from the scene, which allowed old people like me to keep their Social Security?

California's stem cell program, the CIRM, was a magnificent beginning, and which must be nurtured and protected.

A small incident summed up the situation. It was after one of the board meetings, and we were invited to visit the newly completed Lorry I. Lokey Stem Cell Research building.

It was late at night, the building almost completely dark. But *among the unpacked boxes*, scientists were already at work: wasting no time, beginning the years of effort it would take to bring cures. Their efforts must be protected for the sake of all who suffer chronic disease.

But the California stem cell program — without Bob Klein?

Bob recommended a Canadian, Alan Bernstein, to succeed him. Bernstein helped build (and was founding President of) the Canadian Institutes of Health Research, a research institute very similar to CIRM. He was a patient advocate par excellence, and the first Executive Director of the Global HIV Vaccine enterprise. But because of citizenship requirements, he was not eligible.

That requirement is narrow-minded and should be changed. If California should decide to support another round of funding for the CIRM, I suggest the eligibility for Chair should be open to any citizen of Planet Earth — we want the very best, and I would personally hope we could consider Alan Bernstein. In the writing of this book I have spent time with Canadian researchers and leaders, and all view Bernstein as an inspirational champion.

The race came down to two superb individuals, Dr. Frank Litvack and Jonathan Thomas.

I had no vote in the great decision, only the right to recommend. As a member of the public, I would be allowed three minutes to speak my opinion on who should lead; that was it. But (after my family) the California stem cell program is the most important thing in my life. It would be wrong of me not to give my best input on this decision for the program's leadership.

After studying their biographies, I contacted both men, who were kind enough to allow me to interview them at length, each for about 90 minutes. I asked them every question I could think of, and brought up every objection that might be thrown against their candidacy.

Then I waited. For days, my stomach was in knots as I tried to decide which man to support. I remembered Henry Kissinger's comment, that the most difficult decisions are always at 51/49 percent.

Karen Miner said, "If you are that torn, that conflicted, it means they are both good."

I studied each man's position paper until I could have made his presentation for him. The night before the meeting, I contacted each candidate, and told him my decision.

The two men were strikingly different.

Frank Litvack, with his white crewcut hair, looks like a semi-retired weightlifter, and carries himself very tall — a little intimidating.

Jonathan Thomas is dark-haired, slender, built for the bob and weave of basketball, and is both cheerful and approachable, someone you might approach to ask for directions if you were lost.

Litvack was a heart doctor and a medical entrepreneur, capable of recognizing an idea with potential and making it real; he developed several biomedical companies, including one which owned the medicated heart stent, a tube to help the circulation of a failing heart.

Thomas was a lawyer, investment counselor, and bond expert. He had put together funding for multi-billion dollar government projects, like the highway connection to the Los Angeles harbor.

At the public meeting of the board, each man spoke for about 20 minutes. Both men nailed the delivery of their speech. Litvack inspired; he made me want to get up and fight, to do something for the cause. Thomas assured; he gave me confidence that the financial battles would not be lost.

And then at last, public comment was called for. I read the following statement into the record:

"We are fortunate to have such outstanding candidates. I contacted both, and questioned them.

"In personality, Frank Litvack is a strong leader, explosive, charismatic, fun to be around. His answers to most questions were short, punchy, easy to follow.

"Jonathan Thomas, on the other hand is quiet, not an exciting speaker. But he has a bulldog tenacity, which makes him persuasive. When I asked him a question, he would not only answer the question, but would come up with possible objections to his own answer, and answer those as well. He would never be distracted, but kept coming back to the question again and again, until every objection was met. He is focused, and determined, and he basically wore me down.

"Frank Litvack is a highly successful entrepreneur, the active chair of five companies. This is evidence of his success in the business world — but it may also be a possible negative. With the best will in the world, his energies and commitments will necessarily be divided.

"Jonathan Thomas views the job of chair as a full time one, requiring his 100% effort and commitment, year round. That is to my mind the only realistic assessment of the job.

"The new Chair will face huge financial challenges to our program. I asked both candidates: if the worst happened, and there were no General Obligation Bonds, what will you do?

"Litvack said that without General Obligation Bonds, there was no way to fund the program at its current level, and anyone who said otherwise was blowing smoke.

"Asked the same question, Thomas said he would first implement a short-term plan to stop the bleeding, and then develop a long-term financial plan. He laid it out in a 15-minute answer. I do not pretend to understand it, but it appeared to be a mix of bonds, donations from charities and foundations, revenues from biomedicine, national government contributions, and other sources of funding. He said such fundraising challenges had been a major part of his life for the past 30 years, and he was prepared, if necessary, to do it every day of his tenure. He has shown the ability to raise massive amounts of funding: projects involving literally billions of dollars.

"In entrepreneurship, Frank Litvack is wonderfully successful, having developed, for example, a company which manufactured the medicated heart stent and other valuable products.

"But Jonathan Thomas has also demonstrated the ability to pick a winner in the biomedical world. Many years ago, when stem cell research as an industry was just getting started, he showed foresight by picking just one company to support: Advanced Cell Technology. Jonathan Thomas helped Mike West raise the funding he needed — and the field of biomedicine began.

"In a time when complicated financing may mean success or failure for the California stem cell program, Jonathan Thomas's unique skillset makes him the essential choice."

The board went into closed session for three hours; it seemed like three days.

When they returned, the roll was called; the votes were counted.

By a vote of 14–11, Jonathan Thomas was elected Chair of the California stem cell program.

Bob congratulated Jonathan Thomas on "beginning a journey [he] will never forget".

Old Chair and new shook hands.

There was a long line of people who wanted to speak to Dr. Litvack afterward. Everyone knew this was a giant in the field. I was nervous to approach him, but did anyway, and blurted out that I respected his greatness and I hoped he would help us, because the program really needed him. Fortunately, he was as large in spirit as he was in accomplishment.

The next day was the part I had been dreading: the official transfer of authority.

Bob asked that his wife Danielle Guttman-Klein be acknowledged. Herself an advocate for many causes, and a force for positive change, Danielle is lovely as a young Doris Day. I attended their wedding, and the love just shines through between the two of them.

But it broke me down when Bob read the swearing-in oath, and Jonathan Thomas raised his right hand and swore to uphold the Constitution of the United States and the State of California. I was typing, but I could not see the keys anymore. My stomach was shaking, wracked with hopefully silent sobs — but nobody noticed; most folks were having similar difficulties.

And then Bob gave his last official remarks.

"It has been the privilege of my life, working with this board and the staff of CIRM. We are on a mission for all our families, in our state, our country, and the world.

"With the tremendous outpouring of dedication and effort of all involved, including especially the patient advocate groups, I believe we will be successful beyond our wildest dreams.

"It is the dedication of the people in this room, the staff and the empowered scientists, which will see this dream of California through: to reduce human suffering.

"And so, I thank you," he said, and stopped.

The first action of the new Chairmanship of Jonathan Thomas was to call for an ICOC vote that Bob Klein would receive the title, "Chair Emeritus", to honor his contribution forever. He read a statement into the public record, thanking Bob for the tremendous gift of Proposition 71.

The meeting went on to other things. Jonathan Thomas talked; Bob sat in the audience and listened. I had the urge to tell him he was sitting in the wrong place!

It seemed impossible, but there it was. Bob's hands were off the steering wheel. Nothing lasts forever but the story of the stars. The program Bob built must continue without him.

Jonathan Thomas, Chair of the Board of Directors of the California Stem Cell Program, and patient advocate Don C. Reed.

35 THE LIVER LIST

Sevengill sharks, like the ones the author swam with as an aquarium diver, have a liver 1/3 the mass of their bodies.

Before I saw a shark being dissected, I never thought much about livers. But as the veterinarian's scalpel swept along the seven-gill shark's rough belly — an enormous gray-pink *something* bulged out.

The liver was huge — a third of the shark's body mass. Shark livers contain oil (they were once fished as a vitamin source), serving both as a long-term food source and to give it "lift" in the water.

A human being's liver is only about three pounds, but it is the second largest organ in the body (the skin is the biggest).[1]

Right now, Mary S. and her husband Will are involved in a struggle to find a new liver for her. With permission, their dialogue is reprinted:

MARY: "I knew I was in trouble when I felt an extreme loss of energy, had unexplained weight loss, and suffered intestinal discomfort."

WILL: "It was polycystic liver disease; we were told there were no cures."

MARY: "The organs were damaged by tons of cysts. A liver transplant is not easy."

WILL: "By 2012 we knew we had to do something. A liver resection was attempted. Unfortunately problems from the operation caused the need for a liver transplant. We have been waiting ever since. She has lost about 50 pounds and feels tired and weak, with pain all over her body. While staying in hospitals she has been subject to much testing that is always hard on her. Drugs and treatments often have severe side effects.

"We try to stay positive and God is helping us through this. But we are in a catch-22 situation. On one hand we are told over and over that Mary has to be sicker [...] to qualify for a transplant."

MARY: "But my goal is to get strong enough to survive a liver transplant."

WILL: "There is such a need for liver cadaver donations. When you donate your liver after your death it's forever appreciated. It's as if your life goes on [as] an irreplaceable gift to the receiver and all those by whom they are loved."

So, what do you do if you need a liver (or a heart, kidney, or lung)?

You get on "THE LIST"[2]:

"All liver transplants in the United States *must be listed* with UNOS (United Network for Organ Sharing) before a donor liver can be allocated [...]. Currently, there are 17,000 people waiting for a liver transplant in the U.S. The median national waiting time in 2006 was 321 days [...]."

The Model for End-Stage Liver Disease (MELD) is a system used to determine how desperately a person needs a liver. The system is essentially trying to give priority to the sickest people. But even on the list, it is not easy sailing. Some parts of the country have more liver donors than others. If you live in a state where there are too many applicants, your chances are slimmer.

For Mary and Will, it has been hell. They relocated from Oregon to California to Hawaii to Louisiana and back again, hunting for a matching donor. Will is a retired school teacher, like myself, and they are not rich by any means. But they will not quit fighting.

UNOS is not at fault (although improvements are possible), but the facts are inescapable: there is no good way to divide 5,000 livers among 100,000 people who need them right now.

Like skin, the liver can regrow itself and does so continually, every cell dying and being replaced, regenerating a new liver every five months.

---

[1] The liver is a very large organ weighing up to 3 kilogram in adult humans [Internet] [cited 2015 Feb 12]. Available from: http://www.gesa.org.au/content.asp?id=102

[2] The waiting list [Internet] [cited 2015 Feb 12]. Available from: http://livermd.org/waiting.html

To untrained eyes like mine, a liver looks like a plain piece of meat, but it does many subtle and interactive jobs. It is as complicated as a computer, but with no reset button to push.

Livers balance hormones for physical and mental health so that the muscles stay fit and the mind stays sane. The liver converts food sugar into usefulness and destroys cells when they don't work anymore. Above all, the liver filters poisons. Every moment, a quart to a quart and a half of blood is cleansed by the liver; you never know what is happening until something goes wrong.

When the liver stops working, the poisons accumulate. The body can survive on a partially functioning liver, but not well. We become nauseated, like when we are majorly carsick.

And if the liver fails altogether? We die.

The words "liver failure" bring alcoholism to mind, and it is true that alcohol abuse is a major cause of cirrhosis of the liver — a deadly threat.

But one of the largest single causes of liver failure has nothing to do with drinking.

Non-Alcoholic Steato-Hepatitis (NASH) affects Hispanic-Americans more than any other ethnic group — even if they never drink a single can of beer. Why? We don't know.

Liver disease can be caused by environmental poisons, heart or kidney failure, obesity, diabetes, or sexually transmitted diseases.

According to the American Liver Foundation, *30 million Americans have liver disease* — roughly one in ten.[3]

Fortunately, "only a small percentage of these, probably one percent, have bad liver disease," says liver researcher Dr. Mark Zern of the University of California, Davis.

That is still a lot of people: an estimated 100,000 folks need new livers right now. Their only chance for cure? Remove the sick liver and put in a new one. American surgeons perform five thousand liver transplants a year, and would do more — if they had more livers.

Thanks to CIRM grants and strategic partnerships, scientists are accepting the challenge: people like the aforementioned Mark Zern, Holger Willenbring of the University of California at San Francisco, and Professor Lijian Hui of the Shanghai Institute for Biological Sciences are working together.

As Zern and Willenbring explained, "[…] There are not enough livers available for transplantation for all the people who need them. […] One way to improve this situation is the transplantation of liver *cells* rather than whole organ transplantation. We are attempting to develop liver cell lines […] that will act like normal liver cells. [If these function well, they will] hopefully be adapted to future human clinical trials.

"[…] We will compare human embryonic stem cells with other stem cells to determine […] the most effective cells to transplant [into an injured liver]. People who have liver failure or an inherited liver disease could be treated, because there would be an unlimited supply of liver cells."

---

[3]October is Liver Awareness Month [Internet] [cited 2015 Feb 12]. Available from: http://www.liverfoundation.org/chapters/lam2010

Willenbring is also working with Lijian Hui to develop liver cells by another method called direct reprogramming:

"[The recent] discovery that skin cells can be converted into hepatocytes (liver cells) by transfer of a few genes suggests a promising new source […]. We aim to identify which […] human cell type — skin, blood, or fat cells — can be most efficiently converted into hepatocytes [by] gene transfer. The [efficacy and safety of these] will be rigorously tested in animal models of human liver failure.

"If successful, our project will establish the feasibility of therapy [for] liver failure — with cells derived from the patient's own […] cells."

Three great research centers hailing from the U.S. and China — the University of California at Davis, University of California at San Francisco, and the Chinese Academy of Sciences in Shanghai — cooperating to ease suffering and save lives. Is that not the way the world should be?

36 THE WILL OF CONNECTICUT

Connecticut Senator Chris Murphy, working with Governor M. Jody Rell, developed their state's magnificent $100 million stem cell program.

In Hollywood, most folks know where movies come from. Similarly, in a biomedical state like Connecticut, it seems logical our leaders should know a little something about stem cells.

However, in 2012 when Connecticut Democrat Chris Murphy and Republican Linda McMahon were rivals for the U.S. Senate, one was fully knowledgeable about the state's advancements in stem cell research; the other, not so much.

In 2005, Representative Chris Murphy authored and passed the Connecticut Stem Cell Investment Act, a $100 million program which made the Constitution State one of the stem cell centers of the world. Already, the investment was paying dividends; the tiny state leaped past many *nations* in the life sciences industry. Years on, Connecticut's stem cell research program has benefited greatly. Murphy said, "We have over 100 labs [...] and almost 200 researchers between Yale, the University of Connecticut and Wesleyan, doing embryonic and related stem cell research — it's an amazing, amazing thing that's happened."[1]

---

[1] Becker AL. State showcases local expertise at stem cell symposium. *The Connecticut Mirror* [Internet]. 2011 Mar 23 [cited 2015 Feb 12]. Available from: http://ctmirror.org/2011/03/23/state-showcases-local-expertise-stem-cell-symposium/

How does Connecticut's program work?

Public Act 05-149, an act permitting stem cell research, was approved by the General Assembly and signed by Governor M. Jody Rell on June 15, 2005. According to a report to the General Assembly, the Act appropriated the sum of $20 million "for the purpose of grants-in-aid for conducting embryonic or human adult stem cell research. [In addition, for each fiscal year until June 30, 2015], an additional ten million dollars [shall be disbursed] to the stem cell research fund [...]."[2]

Does Murphy's opponent, Linda McMahon, support embryonic stem cell research? Judging from her campaign statements, the answer is: yes, no, couldn't say, or probably not.

Sometimes, it was yes. As reported on August 27, 2010 in the *New Haven Register*, "Republican Linda McMahon [as a Senator], would support federal funding of embryonic stem cell research. [...] Ed Patru, communications director for McMahon, said [she] supports federal funding for stem cell research for the hope it represents for potential cures for a number of diseases."[3]

Sometimes, it was no. In yet another *New Haven Register* report later on October 17 that same year, "McMahon [stated that she opposed] embryonic stem cell research, despite the state's $100 million commitment to support both embryonic and adult stem cell work."[4]

Sometimes, she couldn't say, "On stem cell research, McMahon's campaign didn't offer any reaction until asked. Even then, her spokesman, Ed Patru, said she supports stem cell research, but he couldn't say whether she would impose any conditions," stated an October 14, 2010 report in the *Connecticut Mirror*.

And her most recent stand? Probably not. On April 1, 2012, McMahon said in the *News Times*, "I believe more that we should use the umbilical cords[5] and the research that we can do from there that aren't embryonic stem cells."[6]

Her answers seem to vary day by day.

---

[2]Galvin JR. Report to the General Assembly: An act permitting stem cell research and banning the cloning of human beings. Connecticut Stem Cell Research Advisory Committee; 2007 Jun 30. 51 p.
[3]O'Leary ME. Connecticut Senate hopefuls support stem cell work. *New Haven Register* [Internet]. 2010 Aug 27 [cited 2015 Feb 12]. Available from: http://www.nhregister.com/general-news/20100827/conn-senate-hopefuls-support-stem-cell-work
[4]Endorsement: Blumenthal for U.S. Senate. *New Haven Register* [Internet]. 2010 Oct 17 [cited 2015 Feb 12]. Available from: http://www.nhregister.com/general-news/20101017/endorsement-blumenthal-for-us-senate
[5]Umbilical cords? Presumably Ms. McMahon meant cord blood, which for a time was offered as an alternative to embryonic stem cells, and as such was popular among the Religious Right. Cord blood is not without value (see Chapter 32), but few scientists are eager to put all their hopes into that one narrow area. As Dr. Larry Goldstein put it, "Stem cell researchers find cord blood interesting, but not [...] as a substitute for human embryonic stem cells."
[6]Vigdor N. Social issues pose tricky litmus test for McMahon. *News Times* [Internet]. 2012 Apr 1 [cited 2015 Feb 12]. Available from: http://www.newstimes.com/local/article/Social-issues-pose-tricky-litmus-test-for-McMahon-3451678.php

As for the Democrat, Chris Murphy — what is his position on stem cell research? One of unwavering support.[7]

Representative Murphy took a strong position in the debate about President George W. Bush's restrictive stem cell policy.

In a strong statement, Murphy said, "100 million Americans are affected by a debilitating or life-threatening disease. Somewhere in this vast universe, a cure for their disease exists. I know it. We all know it. Let's stop putting up man-made barriers to finding that cure, to curing our loved ones [...].

"[In Connecticut] we made a commitment of $10 million a year for over 10 years, and over a very short period of time, Connecticut has become one of the centers of excellence for stem cell research and embryonic stem cell research in the country.

"Stem cell research — the investment in potential cures and treatments to our world's cruelest diseases — must be a national priority."

A business executive, Ms. McMahon was co-founder of World Wrestling Entertainment. She knows the tremendous demands professional wrestling puts on the body, as 300-pound muscular giants leap and crash onto each other. Presumably, too, she knows what happens to wrestlers' bodies as they age.

Arthritis happens — whether brought on by smash-and-crash injury or simple wear and tear.

As many as one in five Americans suffers arthritis today, at a cost of roughly $128 billion annually from medical care and lost wages.[8]

Think of a Thanksgiving turkey leg, the white plastic-looking substance at the end of the bones. That's cartilage, which joints must have to operate smoothly. If torn or worn away, it does not repair itself. Resultant arthritic pain can be crippling.

As a Republican, Linda McMahon may be tied to a position that could literally make curing arthritis against the law. *Page 34 of the Republican National Party's platform*, titled We Believe in America, called for a complete ban on all embryonic stem cell research, public or private.[9]

If that Republican position was imposed on Connecticut, it could stop the work of Dr. Caroline Dealy, who is working toward a cure for arthritis. Dr. Dealy wants to replace worn-out cartilage with new, and one of her methods would be to use embryonic stem cells to make more specialized cells called *chondrocytes* — which in turn become cartilage.[10]

---

[7] Chris Murphy [Internet] [cited 2015 Feb 12]. Available from: http://www.ontheissues.org/house/chris_murphy.htm

[8] Arthritis-related statistics [Internet] [updated 2014 Mar 17; cited 2015 Feb 12]. Available from: http://www.cdc.gov/arthritis/data_statistics/arthritis_related_stats.htm#1

[9] Mick J. RNP platform: ban marijuana, embryonic stem cells, morning after pill. *DailyTech* [Internet]. 2012 Sep 2 [cited 2015 Feb 12]. Available from: http://www.dailytech.com/RNP+Platform+Ban+Marijuana+Embryonic+Stem+Cells+Morning+After+Pill/article26568.htm

[10] Gong G, Ferrari D, Dealy CN, Kosher RA (2010). Direct and progressive differentiation of human embryonic stem cells into the chondrogenic lineage. *J Cell Physiol* **224** (3): 664–671.

Connecticut has developed one of the greatest stem cell programs in the world. As reported in *The Connecticut Mirror*, "The state has committed to [distributing $70 million so far]. Scientists from the University of Connecticut, Yale University, Wesleyan University and the Jackson Laboratory thanked the state legislature for the stem cell initiative, saying it has helped further their research, collaborate with each other, and attract funding from other sources."[11]

What has that $70 million bought so far? The state funding eventually helped them build up to $167 million in grant funding (leveraging their funding to greater grants from federal and other sources), and led to 81 inventions and 36 patents. Yale University, meanwhile, reported that it had spun off 40 biotech companies, many from the Yale University science department, and created lab space for these companies. The state's stem cell funding helped the university attract pre-eminent scientists, said Bruce Alexander, Yale University's Vice President for New Haven and state affairs and campus development. Yale University's stem cell department grew from only two researchers in 2006 to 72 researchers in 2012. In addition, 200 high-tech jobs were created in Connecticut and 132 patents were generated. The state's investment of $6 million in 2010 helped Yale University attract $36 million in grants from the NIH and $2.6 million from private sources.[12]

Democrats began and supported the program — former Democratic Governor Jodi Rell signed the bill into law, while current Governor Dannel Patrick Malloy, another Democrat, continues strong support for the research effort.

As Governor Malloy said to scientists at the Connecticut United for Research Excellence, a gathering supported by patient advocates, "Connecticut supports you. Your work in this industry, your willingness to be involved in this research, meshes well with Connecticut's great historical traditions."

And what are the scientists doing with their research grants?

The program's core lab director is Dr. Ren-He Xu. He has been working with embryonic stem cells since 1999, when he started in the lab with Dr. James Thomson of Wisconsin, who was the first to isolate the human embryonic stem cell (hESC). But state funding was limited there, and Connecticut's research funding attracted Dr. Xu.[13]

Dr. Xu is working on projects which might develop *vaccines against cancer*. For instance, he believes that "vaccination with human pluripotent stem cells [can generate] a broad spectrum of immunological and clinical responses against colon cancer."[14]

---

[11] Merritt G. Scientists tell lawmakers: stem cell initiative producing partnerships, breakthroughs. *The Connecticut Mirror* [Internet]. 2012 Sep 25 [cited 2015 Feb 12]. Available from: http://ctmirror.org/2012/09/25/scientists-tell-lawmakers-stem-cell-initiative-producing-partnerships-breakthroughs/

[12] Merritt G.

[13] Kaplan T. Suddenly, Connecticut is stem cell central. *The New York Times* [Internet]. 2007 Nov 25 [cited 2015 Feb 12]. Available from: http://www.nytimes.com/2007/11/25/nyregion/nyregionspecial2/25stemcellct.html?_r=2&

[14] Li Y, Zeng H, Xu RH, Liu B, Li Z (2009) Vaccination with human pluripotent stem cells generates a broad spectrum of immunological and clinical responses against colon cancer. *Stem Cells* **27** (12):

Program Director Dr. Marc Lalande, who is also Executive Director of Genomics and Personalized Medicine at the University of Connecticut, is working on a severe mental disorder called the Angelman syndrome. Patients with this disorder from childhood cannot speak. They have seizures, endure fits of inappropriate laughter, and require constant care. Dr. Lalande is working with induced pluripotent stem cells (iPSCs) taken from the patient's skin.

Other projects include James Li's modeling of Parkinson's disease using human embryonic stem cells (hESCs) and patient-derived stem cells, the induced pluripotent stem cells (iPSCs). Another by Diane Krause, a principal investigator at Yale University, aims to assist in the use of hESCs and iPSCs to study leukemia. Mark Carter is working on the generation of insulin-producing cells from hESCs. Lawrence J. Rizzolo focuses on blindness.

Team efforts are strong too. Laura Grabel, Janice Naegele, and Gloster Aaron — three Wesleyan University colleagues — bring multiple competencies to the table. Grabel provides the embryonic stem cell expertise, Naegele provides her knowledge with epilepsy models, and Aaron helps with his background in neurophysiology.[15]

Perhaps most exciting is the interaction between public and private enterprise, such as the startup company, IMSTEM Biotechnology, a direct result of Connecticut's stem cell program.

When I spoke with Dr. Xiaofang Wang, IMSTEM's Chief Technical Officer and co-founder, he explained that IMSTEM's plan was to attack multiple sclerosis by using embryonic stem cells to make another kind of stem cell — mesenchymal stem cells. They tried using both embryonic stem cells and also adult tissue-derived stem cells, and found embryonic stem cells to be clearly superior. A lot of cells are needed, so there will have to be multiple "passages" or growth sets to make enough cells. Bone marrow-derived cells, if expanded into several passages, became weak. The embryonic stem cells, however, maintained their properties, guaranteeing consistently good quality.

In two kinds of experiments, one to prevent the disease and one to try and cure it, the embryonic stem-derived cells chosen by IMSTEM were effective. The mice had paralysis in both front and hind legs; after the treatment, they walked again.[16]

For more on the experiments with embryonic stem cells to treat multiple sclerosis, visit http://today.uconn.edu/blog/2014/06/embryonic-stem-cells-offer-treatment-promise-for-multiple-sclerosis/

And when election day rolled around, what was the will of the Constitution State?

Connecticut said YES to Chris Murphy, the man who began the state's stem cell program.

3103–3111.

[15] Mullen J. Report to Governor Malloy and the General Assembly: An act permitting stem cell research and banning the cloning of human beings. Connecticut Stem Cell Research Advisory Committee; 2011 Nov. 36 p.

[16] Wang X, Kimbrel EA, Ijichi K (2014) Human ESC-derived MSCs outperform bone marrow MSCs in the treatment of an EAE model of multiple sclerosis. *Stem Cell Reports* **3** (1): 115–130.

Senator Chris Murphy and Governor Dannel Patrick Malloy, a powerhouse team for stem cell research. (Malloy is seated by the flag.)

37 CHAMPIONS FIND A WAY

As you recall, Geron closed down its stem cell department, which not only laid off more than 60 great scientists but also stopped the paralysis human trials with embryonic stem cells — the work begun by the Roman Reed Spinal Cord Injury Research Act.

It wasn't that we were close to human trials — we were *in* human trials, the first in the history of the world. Two of the four people who got the injections were friends — Katie Sharify and T.J. Atchison — Roman even worked with TJ to set up a paralysis research program in Alabama.

Insofar as they had been allowed to proceed, the trials had been absolutely successful. Their purpose in Phase I was to establish safety, and no one had been harmed.[1]

The California stem cell program loaned the company $25 million to help with the trials — and then Geron pulled out. The state money (plus interest) was returned.

So much work! Under the leadership of Tom Okarma, then-President of Geron, the research had gone through eight long years of FDA tests before being approved for human trials. But new leadership apparently felt that it was costing too much money and closed the stem cell department.

That was a blow.

But the science was strong, and it remained, waiting for someone with the vision and strength to bring it along.

Thomas Okarma, who took embryonic stem cells through eight years of testing.

---

[1] Geron presents clinical data update from GRNOPC1 spinal cord injury trial [Internet] [cited 2015 Feb 13]. Available from: http://ir.geron.com/phoenix.zhtml?c=67323&p=irol-newsArticle&ID=1635760

If any one man can be credited with the beginning of modern biomedicine, that man might well be Michael D. West, founder of Geron, Inc., BioTime, and Asterias Biotherapeutics. He never doubted; he always went ahead.

Mike West was that man. One of the pioneers of biomedicine and the original founder of Geron, he built a company called BioTime, from which came Asterias Biotherapeutics, Inc. Asterias Biotherapeutics was focused on studying regeneration. Many of its top people had formerly worked with Geron, but were laid off when that company quit its involvement with stem cells.

But Asterias Biotherapeutics bought the stem cell properties from Geron, and suddenly the world was bright again. The science had never been in question, only the money to finance the trials.

On May 30, 2014, the California stem cell research program awarded $14.3 million to Asterias Biotherapeutics to bring embryonic stem cell treatment closer to paralysis cure.

Remember how it works? When the brain sends a message through a healthy spine to the body, the nerves pass it along. But, like electrical wire, the nerves must be insulated, or the message will short-circuit. When a spine is damaged, so are the individual nerves. Trauma bashes them, so they lose a fatty acid called *myelin* they need to provide insulation.

To make that myelin, embryonic stem cells are turned into oligodendrocytes. Making myelin is what they do. Injected into the spine, the "ollies" reinsulate the nerves.

Several quite wonderful things can happen after that.

First, the amount of damage to the spine may be lessened. In a spinal cord injury, only half the damage is done by the physical accident; the trauma (injury) is just the beginning. The body's immune system literally rips into the wound, enlarging the cavity and doubling the damage. But with the new invention, this secondary damage (and some of the accompanying paralysis) might be lessened. After the stem cells were put in, the size of the wound was apparently reduced.

Second, a wave of nutrients floods the damaged area, serving as a nerve fertilizer, which may help in the regenerative process.

And as the nerves reinsulate themselves, abilities lost may be partly recovered.

I had seen it work. On March 1, 2002, at the Reeve-Irvine Research Center, I held in my hands a formerly paralyzed rat named Fighter. Videos showed her dragging her body like luggage. But now, thanks to stem cells, she could scamper — tail high! — across the purple plastic swimming pool, her play area. It wasn't perfect but it was motion, and pretty quick at that.

The human trials could begin again, with increasing doses: double-checking for safety, and trying for improvement.

A great circle had closed, bringing scientists, patients, advocates, and biomedicine back into the loop again.

Because of that cooperation, there may come a day when paralysis is no longer described as a life sentence.

Photo 1 The man himself, Nobel Laureate Paul Berg, famous for his DNA efforts, was one of the first to appreciate the significance of Prop 71.

Photo 2 President Bill Clinton was the first U.S. President to authorize embryonic stem cell research.

Photo 3   Liz DeLaperouse and David Eagleton: Missouri warriors.

Photo 4   David Jensen: King of the critics: Agree or disagree, Jensen's weblog, the California Stem Cell Report, is must-read material.

Photo 5 Author Don Reed receives the first CIRM advocacy award from Bob Klein.

Photo 6 Reed Family: Terri, Katherine, Don, Roman II, Jason, Gloria, Roman, Desiree Reed-Francois, Jackson Reed-Francois (husband Josh Francois was in Tennessee that day) — Gloria and Don's 45th wedding anniversary.

Photo 7 Author Don C. Reed "walking" great white shark in previous job as diver for Marine World.

Photo 8 True Colors Club: Students at Don Reed's junior high school put on fund-raisers for Christopher Reeve: Here is cast and crew of one of the author's plays, LEGEND OF WING CHUN.

Photo 9 Working to build the Alabama Institute of Medicine: Roman Reed, Tori Minus, Hans Keirstead, Chris Drummond and an un-named friend.

38 DISENFRANCHISE THE DISABLED?

Voter suppression laws may strip away the rights of disabled voters.
Look up your state in the "Map of Shame" to find out if your rights have been protected or suppressed.

Are you one of America's 35 million voting-aged citizens with a disability?[1]

If so, watch out; you may have just lost your right to vote.

Why would any politician stoop so low as to deny the vote of a disabled citizen? Perhaps because disabled individuals may need government assistance programs, paid for by taxes — and support politicians who think their way.

Democrats use government to solve problems, even if that means an occasional tax increase on the wealthy. But Republicans see government itself as the problem. In the words of Grover Norquist, author of the "Taxpayer Protection Pledge", they want to shrink it "small enough to drown in a bathtub". This pledge, which 95% of all Republican

---

[1] Disability Rights California. Voters with disabilities [pamphlet]. California: Disability Rights California; 2014 Sep.

Congress members signed in 2012, is a promise never to allow more taxes — even on billionaires or multinational corporations.²

Because they will not support any increase in taxes, Republican leaders are less likely to support stem cell research, poverty relief, medical care, or anything needed by the disability community.

But the disability community is huge. According to a comprehensive report released by the U.S. Census Bureau, about 56.7 million people, or 19% of the population, had a disability in 2010.³ That's roughly one in five Americans.

Of that number, about 35 million are eligible to vote. Add that to the rising tide of minority, youth, elder, and low-income voters, and Republicans have a problem.

To win, they must discourage Democrats from voting, one way or another.

If enough Democratic voters are cutoff from the voting booth, Republicans can still win, even if the majority of the country disagrees with their policies.

Voter suppression today is not as open as it was in Jim Crow days, when African-Americans were systematically denied their democratic rights, but the effect is similar. If people are not allowed to vote, they become slaves again.

There are numerous methods to block the vote. These include shortening the time allowed for voting (so lines stretch block after block and people cannot reach the polls before they close), denying same day registration, shortening the time allowed for absentee balloting, and putting near-impossible restrictions and penalties on voter registration groups.

Most of all, progressive votes are killed by requiring a government-issued voter ID with a photograph on it, a harmless-looking plastic card.

Right now, consider what government-issued photograph ID card do you have in your possession? Probably the same one as I do — a driver's license.

But what if you are among the approximately 3.4 million *blind* Americans?⁴

The blind cannot drive. And what about America's 5.6 million paralyzed people? Some can drive, but many cannot. Do *non-drivers* have driver's licenses?

But, Republicans counter, there are other forms of government-issued voter IDs, like passports and gun permits.

But if a disabled person is too poor for international traveling, is he or she going to spend the money to purchase a passport? And blind citizens do not do a lot of shooting. Many of the poor do not own a car and therefore have no driver's license — should they be denied the right to vote? What about college students? If they are from out of state, they will not have a local address on their driver's license — should they removed from the eligible list? What about senior citizens — will they have a license if they are no longer safe to drive?

---

²Grover Norquist. *Wikipedia* [Internet] [updated 2015 Feb 7; cited 2015 Feb 13]. Available from: http://en.wikipedia.org/wiki/Grover_Norquist

³Nearly 1 in 5 people have a disability in the U.S., Census Bureau reports [Internet]. 2015 Jul 25 [cited 2015 Feb 13]. Available from: http://www.census.gov/newsroom/releases/archives/miscellaneous/cb12-134.html

⁴The burden of vision loss [Internet] [updated 2009 Sep 25; cited 2015 Feb 13]. Available from: http://www.cdc.gov/visionhealth/basic_information/vision_loss_burden.htm

Indeed, *according to the League of Women Voters, as many as 11% of eligible voters do not have government-issued photo IDs.*[5]

Eleven percent? *Twenty-one million voters* with their participation at risk? This is an assault on democracy.[6]

Across the country, Republicans are imposing voter-suppression laws which may disenfranchise literally millions of Hispanics, African–Americans, the elderly, students, the poor, and the disability community[7] — in other words, Democrats.

The New York University School of Law's Brennan Center for Justice found that "state lawmakers across the country introduced at least 180 restrictive voting bills in 41 states", with 19 states ultimately passing 27 measures.[8]

If the law in your state has been changed, you may not know it until you show up to vote and find a nasty little surprise waiting for you, even if you have voted many times before.

"Where is your government-issued photograph ID? Don't you have a driver's license?"

But, Republicans counter, voter IDs are given free by some states, even those with voter-suppression laws!

Read the small print. First, getting those "free" voter IDs may involve hidden fees and require documents you do not have, like a birth certificate. If you were born at home, for example, you will not have a birth certificate. To obtain needed documents means travel, expense, and time off from work. Among the working poor, many (including disabled citizens) cannot afford to take time off from work to locate documentation, and even if the documentation is acquired, it must be taken to the Department of Motor Vehicles (DMV) to register.

What if the nearest DMV office is several hours' drive away? In Texas, one-third of the counties have no DMV at all.

Visit the Map of Shame website, http://www.lawyerscommittee.org/page?id=0044#wall of shame/, to see if your state has had its voting rights diminished by voter ID requirements. This infographic at http://www.aflcio.org/Multimedia/Infographics/Map-of-Shame-Voting-Rights-4-11-12 also provides a quick overview of vote suppression legislation by state.

What excuse do Republicans give for such voter-suppressing laws?

They claim voter fraud is a major threat, with people pretending to be someone else so they can vote two or more times.

---

[5] Voter ID [internet]. 2012 Oct 15 [cited 2015 Feb 13]. Available from: http://www.brennancenter.org/analysis/voter-id

[6] Gerard L. ALEC's voter ID laws work to overturn hundreds of years of progressive moves to broaden democracy. *AlterNet* [Internet]. 2012 May 7 [cited 2015 Feb 13]. Available from: http://www.alternet.org/story/155307/alec%27s_voter_id_laws_work_to_overturn_hundreds_of_years_of_progressive_moves_to_broaden_democracy

[7] Ballot [Internet] [cited 2015 Feb 13]. Available from: http://craigconnects.org/voter-protection-infographic

[8] States with new voting restrictions since 2010 election [Internet] [updated 2014 Nov 12; cited 2015 Feb 13]. Available from: http://www.brennancenter.org/new-voting-restrictions-2010-election

Is this accurate? Hardly. In some states mandating voter ID legislation, voter fraud has NEVER occurred.[9] It is difficult enough to get people to vote once, let alone twice — not to mention there are jail sentences waiting for anyone stupid enough to try such a thing.

On the federal level? Former President George W. Bush put the Attorney General to work on the issue of voter fraud. Despite years of search they found... nothing. Between 2002 and 2007, a major probe by the Justice Department failed to prosecute a single person for going to the polls and impersonating an eligible voter, which the anti-fraud laws are supposedly designed to stop.[10]

Indeed, according to an infographic by Craigslist founder Craig Newmark, between 2000 and 2007, there were 32,299 UFO sightings in the U.S., 352 deaths caused by lightning, but only nine cases of voter impersonation — a negligible number.[11]

Voter fraud is essentially a non-existent crime.

What is real? Something far more sinister.

David and Charles Koch, the billionaire brothers who bankrolled the Republican Tea Party, funded a systematic campaign in 2011 that introduced legislation designed to impede voters at every step of the electoral process in 38 states.

So what do we do, we in the disability and patient advocate community?

We exercise our rights. We speak up against voter ID laws and other means of voter suppression. Above all, we vote! Americans with a disability and their families must register and vote.

As my paralyzed son, Roman Reed, asserted, "Of course we are going to register and vote. That is our right and our duty. But it is not enough just to vote. We must also reach out to our friends in the community."

Enable the disabled! Wheelchair warriors, spread the word: ask your friends if they are registered to vote. Each one, reach one. Do you have your ride to the polls figured out? If not, plan your ride ahead of time. If you already have a ride, consider sharing it with someone less fortunate.

Reassure friends with disabilities that they are welcome at the polls. National law requires that voters with a disability be assisted. If a person cannot get down from the car to access their voting rights, curbside voting must be brought to them. If help is not provided, ask for the precinct captain. If he or she won't make things right, contact your local news outlet — or even your local Democratic organization. Don't endure; cure!

Here is a non-partisan number for voting questions: *1-866-OUR-VOTE*. When you call, they will ask what state you wish to access, and then transfer you to someone who will tell you exactly what needs to be done to register and vote in your state. For more information, visit http://www.866ourvote.org/.

Remember those who went before, who worked and fought and bled and died for the right to vote. Honor their sacrifice. Because of them, we have a chance to make things better.

Do not be cheated of your right to vote.

---

[9] The "voter fraud" myth debunked. *Rolling Stone* [Internet]. 2012 June 12 [cited 2015 Feb 13]. Available from: http://www.rollingstone.com/politics/pictures/the-voter-fraud-myth-debunked-20120612
[10] Berman A. The GOP war on voting. *Rolling Stone* [Internet]. 2011 Aug 30 [cited 2015 Feb 13]. Available from: http://www.rollingstone.com/politics/news/the-gop-war-on-voting-20110830
[11] Ballot.

39 ADVENTURES IN INTELLECTUAL PROPERTY

Jeanne Loring: As scientist and advocate, she is everywhere.

One of the greatest movies ever made was *The Adventures of Robin Hood* by the Warner Brothers in 1939, starring Errol Flynn. I grew up with Robin Hood and Maid Marian and Friar Tuck and Prince John and Little John and — how did they shoot the arrows into those guys in the movie? It looked so real!

The sword-fighting I could understand. When Errol Flynn fought Basil Rathbone in the great sword battle, it was a trained athlete versus a natural. Rathbone, soon to become famous as detective Sherlock Holmes, had competed in fencing at the Olympic Trials level. Errol Flynn was a terrific athlete, though untrained. They also had the coaching of Fred Niblo, a stunt coordinator and champion swordsman, to direct their glorious fight.

But the arrows? Previously, when an arrow was shot into a stunt man, it was a hollow shaft, slid along a fishline attached to the "wounded" one's shirt. Someone offstage would throw the arrow up the string. The apparently deceased actor would express discomfort, grab the arrow, and fall down. But in Robin Hood, the arrowed victims actually looked shot… and not to mention the famous part when the hero splits the arrow? More in a moment.

But first, consider the problem of intellectual property for stem cell research, and meet some people with white hair.

If you visited the meetings of the CIRM in the summer of 2010 (join us — go to http://www.cirm.ca.gov/ and click on "Upcoming Meetings"), it might have been confusing at first, if you did not know the characters.

Whenever the subject of intellectual property was scheduled, three people would always be there: John Simpson, Mary Maxon, and Ed Penhoet.

Ed is an interesting man with bushy eyebrows and friendly as a red-glowing heater on a winter's day. Ed is like a favorite uncle, and it is easy to miss the depth of his intellect. Co-founder of Chiron Corporation, one of the earliest biomed companies, he served as Chief Executive Officer for 17 years, as well as becoming the Dean of the School of Public Health at University of California, Berkeley, and President of the Gordon and Betty Moore Foundation![1]

The other day, an argument occurred between ICOC members Jeff Sheehy, Francisco Prieto, and representatives of the biotech industry.

The argument was one that had come up before: if an invention is made on CIRM (state) dollars, should it be available to poor people cheaply?

The biotech people were arguing that it would be impossible for them to set up an "access program" (i.e., cheap medicine for the poor) and Jeff Sheehy snapped:

"History does not support your assertion!"

He meant that the big pharmaceutical organizations like Merck routinely make low cost or free medications available for low-income citizens, and biomedicine should do the same.

Dr. Francisco Prieto is a small man, physically. He wears glasses, is soft-spoken and mild-mannered — he reminds me of an actor named Wally Cox, who played a meek television personality, Mr. Peepers, long ago. What the world did not know about Wally Cox was that Mr. Peepers was a weightlifter and had a black belt in judo.

It is not advisable to get between Dr. Prieto and the well-being of his diabetes patients.

"You guys seem to want grants with no strings attached — that's not going to happen!"

The biotech people argued (and it made sense): how could they have an access program designed before they even knew what the product was and how much it would cost, amongst other issues?

I thought, oh no, here it comes; major battle, and another slowdown for stem cells.

Both sides were right. We need cooperation with biotech, or improvements in the lab will not reach the public. We also need to be sure the patients have access to the medicine. We were caught between a rock and a hard place.

And then Ed Penhoet spoke up.

---

[1] Gordon and Betty Moore Foundation names Edward E. Penhoet President [Internet]. 2004 Jul 14 [cited 2015 Feb 14]. Available from: http://www.moore.org/newsroom/press-releases/2004/07/14/gordon-and-betty-moore-foundation-names-edward-e-penhoet-president

"Why don't we have it where you (biotech) design your own access program, and have it approved by this board — before your product goes on the market? That way, you'll know exactly what your product is, and no surprises either way."

It got very quiet in the room. I looked from corner to corner, protagonist to protagonist.

Everybody was nodding.

A bullet had been dodged.

And the white hair folks?

Dr. Mary Maxon, PhD, Deputy Vice Chair at CIRM, is way too young to have that glorious mane of snow-white hair. She looks like someone who would be asked for her ID card if she went to buy a bottle of wine for a dinner party. I asked her about it one time, and she said her hair had been white since she was a teenager.

Dr. Maxon writes the CIRM intellectual property policy. She attends every meeting, carries home endless stacks of legalese documents, studies them all until she understands, and then brings together what everyone is trying to say. Her Christmas vacation was writing the intellectual property policy for non-profit institutions. And all the while knowing that every placement of every comma would be argued over by everyone, rewrite after rewrite, just ahead. Can anybody say unpaid overtime?

The other person with white hair is John Simpson. Now, John comes by his white hair honestly; he is not perhaps as old as I am, but comfortable in that neighborhood. Simpson used to be a newspaperman for a major news chain, and he knows how to find information and put his thoughts together with impact.

But now he worked for the Foundation for Taxpayer and Consumer Rights as an effective spokesman, and I always got a little concerned when he stood up for public comment.

John's organization appears to want every tax nickel to be continually justified. I worried (and I still do) that the hunt for short-term financial returns might slow down long-term progress. If we went for the quick buck, that meant doing what was safe and repeating what was already being done. Our goal was bigger than that. We wanted the answers to chronic disease, and that is a difficult process which is unlikely to yield an immediate profit.

In a democracy, sigh, every side has a right to be heard.

But lately, John has been talking about something that made my ears perk up: it was about the patents on stem cell research.

What if every stem cell in America was owned by one company? Anybody wanting to use stem cells — every researcher company, even the government; everybody — would need to buy or rent from that one company.

That is the position taken by the Wisconsin Alumni Research Foundation (WARF), which funded the research by Dr. James Thomson, who was the first to derive a usable line of human embryonic stem cells. WARF feels they own it all: not just the process, but every embryonic stem cell in the world. This could be hugely damaging to the research.

It must be said that the Wisconsinites have a side to make as well, and they are decent folks whose foresight in funding the research advanced it enormously.

I figured the court battle would go on for decades, and there was not much that could be done about it.

But John Simpson challenged WARF.

His group, and another one, the Public Patent Foundation, went to law and petitioned the U.S. government to "revoke three University of Wisconsin patents they claim are driving scientists overseas and could hinder California's $3 billion stem cell research institute," as reported by the *Mercury News* on July 19, 2006.[2]

Wow.

I doubt that John Simpson thinks of himself in terms of Errol Flynn and old movies.

But to take on one of the single most important problems facing stem cell research today?

That was like when Robin Hood walked into the king's castle, dumped a deer on the table, defied the law, and said, "This will not do. There must be change."

See why you have to come to these meetings?

It's good stuff.

P.S.:

When Robin Hood splits the arrow?

The movie makers hired legendary professional archer Howard Hill, put him in costume, and actually got him to do the legendary shot. He aimed at the arrow already in the bull's eye — and splintered it, doing it several times. But the arrow would not split exactly end to end, as was wanted, and at the end a hollow arrow method was employed.

But there was nothing fancy about those poor soldiers who got shot. Under their shirts, each of the stuntmen wore a wooden board like a kitchen chopping block.

Howard Hill shot them in the wood. It felt, I am told, like being slammed in the chest by a baseball bat.[3]

---

[2] Johnson S. Groups urge revocation of stem-cell patents. *San Jose Mercury News* [Internet]. 2006 Jul 19 [cited 2015 Feb 14]. Available from: http://www.consumerwatchdog.org/story/groups-urge-revocation-stem-cell-patents

[3] The Adventures of Robin Hood (film). *Wikipedia* [Internet] [updated 2015 Feb 8; cited 2015 Feb 14]. Available from: http://en.wikipedia.org/wiki/The_Adventures_of_Robin_Hood_%28film%29

40 SPARTACUS FIGHTS BACK AGAINST STROKE

Kirk Douglas starred in the movie, SPARTACUS, the slave who fought back against Rome. But he also battles on against the effects of stroke, the number one cause of disability in the world.

Something metallic made gentle circles on my neck. I heard a sound like dolphins chirping. It was ultra-sound, to see if there was a blockage in the arteries of my neck.

I closed my eyes. I had just had a stroke.

Remember the Academy Award-winning movie, *Spartacus*, about the slave who fought back against Rome?

Kirk Douglas played Spartacus. A natural athlete who always trained hard, Douglas was in peak physical shape for the film: agile, swift, every muscle defined. There was one scene when he had his shirt off for gladiator training, and his back was chiseled perfection.

After the movie, Douglas stayed in top shape for many years, jumping around as if age could not touch him — and then the stroke hit like a lightning bolt.

When he gave interviews on television he could joke about it: "Now my sons really have to listen to the old man!" But it hurt to see him struggle to shape each word, and to see the devastation of that once magnificent body.

Spirit undimmed, he remained Kirk Douglas. Even now, approaching 100 years of age, he still exercises daily (with an 88-year-old trainer!), writes books, and is active in every way.

But if stroke could do that to him…

Every year, 800,000 Americans have a stroke; one in five of those will die of it. Seven million currently struggle with the paralyzing after-effects: physical, mental, or both.

Stroke, said the CIRM website, "is the number one cause of disability, the second leading cause of dementia, and the third leading cause of death in adults."

Disability, dementia, death… what cheerful words!

Mine was a very mild stroke (a Transient Ischemic Activity, or TIA for short). It was only speech that was interfered with, and that interference happens in a very odd way. When I spoke, it was like seeing a print sentence (like this one) in my mind. I had to "write" the words mentally before I could say them, laboriously telling myself which word went where. This lasted about a week and then went away. There was no noticeable permanent damage. Maybe I have to concentrate a little more to find and pronounce the words, but Gloria says she cannot tell the difference.

And what did the doctors recommend, to prevent a second and more dangerous stroke? "Take a baby aspirin everyday," they said, "To thin the blood". There was also a pill called *statin*, which did about the same thing, and I am taking that as well.

A stroke means damage done to the brain; when the blood flow is blocked off, cells begin to die immediately.

You need to unclog the blockage immediately using a thromboplastin, which is a clot-buster.

Otherwise, part of the brain will shut down. You could lose speech or thought, and your body could become paralyzed. Serious stroke damage leaves an actual hole in the brain where the cells had died. Whatever activities those brain cells were in charge of become diminished or gone.

But what about stem cells? Could they rebuild the missing portion of the brain after a stroke? Or make the remaining cells work better, compensating for the loss?

The Chair of Neurosurgery at Stanford University, Dr. Gary Steinberg, wanted to find out.

His goal is to use embryonic stem cells to improve recovery in the weeks and months following a stroke. He would first mature the embryonic stem cells into the kind of neural stem cells that are normally found in the brain. Then, he would transplant the new cells into the damaged brain.[1]

The stem cells, he hoped, would migrate to where they were needed.

Steinberg was building on more than two decades of struggle — and some remarkable success.

---

[1] Stroke fact sheet [Internet] [cited 2015 Feb 14]. Available from: https://www.cirm.ca.gov/our-progress/disease-information/stroke-fact-sheet

In a previous experiment, Steinberg had transplanted stem cells (specially treated mesenchymal cells) into the brains of 18 long-term stroke survivors.

Although the test was only intended to check for safety, many participants achieved improvement, a reduction of their weakness or paralysis. Furthermore, two of the patients (both women, one 33 and the other 71) made amazing progress.

As described by Steinberg, "They were very disabled. The 71-year-old could only move her left thumb, [but she's] walking now."

The other woman had severe language problems and could barely pick up her arm. The day after the surgery, she could move more strongly than since her accident, and a year later she could walk even faster and talk better.[2]

With scientists at University of California, Los Angeles as well as his home team, Steinberg applied for — and received — a CIRM team grant of $20 million. Because of Proposition 71, California is challenging stroke.

Spartacus would have been proud.

P.S.:

University of California, Santa Barbara stem cell scientists Zhuojin Xu, Lisa Conti, and Alison Blaschke received fellowships from the Kirk Douglas Foundation to attend the World Stem Cell Summit meeting. The Center for Stem Cell Biology and Engineering is grateful for the Foundation's generosity.[3]

---

[2] Goodman B. Stem cells show promise for stroke recovery. *Health Day* [internet]. 2014 Apr 7 [cited 2015 Feb 14]. Available from: http://consumer.healthday.com/cardiovascular-health-information-20/misc-stroke-related-heart-news-360/stem-cells-show-promise-for-stroke-recovery-686541.html

[3] Douglas Foundation sponsors UCSB researchers at World Stem Cell Summit [internet]. 2011 Dec 15 [cited 2015 Feb 14]. Available from: http://www.stemcell.ucsb.edu/news/2011/

41 DIABETES GOING DOWN?

Dr. Francisco Prieto's worst day is having to prescribe the amputation of a foot for a type 1 diabetes patient.

My cousin by marriage, Leroy Gallegos, was an example of the good one person can do. But I did not understand the depth of his contribution until I attended his funeral, in Oakland, California.

Probably 90% of the audience were teenagers. Some looked rough, and I would guess there were gang members present. But everyone was dressed in their best, whatever color that might be, and the funeral had standing room only.

Leroy had volunteered most of his adult life to organizing youth sports, so the kids of his neighborhood would have somewhere to go, something to do. When he could not raise the money for supplies, he paid for them himself. "Anything for the kids," might well have been his motto.

Tonight those kids were there — accepting no adult assistance — to express their thanks and to honor him. They had hired the church hall, and brought mountains of authentic Mexican food.

Diabetes killed my cousin Leroy. He had had three toes amputated, but even that was not enough. He went into a coma, and then he died.

This chapter is dedicated to Leroy Gallegos, the good that he did with his life, and the hope that a cure will be found for his devastating disease.

David and Jeff Bluestone: As scientist and advocate, father and son use different strengths to bring hope to the chronically ill.

The sheer scope of the problem is hard to imagine. Twenty-nine million Americans have diabetes, or roughly one in eleven. Of these, five percent have type 1 (juvenile onset) diabetes. Diabetes kills or contributes to the death of 231,404 Americans a year.[1]

The financial cost is staggering: roughly $245 billion a year, and rising.[2]

But against that are warriors: the California stem cell program, JDRF (formerly the Juvenile Diabetes Research Foundation), and one small biotech company, called Via-Cyte, Inc.

The odds are steep, but we must win.

As the 2014 National Diabetes Report puts it, "Diabetes […] is associated with … heart disease and stroke, blindness, kidney failure, and lower-limb amputation."[3]

The devastation is so common as to almost be taken for granted.

My Uncle Ben was one of the kindest men I ever knew. With my cousin Tommy Snyder, he financed my first computer, to help me write my first book. But diabetes destroyed his kidney function, so he had to have his blood washed by a machine. Dialysis mechanically filtered his blood until a painful and untimely death.

Diabetes: Dr. Francisco Prieto of the California stem cell program board is a gentle man with glasses and a ready smile. His worst moment was having to prescribe the amputation of a patient's foot, because of diabetes.

---

[1] Infographic: a snapshot of diabetes in America [Internet] [updated 2014 Jun 11; cited 2015 Feb 14]. Available from: http://www.diabetes.org/diabetes-basics/statistics/cdc-infographic.html
[2] General diabetes facts [Internet] [cited 2015 Feb 14]. Available from: http://jdrf.org/about-jdrf/fact-sheets/general-diabetes-facts/
[3] National Diabetes Statistics Report. National Center for Chronic Disease Prevention and Health Promotion (US); 2014. 12 p.

And what if the threat was to our children?

Here is a recount of an ordeal faced by Tom Karolya[4], reprinted with permission:

"I entered the hospital, and down the long corridor I saw my wife Jill standing outside a doorway. I saw my two-year-old daughter Kaitlyn's jacket hanging down from Jill's [hand. We] walked into the ICU room […].

"Kaitlin was in a crib-like bed, machinery, tubes, and personnel all around. She had been poked, prodded, and handled. Her color was horrible, she could not even cry any more. I leaned over and pushed her hair out of her eyes. Her eyes were lit with fear. I kissed her.

"She tried to smile. She was … only two. She rolled her head as if to show me everything around her and everything she was hooked up to including the IV in her arms. She looked at them and then at me […]."

"Daddy."

"Yes, honey."

"Fix."

A leaf-sized organ, the pancreas helps with digestion by processing sugar. One old diagnosis was to taste the patient's urine. Sweetness meant diabetes.

And if the pancreas failed? For thousands of years, diabetes meant death.

But the scientists fought back.

A Canadian champion, Dr. Frederick Banting, and a medical student, Charles Best, obtained lab space from Professor John Macleod of the University of Toronto.

They removed a dog's pancreas, squashed it, froze it with salts, and liquefied it, making insulin. As long as it lasted, they could keep that dog (her name was Marjorie) alive.

Before then, the only treatment for diabetes was a near-starvation diet so strict that people sometimes died of the food restriction — and at best they only gained a year or so.

But now there was hope. Using insulin from cows and pigs, more serum was made.[5]

A boy, Leonard Thompson, was on the verge of death, down to just 65 pounds — but the insulin brought him back.

And then — one of the greatest moments in medical history.

Banting and Best went into the diabetes ward. Fifty patients lay in coma, their families gathered around them, helpless, waiting for their loved one to die.

Banting, Best, and James Collip, a medical student, went from bed to bed, injecting an entire ward with the new purified insulin extract. Before they had reached the last dying child, the first few were awakening from their coma, to the joyous exclamations of their families.[6]

---

[4]Karolya T. Diabetes? If life ONLY sucks, only WE can change it [Internet]. 2013 Sep 30 [cited 2015 Feb 14]. Available from: http://diabetesdad.org/2013/09/

[5]Insulin. *Wikipedia* [Internet] [updated 2015 Feb 14; cited 2015 Feb 14]. Available from: http://en.wikipedia.org/wiki/Insulin#History

[6]Discovery of insulin [Internet] [cited 2015 Feb 14]. Available from: http://www.medicalnewstoday.com/info/diabetes/discoveryofinsulin.php

Banting and Macleod received the Nobel Prize. The award money was shared with Best and Collip. Later, another person considered equally deserving was Nicolae Paulesco of Scotland, who discovered much of the science involved in diabetes but received little credit.

The patent for insulin was sold to the University of Toronto for 50 cents — and that institution shared the knowledge with the world.

A pioneering biomedical company, Genentech, produced the first synthetic insulin, and that company's owner, Eli Lilly and Co., made it widely available.

So now we can keep a diabetic person alive — at the cost of multiple injections of insulin every day, as well as stabs in the finger to test the blood. The popular "pump" helps to systematize the process, but still a diabetic must constantly monitor every mouthful of food — how much sugar does it have? — and balance that against exercise and insulin.

Make a mistake, and he or she goes into coma.

And the threat of side effects has not gone away.

Enter ViaCyte, Inc., a small California biomedical company.

With grants of $39 million from the California stem cell program, as well as funding from JDRF, the world's leading funder of diabetes research, and backup from scientists at University of California, San Francisco, ViaCyte is taking on type 1 diabetes.[7]

They are using a device half the size of a credit card. This would be implanted under the skin. Called "an encapsulation medical device", Encaptra, it contains cells designed to replace those taken by the disease.

The scientists use embryonic stem cells, differentiating them into cells called "progenitors", one stage before making insulin. These progenitors are inserted into the Encaptra drug delivery device. Once in the body, the cells mature — and produce insulin.

The "credit card" is an amazing device. Porous as a tea bag, it has holes so small the body's attack cells (the immune system) can't get in to kill the good cells, but the life support nutrients can enter — and the insulin gets out.

ViaCyte's scientists tested miniature versions on diabetic mice, which were made well.

If it works on people, the world changes, and not only for diabetics and their families.

If we can cure just one "incurable" disease, the floodgates to research funding will open for more biomedical solutions.

This fight connects many giants in our field.

For instance, a friend and co-worker is David Bluestone, key to the youth campaign of Proposition 71.

His father, renowned diabetes researcher Jeff Bluestone, is the Chief Academic Officer of the University of California at San Francisco. Provider of much of the scientific expertise behind ViaCyte. Dr. Bluestone is the kind of man you want beside you in the fight. Not only is he a top-level scientist, but he gave up one of his kidneys to save his father's life.

---

[7] ViaCyte, Inc. [Internet] [cited 2015 Feb 14]. Available from: https://www.cirm.ca.gov/our-progress/institutions/viacyte-inc

And without CIRM, there would never have been that $39 million in research funding.

I spoke with Bob Klein about it recently. Father of a diabetic son, Jordan, and a massive fundraiser for the cause, Bob has seen pretty much everything there is when it comes to diabetes.

What did he think about the ViaCyte project?

"One of the most important and promising projects CIRM has ever funded," he said.

Was it appropriate to use the word "cure?". I asked ViaCyte's Chief Scientific Officer, Kevin D'Amour, with regards to the goal of their project.

Like any good scientist, he was cautious, but said, "Our product has the potential to provide patients with what amounts to a cure. The first human trials [just begun as this is written] will be conducted in patients with the disease; while demonstrating safety will be the primary endpoint, we will also be accessing efficacy, all in accordance with FDA requirements."[8]

And then he added words every patient advocate might echo:

"We have already cured thousands of mice; it is time we started curing some people."

---

[8] ViaCyte human trials imminent [Internet]. 2014 Aug 20 [cited 2015 Feb 14]. Available from: http://www.thejdca.org/wp-content/uploads/2014/08/Viacyte-Flash-Report.pdf

42 JAMIE THOMSON, OR, HOW DO YOU FOLLOW AN ACT OF GENIUS?

Jamie Thomson worked with the invisible to achieve the impossible: to bring cure closer for those who have been told: "there is no cure." The world owes him a debt of gratitude.

In 1998, Wisconsin's Jamie Thomson isolated the human embryonic stem cell (hESC), one of the pivotal achievements in medical history. Suddenly it seemed possible to repair — or regrow — any part of the human body. Language did not hold enough superlatives to describe the possibilities.

Others worked toward the same goal at the same time — Ariff Bongso of Singapore, John Gearhardt of Massachusetts, Gail Martin of the University of California, San Francisco — but Thomson brought everything together. (His contributions actually started much earlier, when he isolated ESCs from a rhesus monkey, considered impossible at the time.)

Time passed, and in 2007 the world got excited about another new stem cell development, the induced pluripotent stem cell (iPSC). Shinya Yamanaka won the Nobel Prize for his iPSC research, and deservedly so. The iPSCs were easier to work with, and had no ideological hassles.

ESCs were made from thrown-away fertilized eggs — frozen leftovers from in vitro fertilization procedures. But iPSCs were made from just a patch of skin. Add some genes with names like Oct-4, Sox-2, and Nanog, and the skin tag could be turned into something resembling an ESC.

The religious conservatives liked iPSCs because it was not technically an embryo (although they did not seem to notice that it could conceivably become one), and the scientists and patient advocates liked it because it might be able to do what an ESC could do.

There were problems with iPSCs, of course. The genes that transformed the skin tag were the same ones that made cancer. Also, unlike ESCs, iPSCs took considerable time to make, sometimes up to six months. If you were dying of liver failure and needed a replacement, you might not want to wait half a year. Still the problems seemed fixable, and the excitement was huge.

But while Dr. Yamanaka got the lion's share of the credit for iPSCs, Jamie Thomson was also working on the same project. In fact, he had written up the results of his experiments and had them in print at almost exactly the same time as Yamanaka. If Thomson had used mouse cells, which reproduce far more rapidly than the human cells he was using, he might have beat Yamanaka to the finishing line, so that it could have been him standing up in Stockholm, receiving science's ultimate accolade.

But when I interviewed Dr. Thomson recently, he said he was not going to spend his time being irritated about it. He had other things to do with his life.

At this, my ears perked up. Other things? What possibly could be on the same level as his previous work? To lead not one but two revolutions in biomedicine — surely that was enough? Most scientific breakthroughs are done by young men and women. This is another reason why it is so crucial that research grants be made available for youthful scientists. Most scientists have done their best work before they reach 50.

I began by checking the vital statistics.

Was he still Director of Regenerative Biology at Morgridge Institute for Research in Madison, Wisconsin, as well as having professorships at both the University of Wisconsin, and the University of California, Santa Barbara?

Yes.

And was his company Cellular Dynamics International still going strong?

Yes.

What did he do there?

He took a deep breath. It occurred to me that he was reportedly not fond of being interviewed.

"We make derivatives of iPSCs."

"For drug discovery and toxicity testing?"

All of a sudden he started talking faster than I could type.

Apparently, the FDA does not test every new drug or chemical that comes on the market. They do not have the budget or the staff. So when you or I step into our bathroom or kitchen, it is *terra incognita* — filled with thousands of untested and potentially harmful substances we may use. In Europe, the countries are demanding much more in terms of product safety, he said, referring to a program called REACH[1], which classifies many thousands of new products.

Stem cells could speed up drug and product testing, he said.

---

[1] http://echa.europa.eu/web/guest/regulations/reach

Thomson was a trained veterinarian, and had once wanted to become a marine biologist. One reason he has an office (and a professorship) at the University of California, Santa Barbara was that it was beside the ocean. In a previous interview, he stated, "It would be cool to get a stem cell line from a blue whale!"[2]

Naturally, I brought up my career as a professional diver, 17 years with sharks and dolphins in the giant aquarium tanks of Marine World in Northern California. He was polite but seemed uninterested, so I let that drop.

Was it true he had an uncle who was into rocket science?

"That was Uncle Jack," he said, and I could hear the smile in his voice, "He took me to NASA and showed me the work he did, arc-welding pieces of metal together to build a rocket ship. I remember him whispering, 'They actually pay me for doing this!' I wanted a job I could love like that."

I told him I appreciated him always giving credit to Dr. Junying Yu for her contribution to reprogramming research.

"She never got the credit she deserves," he said, clearly irritated, "She worked her butt off on the iPSC effort, and it went almost unnoticed."

Junying Yu, a name to remember.

What did he think about the California stem cell program?

"California deserves credit for stepping up to the bat, answering an unmet need. But if a scientist has a good idea, his ability to get a grant should not depend on his zip code."

I agreed with him on both points. If Washington had spent an amount of money on stem cell research commensurate with the size of chronic disease, there might have been no need for Proposition 71. But such was not the case. Federal funding of medical research has been flat for many years: no increases for the greater number of people with chronic illness, not even keeping up with inflation. And in Thomson's own state, the inventor of ESC research could not get a nickel of state funding. To this day, his research is privately funded.

"The current governor of Wisconsin has never been a friend of stem cell research," he said, in a colossal understatement: Scott Walker was recently revealed to be a supporter of personhood, which would almost certainly criminalize embryonic stem cell research.

http://www.nytimes.com/2015/02/23/us/politics/scott-walker-hardens-tone-on-social-issues-to-woo-christian-conservatives.html?_r=0

But something troubled me. A recent statement had been attributed to him in the conservative magazine *Financial Times*, which said, "He (Thomson) does not foresee a future for embryonic stem cells in mainstream research."[3]

---

[2]Baker M. James Thomson: shifts from embryonic stem cells to induced pluripotency. *Nature Reports* [Internet]. 2008 Aug 14 [cited 2015 Feb 14]. Available from: http://www.nature.com/stemcells/2008/0808/080814/full/stemcells.2008.118.html

[3]Cookson C. Reluctant star of stem cell research. *Financial Times* [Internet]. 2014 Feb 28 [cited 2015 Feb 14]. Available from: http://www.ft.com/intl/cms/s/2/0a12441a-9f47-11e3-8663-00144feab7de.html

Could this be?

I brought up the subject in a roundabout way, briefly citing examples of what seemed to me to be incontrovertible progress in ESC research:

1. Blindness: In hESC safety trials, legally blind patients have recovered vision, although only given minimal dosages.

    "Human embryonic stem cells [...] improved the vision in more than half of the 18 patients who had become legally blind because of two [...] currently incurable eye diseases."[4] (The two eye diseases are macular degeneration, a common source of adult-onset blindness, and Stargardt's disease, which threatens young people.)

2. Diabetes: There were two major ESC breakthroughs this month. ViaCyte's human trials began for a device which goes under the skin and dispenses ESC-derived insulin, as well as a separate ESC pathway (by Dr. Doug Melton) to make the beta cells needed by diabetics.

    "As San Diego's ViaCyte was in the midst of launching the first FDA-approved embryonic stem cell clinical trial for diabetics last week, Boston's Harvard University reported that cells made from embryonic stem cells 'cured' diabetic mice [...]."[5]

3. Alzheimer's disease: For the first time, scientists were able to study the terrible memory loss condition in a Petri dish. Before this, there was no way to follow the disease through its life cycle, or even to know when Alzheimer's disease began.

But now? "Group members used human embryonic stem cells [...] and grew them with a mixture of chemicals that made them turn into neurons. [Those neurons were given] Alzheimer's genes and [were grown in] petri dishes."[6]

Dr. Thomson did not directly respond to the *Financial Times* statement. However, he did say two things. First, he thought that "the ViaCyte people got it right" in their use of ESCs for diabetes therapy. Second, he went on to question if the Alzheimer's disease modeling had actually been done with ESCs, as the *New York Times* reported, suggesting I should check out the original paper[7], published in the scientific journal *Nature*.

I purchased a copy online (cost me $32!) and read it.

---

[4] Stein R. Embryonic stem cells restore vision in preliminary human test. *Shots* [Internet]. 2014 Oct 14 [cited 2015 Feb 14]. Available from: http://www.npr.org/blogs/health/2014/10/14/346174070/embryonic-stem-cells-restore-vision-in-preliminary-human-test

[5] Fox C. Embryonic stem cells in trial for diabetes. *Bioscience Technology* [Internet]. 2014 Oct 16 [cited 2015 Feb 14]. Available from: http://www.biosciencetechnology.com/articles/2014/10/embryonic-stem-cells-trial-diabetes

[6] Kolata G. Breakthrough replicates human brain cells for use in Alzheimer's research. *The New York Times* [Internet]. 2014 Oct 12 [cited 2015 Feb 14]. Available from: http://www.nytimes.com/2014/10/13/science/researchers-replicate-alzheimers-brain-cells-in-a-petri-dish.html?_r=1

[7] Choi SH, Kim YH, Hebisch M (2014) A three-dimensional human neural cell culture model of Alzheimer's disease. *Nature* **515**: 274–278.

The paper said, in dense scientific language, "These FAD lentiviral constructs were transfected into ReNcell VM human neural stem (ReN) cells." This meant, I thought, that the scientists put the Alzheimer's stuff into the hESCs.

Just to be sure, I contacted Dr. Rudolph Tanzi, the scientist who had done the groundbreaking research, and he was kind enough to write back.

He said, "We used a human embryonic stem cell line that has been immortalized and sold commercially — RenVM cells."

So where did that leave us? Dr. Thomson had previously stated that hESCs are "a pain" to work with. After his breakthrough paper was published, only "about 50 labs" started using ESCs. But after the publication on iPSCs, *thousands* of labs — including Dr. Thomson's — went to work on the simpler methodology. It may well be that iPSCs turn out to be the vehicle of choice for most research. If that is the direction the science leads, so be it. I am a patient advocate, not a scientist; but if 20 years of advocacy have taught me anything, it is that science is difficult, complicated, and unpredictable. iPSCs may be a great tool, but nothing substitutes for the entire toolbox!

The International Society of Stem Cell Researchers (ISSCR) calls ESCs "still the gold standard" against which everything else will be judged.

To my delight, on October 27, 2014, I received the following brief email from Dr. Thomson:

"Human embryonic stem cells remain the gold standard to which induced pluripotent stem cells are compared, and they will be used for a long time. However, the ability to choose a specific genetic background is a significant advantage of iPSCs for some applications."

It is the scientists who should make that decision, project by project. They need the freedom to go forward with the best science available and not be crunched under by politics. May their options remain forever open, and may they have the funding that is necessary to insure scientific freedom — no scientist is free if his or her research is shut down.

And finally, the big question: after ESCs and iPSCs, what might be a third great breakthrough for Dr. Jamie Thomson? The answer was breathtaking. He was working on nothing less than the timing of the lifespan of the cells.

Imagine a little clock inside every cell of your body, telling it when to grow, expand, multiply, and die. What if you could speed up or slow down that cellular "clock"? If you wanted to make a replacement organ, like a liver or a heart, you might want the cell growth to speed up, taking days instead of months. But for a cancer stem cell? We would want that dangerously aggressive cell to slow down, or even stop, so the patient would not die. Changing cellular clocks might even affect the aging process itself, so we could stay younger longer!

And as for Jamie Thomson's work on both embryonic and induced pluripotent stem cells? I could only repeat what I had told him years before, when we met at the World Stem Cell Summit:

"When my paralyzed son walks again, I will always believe he took his first steps in Wisconsin."

43 SHERLEY V. SEBELIUS

Kathleen Sebelius is perhaps best known for her bold and determined defense of America's national health program. But the former Governor of Kansas also led the fight in a crucial battle for scientific freedom ...

On August 23, 2010, Federal Judge Royce Lamberth signed a decree — *shutting down America's funding of embryonic stem cell research*.

Here is the key sentence of Judge Lamberth's decree:

"ORDERED: That defendants [are enjoined from] *funding research involving human embryonic stem cells* [...]" (italics mine).

America's federal government was forbidden to fund embryonic stem cell research.

Do you recognize the name Kathleen Sebelius? Few have paid a heavier price in stress and exhaustion for serving our country than Ms. Sebelius, former Governor of Kansas.

The world remembers the website launch of America's first national health program, the difficulties of which were blamed on Ms. Sebelius. Could the program have begun more smoothly? Of course. But nothing big happens smoothly. The important thing about Ms. Sebelius, and the President she served, was that they did not give up. Because of their endurance, millions of children, men, and women now have health care coverage which they did not have before.

But how could America's funding for embryonic stem cell research suddenly become illegal?

A group of religiously oriented individuals had sued to block federal funding of human embryonic stem cell (hESC) research. Plaintiffs included the following:

1. Drs. James L. Sherley and Theresa Deisher were adult stem cell researchers. They claimed that money spent on embryonic stem cell research meant less money for adult stem cell researchers like themselves, thereby damaging them in the pocketbook.
2. Nightlight Christian Adoptions, a company that (for a fee) hooked up potential parents with embryos left over from the in vitro fertilization process. Their complaint? A possible shortage of embryos for their business.
3. Embryos (!).
4. Two couples, Shane and Tina Nelson, and William and Patricia Flynn, customers of Nightlight. They sued because they might want more embryos to adopt.
5. The Christian Medical Association. They claimed "frustration" because allowing hESC research would make them work harder (and spend more money) to spread their message.

To me, it was nonsense from start to finish.

Scientists suing America because they objected to competition? The National Institutes of Health (NIH) gave out its grants on the basis of a selection process — trying to fund the best science. Was that system of competition to be thrown away to benefit two individuals?

As for Nightlight, should profits take precedence over a chance to heal paralysis, heart failure, cancer, and other forms of chronic illness and injury?

And the couples who might want to adopt more embryos — how many did they wish to adopt? Was not 500,000 (estimated number of blastocysts currently frozen and stored) sufficient to their need?

Embryos suing?! Most sperm-egg combinations are fairly quiet, to the best of my knowledge.

As for the Christian Medical Association, if frustration was cause for filing lawsuits, could not any group sue America? I am frustrated by the Republican Party's blockage of medical research — am I allowed to sue John Boehner, Mitch McConnell, and the GOP?

For a moment, it seemed that reason might prevail, and with a speedy outcome. On October 27, Judge Lamberth dismissed the suit for lack of "standing", meaning none of those suing could claim damage.

But the plaintiffs appealed. Ten months later, on June 25, 2010, the suit was allowed to continue — although plaintiffs with standing were whittled down to just the adult stem cell researchers. The U.S. Court of Appeals for the District of Columbia ruled that these two scientists could claim damage.

The case was sent back to Judge Lamberth.

And then came a major mistake. The plaintiffs asked for a "stay" — blocking all federal funding of hESC research until the trial was over — and Judge Lamberth allowed it.

He said that stay "merely continued the status quo". He was wrong. The status quo included the NIH funding, which by his order was suddenly forbidden to fund hESC research. Every dollar the U.S. Government was spending (or might spend) on the new research was now blocked — even the limited research allowed by the Bush policy.

That was the background on August 23, 2010.

As I said in an interview with *ABC News*, "Every delay is cure delayed, cure denied. We have got to get this political nonsense out of the way."

Sean Morrison, then-Director of the Center for Stem Cell Biology at the University of Michigan, said, "If the ruling [is not] lifted soon, this will do irreparable harm to the field. [...] The judge made a serious error [...]; this ruling would have blocked even the more restrictive Bush administrative policy, [doing] greater harm to the human embryonic stem cell researchers than any policy ever enacted. [This is the] most extreme interpretation of federal law that we have seen yet."[1]

If a scientist had won a grant and was working, he had to go home. If that scientist had lab assistants, they lost their jobs.

The plaintiffs' lawyers included people like Thomas G. Hungar of Gibson, Dunn & Crutcher, and Steven Aden of the Alliance Defense Fund; serious warriors for their cause.

But our side was pretty tough too. Working with Kathleen Sebelius were Eric R. Womack, Kyle Renee Freeny, and Joel L. McElvain, of the Department of Justice.

Our side filed a request that the "stay", (the stoppage of research) be set aside.

The opposition filed arguments demanding that the stay remain.

Here were the three main arguments:

1. Were adult stem cell researchers Sherley and Deisher unfairly damaged by the Obama stem cell policy, and if so, should the program be shut down to accommodate them?

2. Was America's stem cell doctrine in conflict with the Dickey-Wicker Amendment, which said the Feds could not fund research which threatened or destroyed an embryo?

3. Before the new stem cell doctrine was put in place, a public comment period was allowed. An estimated 50,000 letters (many virtually identical) were sent.[2] If a majority of the letters opposed the Obama stem cell policy, did that make it illegal?

From last to first, the plaintiffs argued that the government had entered the public comment period "with a closed mind", unfairly ignoring the negative comments of those who wished to stop the program. But many of these letters were "not responsive". If the question was, how could embryonic stem cell research best go forward, answers that supported shutting it down were off-topic. If someone asks, "Should we go to the movies or bowling?" and someone answers, "Pizza," that is off-topic, although it may be desirable.

---

[1] Hutchison C. Scientists outraged by block on stem cell research. *ABC News* [Internet]. 2010 Aug 25 [cited 2015 Feb 15]. Available from: http://abcnews.go.com/Health/WellnessNews/scientists-outraged-block-stem-cell-research/story?id=11469249&singlePage=true

[2] Listing of Comments on draft NIH human stem cell guidelines [Internet] [cited 2015 Feb 15]. Available from: http://grants.nih.gov/stem_cells/web_listing.htm

As for Dickey-Wicker, that forbids funding for anything that endangers or kills an embryo. The plaintiffs defined an embryo as the union of sperm and egg, instead of what the dictionary listed as the most common definition, which was:

"*embryo*: The developing human individual *from the time of implantation* to the end of the eighth week after conception" (emphasis added).

It was like the religious argument from the Middle Ages, "How many angels can dance on the head of a pin?"(That argument made no sense either — if angels were spirits, they had no physical mass, and therefore an infinite number could fit on the head of a pin.) But foolish or not, the outcome could have very serious consequences.

Harriet Rabb (then-Chief Counsel at the Health and Human Services Department) for President Bill Clinton, argued that since stem cell research involved stem cells, it did not violate Dickey-Wicker. Stem cells are stem cells, not embryos. That argument satisfied three Presidents — Clinton, Bush, and Obama — as well as legislators who twice passed the Stem Cell Research Enhancement Act (S5, Castle/Degette).

Finally, were adult stem cell scientists Sherley and Deisher damaged by the stem cell doctrine, and if so, must the NIH shut down all hESCr funding to avoid inconveniencing these two?

Dr. Deisher had never even applied for a NIH grant to study stem cells; Dr. Sherley had one application for a grant in the process of being considered. Both were perfectly eligible to apply for grants — but where was it written that they were promised success? In terms of actual dollars granted, the odds were on their side anyway. The NIH has awarded many more grants for adult stem cell research than for embryonic stem cell research. In 2009, for example, hESC research received approximately $120 million in federal grants. In contrast, non-embryonic stem cell research got $340 million in grants; nearly triple the amount.[3]

On September 9, 2010, the U.S. Court of Appeals voted 2–1 to overthrow the stay of hESC research. The issue was not decided, but the research could go forward.

It was the turning of the tide. When the case was returned to Judge Lamberth, he dismissed it.

Plaintiffs then appealed to the U.S. Court of Appeals, where they were defeated by unanimous vote. They even went so far as to appeal to the United States Supreme Court, perhaps the most conservative in modern history, but it refused to even hear their argument.

"We couldn't be happier that this frivolous, but at the same time potentially devastating, distraction is behind us," said Doug Melton, scientific director of the Harvard Stem Cell Institute in Cambridge, Massachusetts.[4]

We had preserved again our right to fight disease.

---

[3] Stem cell information [Internet] [updated 2014 Apr 1; cited 2015 Feb 15]. Available from: http://stemcells.nih.gov/research/funding/pages/Funding.aspx

[4] Wadman M. High court ensures continued US funding of human embryonic-stem-cell research. *Nature News* [Internet]. 2013 Jan 7 [cited 2015 Feb 15]. Available from: http://www.nature.com/news/high-court-ensures-continued-us-funding-of-human-embryonic-stem-cell-research-1.12171

44 THE WOMAN WHO WOULD NOT BE SILENCED

Jan Nolta directs the UC Davis stem cell program.

The first I heard of Huntington's disease was December, 2007, when Frances and Margie Saldaña, mother and daughter, spoke at the public meeting of the California stem cell program.

Margie was a beautiful young woman, formerly a model who had worked in movies, but now her body swayed, moving against her will. Her wrists would rise till they almost touched her chest; she would force them back down, like triceps pushdowns on a weight machine.

When it was her turn to speak, it was with great effort, and she repeated herself over and over: she had two children, two children, two children, she said.

When she signed her son up for kindergarten, the office staff started whispering, and sent for a police officer. He asked Margie if she had a problem, and she showed him a letter from her neurologist stating that she had Huntington's disease. The officer took her keys away, at which point the children became terrified and started crying. The officer treated Margie sneeringly, as if she was a criminal or was on drugs.

"He took my keys away," said Margie, "He took my keys... and he would not give them back."

Frances, Margie and Marie Saldana devastated by Huntington's disease, but not defeated.

But her advocate Mom would not accept the injustice. Contacting the chief of police, she pushed for an in-service to be given to the entire department — by a police officer whose wife also had Huntington's disease.

Margie's father, Hector Portillo, also had the disease. It was the cause of his death, when a car struck him as he walked too awkwardly across an intersection.

Margie's sister Marie, also afflicted, was put into an institution. Her roommate was an elderly woman — the driver of the car that had hit Hector! By the time Frances figured this out, Marie had become close friends with the woman. Frances did not want to spoil her daughter's happiness, and Marie died never knowing that her friend had accidentally killed her father.

Margie's husband, Craig, was a man to respect. Often in situations of chronic disease, one spouse (usually the husband, I regret to say) will leave. But Craig stayed. When his wife was diagnosed with Huntington's disease, he provided care for her at home.

Frances' son Michael, now 32, wanted to become a chef, and worked in the field of creative cuisine until the loss of body control from Huntington's disease cost him one job after another. When he could no longer drive, he rode a bike to work, even after being hit twice by cars. Emotionally, he began to lose control. Finally, he flew into a rage, demanding $50,000 from his Mom. He wanted to start a cooking company, he said, and left when she would not comply.

He was found a thousand miles away, sitting in front of a convenience store in Seattle, Washington. He had sat there all day, without his medication and robbed of his money and luggage.

When the owner of the store asked him why he was sitting there, he said he was waiting for his mother to wire him some money, although he had not actually contacted her. By sheer good luck, the owner was familiar with Huntington's disease, and asked for his mother's telephone number. Working through long distance, Frances was able to arrange for an ambulance.

When he got home, all Michael said was, "Hi, Mom. I'm hungry."

As this is written, Michael has been in a board-and-care home for eight years. He is now unable to speak, walk, or chew his food.

When I called Ms. Saldaña for an interview, I asked first about Margie, remembering her advocacy at the meeting years before:

"She died four months ago," said Frances, "We buried her on Valentine's Day."

I burst out with a curse word, and instantly regretted.

But all she said was: "We have to beat this thing."

In 1995, "there was really nothing available in Orange County, California for Huntington's disease families," said Frances Saldaña.

She got a job at University of California, Irvine, where research on Huntington's disease might be done. She worked there all day at her regular job, and used her spare time to contact anyone who might help in the fight.

One of the scientists she met at University of California, Irvine was a neuroscientist, Dr. Leslie Thompson.[1]

Dr. Thompson had been part of the worldwide effort to find the gene that causes Huntington's disease. She worked with a team of scientists and clinicians under the leadership of Dr. Nancy Wexler, whose mother died of the disease. Dr. Wexler's family started the Hereditary Disease Foundation. The team went to South America to work with families around the Lake Maracaibo area of Venezuela where many individuals within the

---

[1] UCImpact: stem cell research. *UCI News* [Internet]. 2009 Dec 17 [cited 2015 Feb 15]. Available from: http://news.uci.edu/features/ucimpact-stem-cell-research/

community share the same gene mutation that causes Huntington's disease, thereby collectively suffering from the disease.

One day, Dr. Thompson came back to the lab to find that the person in charge had died, and she suddenly had to fight for funds or watch the research die passed away.

Fortunately, there were organizations which helped the lab get back on its feet. Also wanting to help was Frances Saldaña. Teaming up with friends Jean Abdala and Linda Pimental, whose families also suffered from Huntington's disease, Ms. Saldaña "founded HD CARE to raise awareness and funds for UC Irvine research and clinical care, [raising an estimated] $1 million [...]."

"Frances is a tireless promoter of HD research and care and her enthusiasm is infectious," Dr. Leslie Thompson told *UCI News*, "Her impact is profound." She added that Frances inspired her everyday.[2]

Dr. Thompson has a framed picture of the Saldaña family on her desk, an ever-present reminder of why the fight must go on. As a hardened 25-year veteran, she and fellow researchers J. Lawrence Marsh, Joan Steffan, and Malcolm Casale keep University of California, Irvine in the struggle.

Frances raises awareness of Huntington's disease in creative ways, inviting scientists to speak at fun events such as chocolate-jazz-and-wine festivals (one raised $70,000) and a *Huntington's ballet*.

Frances explained, "The Huntington's ballet was performed at the Trevor School of the Arts in 2007. I worked on this project with Professor Donald McKayle. Titled *The Long Lonesome Way*, I wrote the script and McKayle choreographed it as a tribute to Woodie Guthrie, who had Huntington's long before it was understood. Leslie Thompson and Dr. Larry Marsh took McKayle to the Huntington's Disease Lab to show him the movement of mice afflicted by the disease."

"Todd Barton, resident composer at the University of Oregon, put the sequence of the HD protein to music, which was combined with Woody Guthrie music.

"Now I want to get that done for all the related neurological diseases, for a production this coming school year. The production brings together true science technology and art [...] with a main dancer interpreting [a Huntington's disease] patient, about 20 dancers interpreting the proteins and neurons, DNA, etc., and a [large screen video] in the back showing at a microscopic level the actual Huntington's disease proteins, and the damage they do [...]."

Why is this important, aside from the plain humanity of wanting to ease suffering?

"In the 14 years that Michael has been receiving government assistance since being diagnosed as permanently and totally disabled, he has received over $1,000,000 in benefits — not counting his medicines and ER visits and hospital stays and ambulance fees," said Frances of her son.

I looked up Dr. Thompson's grants at the California stem cell program. One was labeled "A hESC (human embryonic stem cell)-based Development Candidate for Huntington's Disease". It was funded $3,539,536. More than three million dollars...

---

[2] A cause close to her heart. *UCI News* [Internet]. 2013 Apr 1 [cited 2015 Feb 15]. Available from: http://news.uci.edu/features/a-cause-close-to-her-heart/

but what was its status? Had it failed, or was it dead? A split-second crisis occurred as I held my breath, located its status, and read — ACTIVE.

As described in the abstract, "[Huntington's disease] is a devastating degenerative brain disease [...] that inevitably leads to [death. It] has a 50% chance of being inherited by the children of the parents. Symptoms [...] include uncontrolled movements, difficulties in [holding down a job, and] severe psychiatric manifestations including depression.

"Current treatments do not change the course of the disease. [hESCs] offer a possible long-term treatment [...] that could relieve the tremendous suffering [of] patients and their families [...]".

"In the first year, the team [developed methods to turn] hESCs into neural, neuronal, and astrocyte precursors to be used for transplantation [and] determined the correct cells to use [...]. Once completed, the cell giving the greatest protective benefit will be transplanted into mice [to validate the] approach for human [use]".

"Transplantation of neural cells provided [benefit to a mouse]."[3]

I called up Dr. Thompson. What kind of cells would she be using?

"For the disease in a dish studies, we used induced pluripotent stem cells. But for the actual transplantation, we will be using embryonic stem cells."

So what happens next?

"We continue the mouse work. If that goes well, then we talk to the FDA about human trials. Everything has to work down to the smallest detail. At first, we did not even know which cells to work with. The field is so new. We had to be sure not to create tumors. We changed cell types several times. It took years to figure out. Much of what we now know is thanks to the incredible body of knowledge brought about by CIRM."

Cooperation.

Dean of Medicine Dr. Claire Pomeroy, later a member of the board of directors of the California stem cell program, was strong in the fight for Huntington's disease research. She worked closely with David Richman, Chair of the Neurological Department. They had just hired a neurologist Vickie Wheelock, PhD, and Richman said to her, "You've got to meet the Huntington's families!"

Wheelock convinced University of California, Davis to establish a Huntington's disease clinic in 1997. With the help of the patient advocate community (which included stalwarts like Les Pue and Katie Jackson), money was found.

Dr. Wheelock contacted her friend, nurse practitioner Teresa Tempkin, and suddenly University of California Davis was a center of excellence for Huntington's disease, serving 350 families.

And then there was a scientist named Jan Nolta.

"I was the little girl who at the age of five gathered up mosquito fish," she said. The fish were used as fertilizer for the rice fields. Eventually, the fields dried up and the fish

---

[3] A hESc-based development candidate for Huntington's disease [Internet] [cited 2015 Feb 15]. Available from: https://www.cirm.ca.gov/our-progress/awards/hesc-based-development-candidate-huntingtons-disease

died, but Jan rescued some of them and noticed how, when they multiplied, the fish could be bred for color.

She worked her way through college as a waitress at Denny's. When business was slow, they let her do her homework at the counter, she mused, smiling.

She loved science, and spent two decades developing her skills. She worked with adult stem cells in Missouri (the only kind legally available there), and was fascinated by one particular kind of cell taken from the bone marrow. Called *mesenchymals*, these acted like microscopic emergency vehicles, zipping around in the body, working to heal sick cells.

And when at last she came to University of California, Davis, Vicki Wheelock asked her, "Have you ever thought of using stem cells to fight Huntington's disease?"

"I have spent 20 years using stem cells on every part of the body *except* the brain!" said Nolta.

But the best scientists in the world can do nothing without money.

Nolta went to work on a grant to be submitted to the CIRM. It took two years, but it worked. With $20 million from the CIRM facilities grant, University of California, Davis was able to raise an additional $40 million — and converted a fairgrounds hall (like a barn) into stem cell research headquarters, later named the Institute of Regenerative Medicine.

Dr. Nolta earned a $3 million grant from CIRM to work with mesenchymal cells, her favorite. Could these help fight Huntington's disease?

Huntington's disease destroys a sort of brain fertilizer known as brain-derived neurotrophic factor (BDNF), which is a type of protein. Without BDNF, two very important organs in the brain, the striata, shrivel away.

Without these two small organs, the whole nerve system goes crazy. The abilities of the body and mind are stolen. People with sick mesenchymal cells cannot heal themselves.

But what if healthy mesenchymal cells were put into the brain? Could they produce new BDNF and heal the striata?

Tiny holes would be bored in the sides of a person's skull, following which a syringe would inject healthy mesenchymal stem cells (MSCs). Then, inside, each MSC does something very similar. It sticks out a nanotubule, inserting it into a sick cell. It injects a jolt of liquid BDNF which Huntington's disease destroys.

It worked in mice. But humans are a whole bunch more complicated.

Could this be a cure? "No," said Dr. Nolta, "But it buys us time."

She and Dr. Wheelock agreed that embryonic or induced pluripotent stem cells will be needed for the actual cure.

How does one measure improvement? Patients will be examined at six-month intervals, both before and after the cells are injected. An "arc of progress" (as the downward trend is called) demonstrates that the disease is being fought.

But nothing could happen without the funding.

Wheelock and Nolta applied for CIRM funding, and I was in the room when the governing board made the decision. The room was filled with what seemed like half of Sacramento — Jan Nolta, Vickie Wheelock, Les Pue, Katie Jackson, Teresa Tempkin, Stacy West Brookhyser, and many more.

Physically absent but there in spirit was Kenneth Serbin, a history professor at the University of San Diego. He is known in Huntington's disease circles as Gene Veritas, a pseudonym meaning "the truth in my genes" because he carries the disease's mutation. His mother died of Huntington's disease in 2006. A terrific writer and editor of the newsletter for the San Diego chapter of the Huntington's Disease Society of America, he was a strong supporter of Proposition 71. (Check out his blog, *At Risk from Huntington's*, at http://curehd.blogspot.sg/.)

(Note: One highly respected advocate asked for personal reasons not to be mentioned in the aforementioned list of people present in the room. Naturally, that request is respected, though with sadness, for she is a tremendous worker. If you know Huntington's disease advocacy, you will know who she is — and she deserves a hug.)

This was Huntington's hour. I spoke, the board spoke, and everybody who had something to say spoke. And then there was nothing but the vote: thumbs up or thumbs down.

Huntington's disease advocate Katie Jackson wrote:

"[…] As the voting started, I felt the emotions in the room escalating [and then] the votes came in; every single ICOC member voted YES. When the last yes came in, confetti streamed over our heads. We all jumped up and down with joy, ran around and hugged each other […]."

"Our deepest appreciation goes to the ICOC members who voted yes, to Robert Klein who authored Proposition 71, to the voters of California, [and] to all who advocate for Huntington's [disease, including] the most important person at UC Davis School of Medicine, Dean Dr. Claire Pomeroy for bringing Dr. Jan Nolta and Dr. Gerhard Bauer [to UC Davis]."[4]

Nineteen million dollars would go to Vicki Wheelock and Jan Nolta to try and realize the MSC method to attack Huntington's disease. The money would be given over a period of four years, and was subject to many stringent conditions in the form of "go/no-go" progress requirements.

But if the science is right, the money is there.

When Huntington's disease is cured, it will be forgotten like the nightmare it is.

But we should remember those who fought so hard: the scientists and patient advocates, including Margie Portillo Hayes, the daughter of Frances Saldaña. For all who saw Ms. Hayes struggling to speak, refusing to allow the condition to silence her, fighting in the only way she had left —

A part of her will remain in our hearts.

---

[4] Jackson K. Historical vote for HD stem cell research: CIRM awards $18.9 million to UC Davis stem cell program for Huntington's disease clinical trial — a personal story. *The Huntington's Post* [Internet]. 2012 Jul 26 [cited 2015 Feb 15]. Available from: http://www.thehuntingtonspost.org/files/KJ_cirm_metting_FINAL2.pdf

FLASH!

Huntington's disease will be challenged by embryonic stem cells from BioTime, Inc., via its "wholly-owned subsidiary ES Cell International (ESI, Singapore).

Dr. Leslie Thompson of UC Irvine will be leading the charge, thanks to a $5 million grant from the California stem cell program. Working with her will be UC Davis' Good Manufacturing Practices lab, led by Dr. Gerhard Bauer.

United against this horrific disease are Singaporean/American private enterprise, the research departments of two great universities — and the California stem cell program.

—

http://www.businesswire.com/news/home/20150427005320/en/BioTime%E2%80%99s-Clinical-Grade-Stem-Cells-Subsidiary-ES#.VT-oFU10zIU

45 THE GORILLA GYNECOLOGIST, OR, THE PERA-CHEN ANTI-URINARY-INCONTINENCE METHOD

As a younger man, I had a bladder like a camel, and could wait for what seemed like weeks between restroom visits. But as a "male of a certain age", when the need arose, I had little warning; the time to go was right now. The frequency (and intensity) was frightening. What if I had to use the facilities every 45 minutes? Would I spend the rest of my life chained to bathroom needs? In supermarkets, I started noticing the adult diaper section. After my prostate surgery, that interest is no longer theoretical, but practical.

But for seven million women in California, the inability to hold one's water is a far more serious problem.[1]

Urinary incontinence (UTI) has been regarded as "the primary reason for elderly women to be institutionalized".[2]

The danger is in falling. As women grow older, bones grow thin, coordination may diminish, and a rush to the restroom (especially at night) may cause falls and slow-healing fractures.

There is also embarrassment and worry. When will leakages happen? Will others know? (Important note for those who worry about offending others with odor of a wetness problem: a recent bathroom accident has little odor; the smell arrives only if the urine is not cleaned up over a period of hours.) Depression may keep an otherwise healthy woman from leaving the home; some may shrink from medical care. Isolation can ruin a person's life.

Why does UI hit women roughly twice as often as men? Childbirth may damage the muscles around the bladder opening. Also, women live longer than men: in California, females reach an average age of 82 compared to 75 for men.

The problem is huge. In Japan, more diapers are sold for adult incontinence than for babies.[3]

---

[1] Urinary incontinence in women [Internet] [updated 2013 Sep 18; cited 2015 Feb 15]. Available from: http://kidney.niddk.nih.gov/KUDiseases/pubs/uiwomen/index.aspx
[2] Autologous iPSC therapy for urinary incontinence [Internet] [cited 2015 Feb 15]. Available from: https://www.cirm.ca.gov/our-progress/awards/autologous-ipsc-therapy-urinary-incontinence-0
[3] Aquino F. Sales of adult diapers in Japan surpass those for babies. *The Japan Daily Press* [Internet]. 2013 Jul 11 [cited 2015 Feb 15]. Available from: http://japandailypress.com/sales-of-adult-diapers-in-japan-surpass-those-for-babies-1132153/

If a person goes through three to six diapers a day, think of the sheer trash-disposal problems — diaper mountains for landfills.

Nationally, the cost of incontinence exceeds $20 billion a year — comparable to the expense of chronic diseases such as Alzheimer's disease or arthritis.

Surgery is common. In 2008, 20,330 women underwent surgery for UI in California; this extrapolates to 172,500 UI procedures per year nationally.[4] But the relief is often only temporary, since leakage may reappear as women age.[5]

Fortunately, two scientists — and the California stem cell program — are challenging the problem.

Bertha Chen and Renee Reijo-Pera, using stem cells to battle female urinary incontinence, the number one cause of women being institutionalized.

Stem cell researcher Renee Reijo-Pera and urogynecologist Bertha Chen worked together on a research grant from the California stem cell program.

Bertha Chen was the gynecologist for the famous gorilla Koko, of sign language fame.

"We were trying to figure out why she was not getting pregnant; turns out the problem was with the male, not her," said Dr. Chen. In conversation, she noted that the gorilla's reproductive equipment was "very similar to human females, only somewhat larger."

---

[4] Autologous iPSC therapy for urinary incontinence.
[5] Reed DC. The Pera-Chen method: urinary incontinence, and how stem cell therapy may defeat it. Huffington Post [Internet]. 2013 Oct 23 [cited 2015 Feb 15]. Available from: http://www.huffingtonpost.com/don-c-reed/urinary-incontinence-stem-cells_b_4143565.html

The doctor and the scientist combined their skills to try and find an answer to a worldwide problem affecting millions of women.

A healthy sphincter, the shut-off valve of the urinary tract, stays closed because of quietly operating smooth muscles. These keep the urine in during our daily activities. But as the muscle cells age, they weaken and some die; if there are not enough active cells to maintain closure, urine will spill during activity, or even following a cough. Childbirth may over-stretch the opening and also cause damage to the cells. If the opening cannot close securely, the person will wet herself.

One way to diminish the problem is through "bulking". Injections of fat, collagen, or ground-up bone can narrow the opening. But while this lessens leakage, it does not restore control. Further, it often requires repeated injections. Other surgeries may enhance the strength of a woman's internal ligaments, but these decrease in effectiveness when repeated, as is typically needed.

But what if young and healthy muscle could be regrown inside the woman herself? Normal self-control could be restored, and the problem would be gone.

The Pera-Chen technique would inject muscle cells grown from the woman's own stem cells into the urethral sphincter. There, it is hoped, younger and healthier muscles will restore function. If successful, the technique might well be useful for other conditions, including diseases of the blood vessels, respiratory tract (e.g., asthma), digestive system (e.g., gastric reflux, fecal incontinence) and others, with potential applications for secondary conditions relating to diabetes, neuro-degenerative disorders and other common health problems.[6]

The cells would be delivered in a way that is "minimally invasive" — without risk of major surgery.

Where will the cells come from? The patients themselves. Nobel Prize winner Shinya Yamanaka developed stem cells very much like human embryonic stem cells through the method of induced pluripotent stem cells which are made from the patient's skin. These cells would be used, with human embryonic stem cells serving as a possible backup.

And the hoped for outcome?

As outlined in Bertha and Renee's grant abstract[7], "Our research overcomes major limitations to provide a ready stem cell-derived target product that we anticipate will provide a safe and effective treatment of urinary incontinence resulting in improved quality of life for a significant fraction of our population."

Renee Reijo-Pera, to the sadness of many, has returned to her home state of South Dakota. She will provide medical care for Native Americans there, which is of course an honorable reason, but still she is missed.

But Dr. Bertha Chen, smiling and determined, continues on.

To read about her progress, go to https://www.cirm.ca.gov/our-progress/awards/autologous-ipsc-therapy-urinary-incontinence-0.

---

[6] Reed DC.
[7] Autologous iPSC therapy for urinary incontinence.

46 TURNING OVER ROCKS: THE BATTLE FOR PARALYSIS CURE

No matter how many times I make the Sacramento run, I always get excited by the Capitol building — white sculptured marble, shining gold dome, the hope of peaceful change.

I made that run a lot, trying to convince state government to support paralysis research for cure. The Roman Reed Spinal Cord Injury Research Act of 1999 had been a tremendous success, helping over 300 scientists and staff as they worked toward paralysis cure. The catch was that it had to be renewed every five years.

The first renewal was easy. In 2004, Roman, Karen and I had made the rounds, tugging the coat sleeves of all the California Assembly folks and Senators. A near-unanimous vote (one vote against it) plus Arnold Schwarzenegger's signature kept the program alive.

Paralyzed in body but not in will, Karen Miner personifies the warrior spirit patient advocates need. Here being honored by Unite 2 Fight Paralysis, Karen has fought for cure research for more than two decades …

Five years later, in 2009, we received a 100% vote of support from Assembly and Senate — but this time there was no funding to pay for the research. It was all very polite. Nothing against the program, we were told; there just wasn't enough money in the General Fund to pay for it.

No money? In the richest state of the richest country on planet Earth?

The problem was that California has an anti-tax law, Proposition 13, which makes it nearly impossible to raise taxes on the rich — except by a 2/3 majority in both houses of the legislature.[1]

In practical terms, a 2/3 majority means every Democrat has to say yes — and at least a couple of Republicans as well. Unfortunately, all Republicans (about 95%) sign the Grover Norquist pledge to never raise taxes under any circumstances.[2] If a Republican violated that no-new-taxes pledge, he or she would be attacked by their own party — and a more conservative candidate would run against them in the next election.

If California was on fire, we would have trouble raising taxes for a new firehose.

In the second-floor visitors' gallery, I could see all the Assembly meeting room. On the far wall, a painting of Lincoln quietly presided. Crystal chandeliers glistened in the soft light. The thought of swinging from them occurred to me, like Tarzan of the Capitol. But the grandeur of the place imposed dignity, even silencing the troops of school kids who shuffled in and out for a five-minute glimpse of democracy.

Twenty feet below, at their rows of double desks, were the elected leaders of California. They would battle out the fates of many bills existing today, each as important to their sponsors as AB 714 (Wieckowski, D-Fremont) was to me and Roman and the paralysis community.

WHONK! A gavel focused attention to the speaker's podium. To the left of acting speaker Nora Campos was a black scoreboard for every Assembly member. Beside each name were two buttons — one green and one red, indicating life or death for the bills: thumbs up or thumbs down.

All 80 assembly members had been visited by Roman, myself, and/or Karen Miner. We divided up the members, floor by floor, room by room. Once there, we queried who our legislative aide was by asking, "Who is in charge of AB 714?" We distributed an overview letter, statistics sheets, and a 58-page bound folder, and asked them for one million dollars to fight against paralysis.

As Karen put it, we were "turning over rocks" like the advocate's motto. Imagine if you had a hundred rocks, and one of them had gold underneath it. If you didn't find gold under the first rock, you would not be discouraged. You would just keep turning over more rocks.

---

[1] California Proposition 13 (1978). *Wikipedia* [Internet] [updated 2015 Feb 6; cited 2015 Feb 15]. Available from: http://en.wikipedia.org/wiki/California_Proposition_13_(1978)
[2] Grover Norquist. *Wikipedia* [Internet] [updated 2015 Feb 15; cited 2015 Feb 15]. Available from: http://en.wikipedia.org/wiki/Grover_Norquist

If you never quit trying, you can only win, or die. And everybody dies — so why not try?

In financial terms alone, Roman's Law had been hugely successful. How many government programs made a profit? In ten years, our small program had spent roughly 15 million California dollars. But *we brought in more than $85 million* from out of state sources like the National Institutes of Health. This was new money for the Golden State.

The need was undeniable: every paralyzed person faced extraordinary expenses, beginning with up to $775,567 in the first year alone, with an additional $138,923 every year thereafter.[3] Three quarters of a million dollars for just the first year of spinal cord injury and over a hundred thousand every year after that — for as long as you lived?

But now that we had the California stem cell program, maybe there was no need for AB 714?

Stem cells were only a tiny part of the paralysis program: only four projects out of 129 involved embryonic stem cells. The rest was the "everything else" which had to be done: fighting pressure sores which could rot flesh down to the bone, easing potentially fatal blood pressure irregularities, taking on bowel and bladder problems — so many chores to alleviate paralysis.

In addition to 175 peer-reviewed scientific publications, the Roman Reed Act had funded:

- cost-saving robotics, which enabled new ways to rehabilitate frozen muscles;
- a helmet which utilized brain waves so that a paralyzed person could operate a computer by the power of thought;
- research tools, such as a new Petri dish *which could sort cells by their electrical impulses*, replacing the traditional cell-sorter machine which cost millions of dollars; and
- ways to constrain the damage of the injury itself, limiting the body's reaction to the trauma.

Everybody liked the program, we were told. But there was no extra money in the General Fund. Be creative, they said. Find a non-tax way to pay for the program. And we did.

2011. Since car crash is the cause of many paralyzing accidents, we proposed a small fee ($3) to every traffic ticket. Seven other states funded spinal cord injury research with driving violations.

Nobody liked the new funding mechanism. Traffic tickets cost way too much. Some folks asked for jail time instead, because they could not afford several hundred dollars in

---

[3] Spinal cord injury facts and figures at a glance [Internet]. 2006. [cited 2012 Feb 15]. Available from: http://www.spinalcord.uab.edu/show.asp?durki=21446

fines. Adding on another $3 seemed cruel. But what were we to do, with no way to raise taxes on billionaires?

Every committee faced was a tough struggle, won by only a couple of votes — and then we came to the finance committee, where our bill was shot down. So we kept turning over rocks.

2012. We lowered our "ask" to as low as we could, just a $1 add-on to traffic violations. I even tried fifty cents — "Pennies for Paralysis" — but was told a dollar was as low as we could go.

Reckless drivers would contribute one dollar to paralysis research. Surely this was not too much to ask? We even added an *opt-out* provision, so a city or a county could choose to not take part.

We picked up momentum. The Assembly Appropriations committee, which had killed it before, took another look. Chairman Fuentes did not approve of the funding mechanism, but as he said, a dollar was so small and the reason was so good — he let the bill go through.

Roman's Law passed both Assembly and Senate. But Governor Jerry Brown did not like our funding mechanism. He vetoed the program and said, "While the purposes of the program were noble, they should be paid for by the General Fund."[4]

We thought so too! For ten years the program had been funded by the General Fund, but now the money had been taken away. What were we supposed to do? But he was the Governor.

2013. We honored the Governor's objections. Doing exactly as he had suggested in the previous rejection notice, we made a straight General Fund request for just one million dollars a year — an amount so small in budget terms, it was described by a Republican aide as "budgetary dust!"

And then came the Assembly vote. What kind of arguments would there be? We had gained the crucial support of Diane Harkey, a biomedicine-supportive Republican; would she prevail for us?

According to the list, 193 bills would be read before ours. Every bill to be considered was named by Speaker Campos. If she then added the words, "the clerk will read", a dozen or so words will be rattled off by a gentleman describing the bill as fast as he could. Then the bill's author would speechify, followed by objections and support. It might take five minutes, or twenty.

"Hey, Dad, having a good nap?"

What, who, where did he come from? My golden son, wheeling his powerchair in next to me.

"I was watching the insides of my eyelids," I said.

"Look who else is here," said Roman.

---

[4]Richman J. Political Blotter: Jerry Brown vetoes bill to hike traffic tickets for spinal cord research. San Jose Mercury News [Internet]. 2012 Sep 21 [cited 2015 Feb 15]. Available from: http://www.mercurynews.com/breaking-news/ci_21602781/political-blotter-jerry-brown-vetoes-bill-hike-traffic

"Hey…" said soft-voiced Karen, so frail, vulnerable — and indomitable.

Morning passed into lunch, and after. They might not make it to our bill today.

And then, at 3:44 p.m., I heard the words: "Roman Reed, spinal cord injury, votes required, 54." Fifty-four green lights and we were headed to the Senate. Fifty-three votes or less, and we had lost again. I expected the usual short silence, a brief as the Assembly members adjusted to the change of subject. Instead, there was a burst of noise and laughter. Was that good or bad?

Bill author Bob Wieckowski made a short friendly speech, followed by Assembly member Travis Williams in support. I waited for the opposition to speak; nobody did. "The clerk will call the roll," said acting speaker Campos. Green lights flashed, as did a running total in lights.

Five votes in support… 7, 12… 23… 36, 38, 42… 53 — there it stuck, one vote short — 64. 64?

"Measure passes," declared Nora Campos. The final vote (with members outside the room) was 68 for and only three against. That was the Assembly. And the Senate? They passed our bill unanimously: 39-0.

All we needed now was the Governor Jerry Brown's signature. In the weeks that followed, Roman and I visited several times with members of his staff, as well as the deputy Secretary of Health Katie Johnson. She knew every aspect of the program. Bob Klein spoke with Diana Dooley, Secretary of Health, as well as the Assistant Secretary.

But one day, as I was driving home from Sacramento, there was a buzz from my cell phone. I pulled off the freeway to take the call. It was Jeff Barbosa, legislative director for Wieckowski. Governor Brown had vetoed our bill[5] — again. The reason given was that the program was administered by the University of California system, and they already had their budget, and it would be unfair to give them any more money.

The only listed opposition was religious — the Conference of Catholic Bishops. Governor Brown had once trained to be a Jesuit priest; did that influence his decision? Was the veto of our small program a nod to a conservative constituency? We had earned the support of essentially the entire legislature, including Lieutenant Governor Gavin Newsom. Why did Jerry Brown insist on killing our bill? To this day, I do not know.

But I am very glad indeed that Proposition 71 was written into the State Constitution, where it could not lightly be undone — by the whim of a politician.

---

[5] Brown Jr, Edmund G. Letter to: Members of the California State Assembly. Available from: http://leginfo.ca.gov/pub/13-14/bill/asm/ab_0701-0750/ab_714_vt_20131005.html

47 IN MEMORY STILL GREEN: THE PASSING OF THREE GIANTS

Dr. Jerry Yang wanted to clone a cow which could give more milk to a starving Chinese village — and he succeeded, first person in the world to do so.

"Put your finger in the water, then pull it out. The hole you leave in the water is how much difference our lives make in the world," a friend of mine once said.

I disagree. As I see it, we all make ripples with our lives, and those ripples spread, affecting other people's lives; whether for good or ill is up to us.

Jerry Yang, Leon Thal, and Duane Roth were very different men. Oddly, the one I never met was the one I knew the best.

Born in the midst of famine in Dongcun, China, Xiangzhong "Jerry" Yang nearly died of starvation before he could walk. "Is he dead?" his aunt once asked of him, his body so emaciated from hunger. But the spark of life was strong, and he survived.

Time went by, and the family could only afford to send one child to school. Jerry's older brother was chosen for the honor. Since both parents were needed to work in the fields, nine-year-old Jerry was put in charge of babysitting his infant brother. The first thing he thought of was to try babysitting AND going to school at the same time, by bringing his brother along. But the teacher said the baby would disrupt the class's learning.

Undiscouraged, Jerry sat *outside the classroom by the window* with his baby brother. He would listen and watch, and also chew sweet potatoes into pulp which served as self-made baby food for his brother. With a Lincolnesque determination to learn, he read every book in the village. His determination impressed local officials, and when a national test was offered, he took it at the age of 17, and triumphed. At last he began formal education. As he grew older, he passed increasingly difficult examinations, eventually winning a scholarship to Beijing Agricultural College.

He was selected to travel to the University of Connecticut for advanced study, took a job at the University, and worked there for the rest of his life.

(To know more about Jerry's incredible story, read the excellent book-length series of articles written by William Hathaway at the Hartford Courant here: http://www.courant.com/health/hc-yang-sg-storygallery.html.)

But he never forgot the hunger. He knew the difference one good milk cow could make for a small village like his: survival. Reading about Ian Wilmut and Dolly, the world's first cloned sheep, Jerry Yang determined to develop — and clone — the perfect cow.

He succeeded. The cow, Amy, was the first cloned farm animal in America.

Jerry Yang established a non-profit company which sent the best cow embryos to farms like the one he grew up in, making it possible for cows to be terrific milk producers and bulls to grow laden with high-protein meat.

His work was so advanced that the FDA relied heavily on it, and they determined that cloned meat and milk was indeed safe for human consumption.

Jerry Yang also wanted to develop a different kind of cloning — therapeutic cloning, or the copying of cells, (no baby-making involved — for the fight against human disease.

He married Xiuchun "Cindy" Tian. He was not romantic, Cindy said of their courtship later, but he was persuasive.

When I first talked to Jerry, I used my baby-talk Mandarin.

"Wo shwo da bu hao," I said, "I speak very badly." He was delighted, responding with a cheerful lie, "Ni de zhong wen hen hao (your mandarin is very good)!"

Jerry eventually had a cancer of the salivary gland, and some of the muscles in his face had been cut away during surgery.

But he kept on working until the last. His life deserves honor. It is my hope that the University of Connecticut will one day dedicate a research grant in his memory; they might perhaps call it the Jerry Yang Courage and Commitment Award.

Leon Thal died February 3, 2007, when the plane he was piloting went down. An expert pilot, Dr. Thal had flown alone seven times across the United States. Flying was integral to his incredibly busy life as well as a great diversion — a way to be at peace, soaring through the sky.

Appointed to the California stem cell program's board of directors by Governor Arnold Schwarzenegger, Dr. Thal did not seem to have much to say, though I noticed the room would get quiet when he did offer an opinion. He was a peaceful sort of person, never in a hurry, always sparing time to talk if you had a question about stem cell research or Alzheimer's disease.

His wife, Donna Thal, was a distinguished professor emeritus at San Diego State University, so I suspect even his dinner conversations advanced the cause. But I had no clue that he was a giant in the battle against Alzheimer's, an incurable disease affecting some 4.5 million Americans.

"In my view, shared by many, Leon Thal was the most prominent scientist in the world in [...] Alzheimer's disease," said David N. Bailey, Dean at the School of Medicine at the University of California San Diego.

Bailey added: "Dr. Thal [...] achieved a remarkable body of research [including] more than 300 peer-reviewed papers. [...] He directed more than $100 million in federally funded research grants, and was a collaborator in many others."[1]

In addition to being the Director of University of California San Diego's Shiley-Marcos Alzheimer's Disease Research Center, he also headed the Alzheimer's Disease Cooperative Study, *a consortium of 80 clinical sites* in the U.S. and Canada.[2]

"Leon is the reason we first started giving in support of Alzheimer's disease," said philanthropist Darlene Shiley, who along with her husband Donald had made a four million dollar gift to support the Shiley-Marcos Alzheimer's Disease Research Center. (The Center was named after Darlene's mother, Dee Marcos, whose Alzheimer's condition was first diagnosed by Dr. Thal.)

Dr. Mark Tuszynski, vice chair of University of California San Diego's Department of Neurosciences and who knew Dr. Thal for 20 years, said, "[He] led us to a greater understanding of what might work, what doesn't work, and where future research has to go."[3]

The final act of Duane Roth's life was consistent with his core beliefs. He was doing a cancer research fundraiser for Pedal the Cause. It was a bike-riding event, and Roth hit an embankment at speed. His protective helmet shattered, and he died in hospital.

Duane was vice chair of the Independent Citizens' Oversight Committee (ICOC), the governing board of the California stem cell program, and I saw him in action many times. He once joked that he was the only Republican on the board; I don't know if that was accurate or not, but if he might have been outnumbered, he was certainly never intimidated.

---

[1] Franz L. Dr. Leon Thal, renowned Alzheimer's expert, dies in plane crash. *UCSD News Center* [Internet]. 2007 Feb 5 [cited 2015 Feb 16]. Available from: http://ucsdnews.ucsd.edu/archive/newsrel/health/thal07.asp

[2] Woo E. Dr. Leon Thal, 62; UC San Diego Alzheimer's expert killed in crash. *Los Angeles Times* [Internet]. 2007 Feb 8 [cited 2015 Feb 16]. Available from: http://articles.latimes.com/2007/feb/08/local/me-thal8

[3] ibid.

I remember one argument we had. Roth was never shy about sharing his beliefs, and we got right into it at the hallway outside the ICOC meeting room. I don't recall exactly what the fight was about, but I listened politely, and then said what I wanted as well. I remember the innate courtesy: he suggested we go outside where we could yell without disturbing anyone. When we stopped arguing, we shook hands, and went back to work.

As the Chairman of Connect, a biomedical support group in San Diego, Roth was a leader in business. Before joining Connect, he founded Alliance Pharmaceutical Corporation and was Chairman of the Board. Before Alliance, he held senior management positions at Johnson & Johnson.[4]

Asked to name a person who inspired him, Duane Roth said, "Jim Burke, past chairman of Johnson & Johnson. I started my career with J&J and was a young sales representative in Iowa when the Tylenol capsule poisonings occurred in Chicago. […] Chairman Burke made the decision to recall 100% of the capsules worldwide to make sure there would be no further risk to any person anywhere. He then led us as we regained consumer trust, when the experts said the brand was dead."[5]

Duane Roth married his high school sweetheart Renee, to whom the following message was sent from Representative Susan Davis:

"What a great […] legacy Duane leaves. While his loss will be felt throughout the community, the impact Duane has made in making San Diego a technology and life science leader, [along with his charitable work] will live on for generations to come."[6]

Roth was vice chair of the Sanford-Burnham Medical Research Institute — and had been slated to become its chair in just two months from the date of his passing. We will never know what great things he might have achieved in that position.

But we do know what he did for the California stem cell program. His leadership role on the board of directors meant a voice for the business side of regenerative medicine. And this was crucial, because without the involvement of the business community, how will products of cure reach patients?

"He was one of the true stewards of the mission, offering countless insights on the role of industry in the world of regenerative medicine, and how best and efficiently to drive therapies through to patients. He was unfailingly a voice of reason and optimism, and always sought ways to make things happen, refusing to take no for an answer," said Jonathan Thomas, Chairman of the Board on the California stem cell program.

Duane Roth, Leon Thal, and Jerry Yang.

Champions all, they were taken from us too soon; we mourn their loss.

But their deeds live on, waves in the tide of cure research.

---

[4] Duane J. Roth [Internet] [cited 2015 Feb 16]. Available from: https://foundation.ucsd.edu/_files/board-of-trustees-bios/_roth_duane.pdf

[5] Duane Roth accelerates innovation with Connect. *La Jolla Light* [Internet]. 2009 Apr 22 [cited 2015 Feb 16]. Available from: http://www.lajollalight.com/news/2009/apr/22/duane-roth-accelerates-innovation-with-connect/

[6] Fikes BJ. Tech leader Duane Roth dies from injuries [Internet]. 2013 Aug 3 [cited 2015 Feb 16]. Available from: http://www.billwalton.com/news/408-tech-leader-duane-roth-dies-from-injuries

48 INVITATION TO MEXICO

With Gloria being Mexican American, I have a strong personal interest in seeing Mexico succeed. So I was delighted to receive an invitation to attend a two-day conference at the Mexican Society of Regenerative Medicine in Tijuana, Mexico.

We were treated like honored guests, and shown only the beauty (which is considerable) of the border city. But though we traveled in limousines and were taken to a luxury hotel, the grinding poverty was visible whenever you looked out a window.

It was a classy event. The speakers included Dr. Juan J. Parcero Valdez, President of the Society; Dr. Edward Holmes, President of the Sanford Consortium for Regenerative Medicine; Dr. Toshio Miki, Director of the Eli and Edith Broad Center for Regenerative Medicine; Dr. Julie G. Allickson, Director of Translational Research at the Wake Forest Institute for Regenerative Medicine; Howard Leonhardt, founder of BioHeart; and others.

But the embryonic stem cell research I supported was against the law in this country, where the Catholic Church holds sway. Priests were in attendance at this event, and they carried themselves like policemen, looking arrogant in their swirling robes.

I was scheduled only to attend, not to speak. But then I heard the presentation of (let's call him) Father X, representing the Catholic Church's anti-embryonic stem cell research position.

His presentation was a professionally prepared slide show: lots of diagrams, the usual pictures of a baby in the womb to supposedly show what embryonic stem cell research was all about, and enough science talk to make it sound authentic, yet not so much as to be difficult — or accurate.

He talked trash about science, impugning both the motives and methods of stem cell research.

I listened, shaking my head side to side. I was a guest, after all, and the hosts had treated Gloria and me with the utmost courtesy. At one point my wife had lost her purse. We figured that was it, gone — but no, the purse was found and returned, all contents intact.

But just in front of me sat two ladies from a biomedical company, and one of them said:

"You know, I always thought we needed embryonic stem cells for research, but after hearing Father X, I changed my mind. Adult stem cells are enough."

That was too much. I found the Master of Ceremonies, and asked for the favor of addressing the audience — he said it was unexpected, but allowed me to do so.

I told the assembled dignitaries that I disagreed 100% with Father X's presentation. I told them about my son Roman who was paralyzed and my hopes that he would one day walk again.

Father X talked about embryonic stem cell research as if it was cloning babies for spare parts. But how can there be a baby, when there is no involvement of the womb? A blastocyst is a dot of living tissue, not a life. You can watch it in a petri dish forever, and nothing will come of it.

"Cloning a baby for spare parts" had (to the best of my knowledge) never been proposed by any scientist in the world. In California, prohibitions of reproductive cloning are in the Constitution. I explained briefly how therapeutic cloning worked — a skin cell added to a human egg, again having nothing to do with babies at all, but with possibilities for cure research.

I told them about Hans Keirstead's work, and the formerly paralyzed rats which walked again after receiving oligodendrocytes derived from embryonic stem cells.

There was hesitant applause afterward, nothing like the thunderous ovation that Father X received.

But as Gloria and I entered the car to be taken to dinner, the priest climbed in beside me, and began to lecture. Personally I see no point in arguments between people of diametrically opposed positions. Neither one of us was likely to change our mind after all these years. I answered him briefly, politely, as I would any hostile question from an audience.

But he wanted to convince me I was wrong. Father X was tall and lean and hawk-faced, with short hair and a close-cropped beard. He carried himself with absolute authority, as if there could be no questioning of his position, no slightest possibility of error.

At dinner there were reserved seats, so I thought we would be separated, but he took the seat beside me instead, continuing the speech and hammering home his many points.

I tried to change the subject, asking him what was his favorite form of recreation, but he answered in one snapped-off word — "sailing!" — and picked up the scolding once more. And around us there was dead silence. Everyone was listening.

Well, all right then. I set down my napkin, turned my chair to face his.

And we got down to it. He came at me like a verbal linebacker.

Did I not know about the wonderful things being done by adult stem cells?

Of course I did. Adult stem cells had a 50-year head start over embryonic stem cells; it would be an embarrassment if nothing had been accomplished. Also, we were not trying to criminalize adult stem cell research; our point was to use the best form of cellular research and therapy available, adult or embryonic — but not block out any for religious objections.

Why did I keep insisting that there were no babies involved? Why was I so hung up on implantation, when life so clearly begun at conception?

Because without implantation there can be no babies. No womb, no baby, I said, reminding him that Webster's definition of embryo included implantation.

His face got redder and redder.

Finally he snapped out what appeared to be his ultimate argument:

"Our Lord Himself began life as an embryo!"

Dead silence. No one in the room was chewing their food; everyone was waiting for the lightning to strike.

But I grew up in the Church. Religious arguments are not new to me. I heard myself say:

"It is my understanding that Jesus was born of a virgin birth, no sperm involved — so how could He have been an embryo, which is the joining of sperm and egg?"

He breathed through his nose, deeply, like a dragon inhaling before a blast of fire. He stood up from the table, abruptly. Gloria said later she thought he was going to punch me.

Then he stalked away in a swirl of robes, leaving his meal. It was a nice piece of chicken, I remember, gravy and yellow rice, fresh green peas — since he had not touched it, I considered eating his lunch, but did not want to be rude.

On returning to the U.S., I wrote a letter to the California stem cell program suggesting an international agreement with Mexico so that research on both sides of the border could cooperate. I was not successful, unfortunately. In the future, I hope that situation can change.

Somewhere in Mexico there is a boy or girl with a mind like the great Carvajal, Hispanic founder of neuroscience. If given education and opportunity, that child could change the world.

But his or her time will never come — if religion is allowed to block the progress of science.

The great Hispanic neuroanatomist, Santiago Ramon y Cajal left a legacy of knowledge and artwork which changed forever the way the world looks at the nervous system.

49 OF PRESIDENTS, AND THE VALLEY OF DEATH

Randy Mills: President, California Institute for Regenerative Medicine.

A tale is told of the elephants' graveyard, where old gray pachyderms go to die and fortunes in ivory lie waiting. It is a myth, of course, though elephants have died of disease, and on rare occasions their bodies have been found grouped together.[1]

In biomedicine, there is a place you will not find on any map. It has no physical location. But it is real. It is the Valley of Death where, all too often, new medicines go and die.

The Valley of Death is the series of tests every new medicine or treatment must undergo, before the FDA can approve it for public use.

These tests, or clinical trials, are long, rigorous, and staggeringly expensive. If the company runs out of money during the trials, the new product or therapy is gone. The Valley of Death has claimed another victim — a potential cure lost to the world.

Ever wonder why medicines cost so much? Look to the Valley of Death.

---

[1] Elephants' graveyard. *Wikipedia* [Internet] [updated 2014 Oct 25; cited 2015 Feb 16]. Available from: http://en.wikipedia.org/wiki/Elephants%27_graveyard

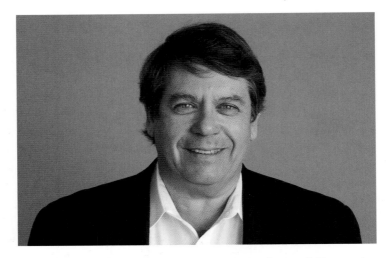

Alan Trounson: Pioneering scientist with In Vitro Fertilization (IVF) procedure, former President of the California Institute for Regenerative Medicine.

Before any new treatment can be sold on the market, the Federal Drug Administration (FDA) asks two crucial questions: is it safe, and will it work? This is right and proper.

Unfortunately, answering those questions may take years of testing through clinical trials and mountains of money.

According to CIRM's President Dr. Alan Trouson, "The average cost of delivery of a new biopharmaceutical drug into practice has been estimated to be [as much as] $3.9 *billion*, including capital costs and the costs of failed drugs."[2]

We need the tests, of course. Without them, unscrupulous outfits might give injections of salt water and say it was stem cells while bilking patients of their life savings.

But if the price of testing continues to skyrocket, valuable new treatments may not even be tried. Chronically ill people will continue to suffer with therapies just out of reach.

"Alpha clinics", an idea under development by the California stem cell program, may streamline these costs of testing without any compromise of safety. When a new therapy was ready to be tested, there would be a pre-established place to go.

Alpha clinics would be hospitals and laboratories, carefully prepared to administer safety tests for new stem cell research methods. Instead of endlessly reinventing the wheel, designing and redesigning tests, we would have centers of expertise in place. Teams of experts would make clear to both researchers and patients exactly what must be done.

There might be half a dozen of these alpha clinic sites, up and down the state. Their locations would be decided through competition, with potential sites bringing their best to the table in the form of equipment, housing, and access to clinicians, patients, or donors.

---

[2] Trouson A, DeWitt ND, Feigal EG (2012). The alpha stem cell clinic: a model for evaluating and delivering stem cell-based therapies. *Stem Cells Transl Med* **1** (1): 9–14.

Each site would focus on a different group of ailments, perhaps heart disease, neurological disorders, diabetes, blindness, immune deficiency, cancer. Like the arms of a starfish, the sites would all connect back to a central management center, which would collect and organize data.

This data would be shareable. In the past, knowledge and information gleaned through clinical trials was often hidden; with more open information, we might avoid repeating mistakes made in the past. (Patient privacy and patent information would of course be maintained.)

A meeting was called by the CIRM on November 14 and 15, 2012.[3] Invited were scientists, patient advocates, doctors, clinicians, cell manufacturers, and biomed and pharmaceutical companies.

I did not understand everything I heard, of course; my brain was swimming with new information. Changes offered were so wide-ranging it seemed that medicine itself — not just stem cell research — could be forever changed, and for the better.

Dr. Alan Trounson is someone who would understand the situation better than I ever will. He explained, "The Alpha Clinics Proposal [will be] a network of cell therapy clinics [...] to provide for the rising number of clinical trials and recognized treatments."

"The network can provide data and a learning base that will assist other studies to progress, provide the opportunity for sound clinical development in major medical centers, and provide support for businesses working in regenerative medicine.

"We need to ensure the centers have the clinicians, facilities, equipment and support staff to ensure the work we are supporting gets to the patients. [...] Independent counselors would assist patients to find the best treatment options [...] rather than seek unscientific and unregulated procedures that have no benefit and are very risky.

"We will work together with the major Californian medical centers to create the network and a hub [to] help with quality control standards, data management, business development, [and] assistance with reimbursement, [amongst other things]."

Unfortunately, Dr. Trounson was retiring. After six years in office, he would be returning to his native Australia to spend more time with his family. He had made his mark. He would be missed — and I had no clue about his replacement. All I could find out was there had been "dozens of candidates", then seven top finalists, and one who really stood out.

On April 30, 2014, we learned who would be the new President of the California stem cell program.

An executive search company, Korn/Ferry, had been hired to do an international talent search. Senior Partner Warren Ross, MD, spoke enthusiastically about the stem cell program and the quality of the people being interviewed.

This was a board decision. The finalists were exhaustively interviewed by the ICOC.

Today, the final selection would be announced.

---

[3] DeWitt ND, Lomax G, Millan M, Feigal EG, Trounson A. Alpha stem cell clinics: delivering a new kind of medicine [Internet] [cited 2015 Feb 16]. Available from: http://www.cirm.ca.gov/sites/default/files/files/about_cirm/Alpha_stem_stem_cell_clinics.pdf

ICOC Chair Jonathan Thomas summed up the situation: "CIRM's portfolio [...] in cellular therapy and regenerative medicine is second to none. We've had tremendous growth over the past ten years in terms of funding research, currently in 39 incurable diseases and conditions. We now reach a time in our CIRM life [...] where we want to place increasing emphasis on project developments getting [into] clinical trials [...]."[4]

Randy Mills, former President and Chief Executive Officer of Maryland-based Osiris, was announced as the new President of the California stem cell program.

His background? Extensive and varied.

For instance, during his 10-year tenure at Osiris, Mills oversaw the first commercialization of a stem cell drug, *Prochymal*, which is used to treat acute graft-versus-host disease in children.[5] The deadly disease is a complication of bone marrow transplants.

Dr. Anthony Atala, Director of the Institute for Regenerative Medicine at Wake Forest University, knew Randy Mills well and had this to say:

"Over the years, what has impressed me most about Randy is the consistent sense of urgency he has toward helping patients in need. Those of us who work in this field know that the potential of regenerative medicine is enormous. However, Randy has demonstrated the unique ability to turn that potential into reality. He will make a great President at CIRM, and will keep the organization focused on developing and delivering stem cell-based technologies that improve patient's lives."[6]

And from the new President himself?

Randy Mills declared, "We are entering a new phase in regenerative medicine, where an increasing number of therapies are heading into clinical trials. It is our mission to do everything possible to accelerate the development of these treatments for the patients who need them."[7]

He had worked with CIRM before as a member of the out-of-state review board, which studies and makes recommendations on stem cell projects being considered.

He had worked his way through college in the emergency room, so he has a strong sense of urgency for the work at hand. Regarding himself, he stressed that he was "patient centered," and said, "We are going to work relentlessly for the benefit of the patients and the people of the state of California."

---

[4]Allday E. State stem cell agency hires new president, C. Randall Mills. *San Francisco Chronicle* [Internet]. 2014 May 1 [cited 2015 Feb 16]. Available from: http://www.sfgate.com/health/article/State-stem-cell-agency-hires-new-president-C-5443427.php

[5] Morin M, Brown E. C. Randal Mills to head California's stem cell agency. *Los Angeles Times* [Internet]. 2014 Apr 30 [cited 2015 Feb 16]. Available from: http://www.latimes.com/science/sciencenow/la-sci-sn-c-randal-mills-to-head-californias-stem-cell-agency-20140430-story.html

[6]C. Randal Mills named as new President of California's Stem Cell Agency [Internet]. 2014 Apr 30 [cited 2015 Feb 16]. Available from: https://www.cirm.ca.gov/about-cirm/newsroom/press-releases/04302014/c-randal-mills-named-new-president-california%E2%80%99s-stem

[7]Sherman N. Former Osiris chief to lead California's stem cell agency. *The Baltimore Sun* [Internet]. 2014 Apr 30 [cited 2015 Feb 16]. Available from: http://articles.baltimoresun.com/2014-04-30/business/bs-bz-former-osiris-20140430_1_cirm-stem-cell-agency-prochymal

His acceptance speech was very short. Add up all the above quotes, and that is pretty much all he said. He acknowledged that brevity by saying in closing, "Well-done is better than well-said."

During the public comment period, I said to him, "You are entering the greatest government program in the world. Research decisions you influence, and executive actions you carry out, will affect the lives of millions, and, in Shakespeare's words, 'echo down the corridors of time'."

"You will be exhausted often, because there is just too much to do. But you will be supported always, not only by a fantastic staff, but also by the patient advocate community."

"On behalf of all who work in the cause of cure, welcome."

50 LITTLE HOOVER AND THE INSTITUTE OF MEDICINE

Jonathan Thomas: Chairman of the Board, California Stem Cell Program.

When I heard that the California stem cell program would be "studied" by the Little Hoover Commission (LHC), I expected trouble. I was not disappointed.

The LHC was named after Herbert Hoover, the Republican President whose policies gave us the Great Depression. The LHC's mission is to make government "more efficient and more effective", which sounds reasonable, at first.[1]

However, it has been my experience that when Republicans talk about making government "more efficient", that generally means less food for the hungry, less books for the library, fewer teachers in the classroom, basically reducing or eliminating programs that help people.

But it was only a study — why would I worry?

(The full study can be found at http://www.lhc.ca.gov/studies/198/report198.html.)

First, a negative study is ammunition for the other side, which does not want us to succeed. If our stem cell program could be discredited, no one would want to copy it.

---

[1] FAQs [Internet] [cited 2015 Feb 16]. Available from: http://www.lhc.ca.gov/about/faqs.html

Secondly, LHC studies often come with suggested legislation, which become *ways to rewrite the program.*

I viewed this as a political attack. The LHC wanted to redo California's stem cell program, overriding the will of the voters. The supposedly unbiased and even-handed "study" gave voice to long-term opponents of the program who clearly wanted to shut us down:

A review drafted by LHC said, "Jesse Reynolds, [of] the Center for Genetics and Society [which opposed Proposition 71 from the beginning] told the Commission that the justification for California's stem cell program has largely dissolved and it is time to re-evaluate its mission."

"John Simpson, [of] Consumer Watchdog, suggested that with anticipated NIH funding for stem cell research, California should reconsider spending the entire $3 billion that voters authorized."[2]

And the LHC's recommendation to "improve" us?

They wanted to "decrease the size of the board to 15 members [...], allowing the Governor to appoint 11 [...]."[3]

Our board would be cut in half from 29 to 15, and of these 15, 11 *would be appointed by the Governor.* Talk about a recipe for dictatorship! What if we had a governor who was against embryonic stem cell research, or just did not want to spend the money? He or she could easily block the program — by appointing anti-research conservatives to the board.

Their reason for suggesting this?

In their report, LHC argued that CIRM's governing board includes representatives from institutions that benefited from grants the committee approved and could never be entirely free of conflict of interest.

The ICOC does include colleges whose scientists have received grants — is that not logical? Where should grant money go, except to where the scientists are?

To prevent "self-dealing", the members are never allowed to vote on — or even discuss — any grant which might go to their parent institution. We need an expert board to understand the incredibly complicated nature of stem cell research funding.

But maybe I am too prejudiced in favor of the program? Fair enough, so listen instead to the words of California State Senator and Senate Majority Leader Dean Florez, a member of the LHC:

"The report recommends reducing the size of CIRM's governing board from 29 members to 15 …, and concentrating the power of appointment in the Governor, who would appoint 11 of 15 members."

"[I am] concerned about the Commission's attempt to shift power over the agency to the Governor. Like the Little Hoover Commission itself, CIRM was designed to be an independent agency. Proposition 71 therefore [shared] appointment authority between

---

[2] Review of the California Institute for Regenerative Medicine. California: The Little Hoover Commission (U.S.); 2009 Jun. 88 p. [Internet] [cited 2015 Feb 16]. Available from: http://www.lhc.ca.gov/studies/198/cirm/Report198.pdf

[3] Review of the California Institute for Regenerative Medicine.

the Governor, the Lieutenant Governor, the Controller, the Treasurer, the Legislature, and UC Chancellors".

"By concentrating appointment authority in the Governor, the Commission would undermine the careful [...] balance struck by Proposition 71. In a controversial area like stem cell research, such a change would threaten the independence that CIRM needs to ensure the success of its mission".

"Finally, I am concerned about the Commission's apparent rush to conclude its report. As one member said at the meeting, five minutes and a sandwich is not adequate time for Commission members to absorb the information that was presented. [...] I am concerned that due to its rush to approve the report, the Commission gave disproportionate weight to CIRM's critics and did not consider a broader range of views on the complex issues that are the subject of the report."[4]

Reading that phrase "the Commission gave disproportionate weight to CIRM's critics" made me want to stand up and cheer — at last, someone said what had been bothering me for so long.

In both the LHC Report and the Institute of Medicine Report, long-term critics of the program are cited as sources of information. Fair enough. Critics like John Simpson, Jesse Reynolds, Marcy Darnovsky, and David Jensen have a right to be heard.

But where were the supporters? Seven million Californians voted in favor of Proposition 71 — why are we not represented in the study? Nearly 100 groups put their good names in support of Proposition 71 — why were their millions of members not represented in a supposedly unbiased study?

I was allowed to speak, but only because I requested it, and only in the public comment section at the very end of the program.[5] Here is a shortened version of my comments:

"Conflict of interest is something very easy to attest, and very difficult to know for sure. [...] For instance, the *Sacramento Bee* has systematically and enthusiastically opposed the California stem cell program, writing literally scores of editorials against it ... Director Drown, you were an editor of that paper. Is that not at least a perception of conflict of interest?

"Additionally, [if a] member of the LHC board is a Republican, a party whose official platform calls for criminalization of embryonic stem cell research — is that not a conflict of interest, as he/she considers [...] an organization dedicated to that research?"

"It is in no way my intent to cast aspersions on the honor and dignity of the Little Hoover Commission, nor any individuals involved: but merely to point out that the perception of conflict of interest can be found everywhere, even here".

"As you consider the future of the California stem cell program, which embodies the hopes and dreams of suffering men, women, and children, here and around the world [...], consider the substance of what is, not the shadows of what might be perceived."

---

[4] ibid.
[5] Reed, Don C. Letter to: Drown, Stuart. Available from: http://www.lhc.ca.gov/studies/198/cirm/ReedNov08.pdf

But CIRM survived the Little Hoover Commission only to face a far more dangerous attack in 2012, from the Institute of Medicine itself.

The study began politely, complimenting CIRM: "[CIRM] enhances California's position as key international hub of activity [by] enriching regenerative medicine everywhere. [It has an] impressive research portfolio [...], exemplary training program, [and] translational projects [...] for industry involvement."

It then recommended radical change, like a doctor saying a patient is healthy, followed by, "Let's operate!"

Their proposed "improvements" would not only overturn the program's structure, but might even violate state law. CIRM, after all, is in our State Constitution.

The study objected to the program's board of directors, the ICOC, which is a 29-member panel of experts appointed by public officials. These board members, the study believes, have an automatic conflict of interest, meaning they could use their votes to benefit the colleges or businesses for which they work.[6]

"They make proposals to themselves, essentially, regarding what should be funded. They cannot exert independent oversight," said Harold Shapiro, chair of the study.[7]

This was wrong, and demonstrably so.

Members of the ICOC can NEVER "make proposals to themselves". They are prevented by law from so doing: board members may neither discuss nor vote on projects for home organizations.

How strict were our standards? CIRM's "conflict of interest rules are modeled on, and in some cases exceed, the standards established by the National Institutes of Health."[8]

Here is what actually happens.

First, the CIRM sends out Requests for Applications (RFAs) on whatever subject is needed (a new way to fight autism, for example) and this goes out to California scientists. The scientists respond with proposals, which are evaluated by the Grants Working Group (GWG), a panel with a majority of out-of-state stem cell experts.

Each project is scored from 1 to 100. The grant proposals fall into three categories: YES, MAYBE, and NO. (Generally speaking, a project is usually approved if it has a score of 70 or better.)

At a public hearing, the projects are listed on a light screen in order of scoring. The proposals are grouped by color as well, green for go, yellow for maybe, and red for not recommended. The board has already read the proposals, which were given to them

---

[6] Committee on a Review of the California Institute for Regenerative Medicine. *The California Institute for Regenerative Medicine: Science, Governance, and the Pursuit of Cures.* Washington DC: The National Academies Press; 2012.

[7] Brown E. Stem cell agency board criticized for conflicts of interest. *Los Angeles Times* [Internet]. 2012 Dec 7 [cited 2015 Feb 16]. Available from: http://articles.latimes.com/2012/dec/07/science/la-sci-stem-cells-cirm-report-20121207

[8] Torres, Art. Letter to: Citizens' Financial Accountability Oversight Committee. Available from: http://www.sco.ca.gov/Files-EO/CFAOC/Item_6G_-_2012_CIRM_Conflict_Policies.pdf

early, as well as being posted (without names) on the public CIRM website. Members are reminded which projects they may not vote.

The chairperson then asks, "Does anyone want to move a proposal up or down?"

In the overwhelming majority of cases, probably 90%, the ICOC agrees with the recommendations.

But if a board member challenges the placement of a grant, then the CIRM's scientific staff will give its opinions. The scientist applying for the grant may also speak. The board makes comments, as do we in the public; it is their one and only chance to do so.

This is America at our best: everybody having their say — the board, the scientists, the patients, and the public; transparent government, out in the open, plenty of arguments but no secrecy.

Unfortunately, this is what the study would destroy. According to Section 3, page 11, "The board [...] should not be involved in day-to-day management. [It] should delegate day-to-day management responsibilities to the President."[9]

What is "day-to-day management"? Nothing less than who gets the money.

Giving grants for stem cell research is not small; it is the whole program. Take that away from the board, and they might as well go home. And who does the study suggest would now make these crucial decisions?

Instead of our 29-member board, interacting with scientists, staff, patient advocates, and the public, only two people would be in charge, one of whom works for the other.

According to Section 4, page 18, "The Senior Vice President and the President [...] decide on a final slate of proposals to submit to the *ICOC for a 'yes' or 'no' vote on the entire slate* [...]. The ICOC [Board] should NOT [...] evaluate individual applications [...]" (italics mine).

A "yes" or "no" vote on the entire slate? If the Board can only make a blanket decision on several dozen projects at once, for all the research projects going forward or none of them, what kind of decision is that? Are they going to say NO, and then block all the research?

Instead of our open process, funding decisions would be made in private; the public would be denied meaningful participation, with our Board reduced to a rubber stamp.

And there was something else I found difficult to believe. Judging by his committee's recommendations, Institute of Medicine's Committee Chair Walter Shapiro did not want people with diseases making decisions!:

"The committee believes that personal conflicts of interest arising from one's own or a family member's affliction with a particular disease [...] can create bias for board members."

Patient advocates — not eligible? These are the people who fought for Proposition 71, the citizens' initiative leading to the CIRM. Having done the work, patient advocates were now to be denied a role in decision-making?

---

[9] Committee on a Review of the California Institute for Regenerative Medicine.

Stem cell scientist Dr. Jeanne Loring asked me to read her letter into the public record:

"Patient advocates are extremely valuable to us researchers. […] With my first CIRM grant, I started meeting patient advocates, and now I can't imagine pursuing a disease-related research project without them. Jeff Sheehy taught me about HIV/AIDS and patient activism, I learned about Parkinson's from Joan Samuelson, autism from Jonathan Shestack; David Serrano-Sewell and Diane Winokur have educated me about MS and ALS. Advocacy makes CIRM-funded research breathtakingly relevant and uniquely powerful to change the course of medicine."[10]

The Board listened to the Institute of Medicine, and offered changes to address the criticism.

"One of the big issues raised by the Institute of Medicine report was that our current governing structure created a perception of conflicts of interest in how we fund research," says Chairman Jonathan Thomas. "While no one has found any actual conflicts, these changes directly address the […] perception."

Translation: No actual conflicts of interest were found, only the possibility that someone might think there was a possibility such conflicts could arise.

Personally, I wanted to keep the program exactly as it was. The critics might never be satisfied with anything less than our extinction. The people had spoken — so let the program do its work!

But cooler heads were on the Board. Chairman Thomas made a bold (if painful) suggestion:

"By having board members who belong to institutions that can get CIRM money *abstain from all votes on funding*, we take even the perception of conflict of interest out of the picture."

Thirteen members of the board of directors just had their right to vote taken away. The institutions which might receive funding from the program would no longer be able to vote on any funding matters at all.

And the patient advocates? Their vote was taken away on the Grants Working Group, which recommends research projects to the board — but those on the governing board will still vote.

How important were these and other changes? In a public e-mail, study leader Harold Shapiro, former President of Princeton, commended the stem cell agency for its response, calling it "very thoughtful and significant, [which] will serve the long-term interests of the citizens of California and the field of regenerative medicine [by] dealing with financial conflicts of interest, [and] enhancing the credibility and integrity of the scientific review process […]."

---

[10] Jensen D. Loring on patient advocates and their role at the California stem cell agency [Internet]. 2013 Jan 28 [cited 2015 Feb 16]. Available from: http://californiastemcellreport.blogspot.sg/2013/01/loring-on-patient-advocates-and-their.html

So why did Jonathan Thomas recommend these major changes, and why did the ICOC vote to back him up?

"We must get past long-standing criticism [...] *that has for too long stolen focus in the media,* [distracting] *from the incredible scientific work that* [CIRM has made possible]."[11]

---

[11] Stem cell agency's governing board proposes dramatic changes in response to IOM report [Internet]. 2013 Jan 24 [cited 2015 Feb 16]. Available from: https://www.cirm.ca.gov/about-cirm/newsroom/press-releases/01242013/stem-cell-agency%E2%80%99s-governing-board-proposes-dramatic

51 STUDYING THE MOON, LOOKING THROUGH A STRAW

Fanyi Zeng as a child with family friend Dr. CC Tan (Tan, Jiazen) considered the father of Chinese genetics.

"The beginnings of stem cell research in China may be traced back to 1963 [...]. Dizhou Tong transferred the DNA from a cell of a male Asian carp to the egg of a female Asian carp, and produced the world's first cloned fish. [...] Chinese scientists developed fish-breeding techniques so powerful that the nation now produces more than half the world's aquaculture harvest."
   http://www.ncbi.nlm.nih.gov/pmc/articles/PMC2435574/[1]

For an outsider, figuring out China's stem cell policies and progress is like studying the moon through a straw. You can only see a little at a time, and it is hard to see how the pieces fit.

---

[1] Liao L, Li L, Zhao RC (2007). Stem cell research in China. *Philos Trans R Soc Lond B Biol Sci* **362** (1482): 1107–1112.

Fanyi Zeng family with Dr. CC Tan, still active in his nineties: *left* to *right*, Mrs. Tan (Dr. Yunfang Qiu), Dr. Fanyi Zeng, Dr. CC Tan, Dr. Yi-Tao Zeng (father) and Dr. Shu-Zhen Huang (mother).

For one thing, there is the language barrier. While most Chinese know English, most Americans do not know Mandarin. It would seem only courteous (as well as practical) to make an effort.

So, for the past ten years, I have spent half an hour a day studying Mandarin. The results, to put it kindly, have been somewhat less than spectacular.

"Wo shuo de bu hao" — I speak very badly!

Once I went to a Chinese movie and tried to follow along. During the entire Mandarin movie, I understood exactly one sentence — "wo bu zhi dao," which means, "I do not know!"

But still I must try, however laughable it may sound. Let me tell you why.

There was a little girl, five years old, a musical prodigy who played the piano on a concert stage. She sang, danced, and even played the trombone. So, would she make her career in music?

But Fanyi Zeng (the Chinese would say the family name first, so she would more accurately be Zeng, Fanyi) always wanted to be a doctor or a scientist like her father, Dr. Zeng Yitou — and her mother, Dr. Shu-Zhen Huang, a well-known biologist as well as an award-winning singer and actor — truly a family dedicated to "helping the cause of life".

Time passed, and Fanyi Zeng studied both music and medicine. She attended UC San Diego, where she majored in biology and minored in music; she then earned a full scholarship for her MD and PhD degrees from the prestigious University of Pennsylvania.

Fanyi Zeng: Continuing the family tradition, researching for cure.

She worked long hours. And sometimes when the body was weary, she would revive herself by playing the saxophone, so haunting noises emerged from the lab at midnight.

Returning to her home country, she followed in her father's footsteps, studying under the legendary Tan Jiazhen, considered the father of Chinese genetics, until he died at the age of 99.

In 2009, a breakthrough…. Japan's great researcher, Shinya Yamanaka won the Nobel Prize for introducing induced pluripotent stem cells.

With Dr. Qi Zhou, her colleague at the Chinese Academy of Sciences, the young scientist helped develop proof of the iPS method's power — by growing an entire mouse (named Xiao Xiao, or "Tiny") from a single skin cell. It should be noted here that iPS is still a new procedure with many unknowns compared to the better-studied embryonic stem cell; Dr. Zeng supports full stem cell research.

She helped in the fight against beta thalassemia, a blood disorder, and worked to develop early diagnoses of deadly diseases like progressive muscular dystrophy, hard to detect until cripplingly advanced.

Her work did not go unnoticed. She received numerous awards, including the First Young Woman Scientist Award from the Third World Organization for Women in Science.

In person, Fanyi Zeng is small, intense, full of fire, a living exclamation mark and Director of the Shanghai Institute of Medical Genetics, at the world-renowned Jiao Tong University.

I met her at one of Bernie Siegel's World Stem Cell Summits. I used my baby-talk Mandarin, to which she responded, "Ni shuo de hen hao" (you speak very well), which is really nothing more than a polite Chinese fib.

We became friends, and when CIRM was developing the Cooperative Funding program and reaching out to China, I made sure she was in contact with leadership folks.

Dr. Zeng and her parents are currently working on a new way "to make medicine (which) could be revolutionary in the pharmaceutical industry… "It is a project that "My parents and I have spent almost 20 years on…. and hopefully it will be close to the final stretch…"

Fanyi Zeng: a name to remember, and a career to watch.

In 2006, then-President of China Hu Jintao stated that "biotechnology is the priority of high-tech industries by which China will try to catch up with the developed countries."

Since then, their Ministry Of Science and Technology (MOST) has funded stem cell research through two main efforts, Program 73 and Program 86: one for basic research on scientific principles, the other for translational medicine, in which the goal is to turn theory into therapy.

(For more information on Program 73, visit: http://www.973.gov.cn/English/Index.aspx.)

Chinese stem cell policies are supportive. Embryonic stem cell research, fetal cell research, and therapeutic cloning for cells is eligible for funding.

But reproductive cloning to make human babies? China practices the "Four No's", as in this one-sentence policy directive from the Ministry of Health, November, 2002:

"Under no situation, under no circumstances, will human reproductive cloning experiments be (1) endorsed, (2) permitted, (3) supported, or (4) accepted."[2]

In 2007, China spent $2 billion on biomedical research and development. In 2012, that number had quadrupled to $8 billion.[3] On stem cells specifically, one estimate cites China as spending roughly $320 million a year on stem cell research. The money came from all sources, public and private.[4]

Most sources believe this amount is increasing sharply. The most recent figure I can find estimates spending at close to $500 million.[5]

The California stem cell program is an international catalyst of cooperation. But if the terms of Proposition 71 require that every CIRM research dollar be spent in California, how do we work with other nations? There is a solution that is simple, and sheer genius. Suppose a scientist in Paris and a researcher in Los Angeles want to work together. Since

---

[2] Liao L, Li L, Zhao RC (2007).
[3] Chakma J, Sun GH, Steinberg JD, Sammut SM, Jagsi R (2014). Asia's ascent — global trends in biomedical R&D expenditures. *N Engl J Med* **370** (1): 3–6.
[4] Khan N. China halts stem-cell trial applications as regulation tightens. *Bloomberg* [Internet]. 2012 Jan 10 [cited 2015 Feb 16]. Available from: http://www.bloomberg.com/news/articles/2012-01-10/china-to-halt-stem-cell-trial-applications-in-effort-to-tighten-regulation
[5] van Servellen A, Oba I. Stem cell research: trends in and perspectives on the evolving international landscape. *Research Trends* [Internet]. 2014 Mar [cited 2015 Feb 16]. Available from: http://www.researchtrends.com/issue-36-march-2014/stem-cell-research/

France and CIRM have signed a Cooperative Funding agreement, California pays our researcher and France pays their scientist — and the project just doubled.

On October 18, 2009, Xiaoming Jin, Director General of International Cooperation of MOST, signed a memorandum of understanding with CIRM President Alan Trounson. According to the memorandum, both sides would contribute to the joint research in stem cells between the U.S. and China. Also present at the historic meeting was the Minister of MOST, Wan Gang.

More than a dozen countries — Israel, Germany, Italy, France, Andalucia, Japan, England, Scotland, Argentina, Brazil, India, Australia, Poland, Scotland, and China — have signed cooperative funding agreements with the California stem cell program. Every country of course has its own approach, and there are still a few (but not many) who do not support the research. Pay William Hoffman's World Stem Cell Map a visit at http://www.mbbnet.umn.edu/scmap.html, and you can see that most of the world has supportive stem cell policies.

Did you know that the Chinese word for America is a compliment? "Meiguo" means "beautiful country", and it gives me joy that in these times of international tension, science still offers possibilities for our nations to work together in peace, for the good of everyone.

Gan xi bao shi fei chang hao de! (Stem cells are very good!)

52 THE GREAT NEBRASKA COMPROMISE

John Garfield

"What are you gonna do? Kill me? Everybody dies."

> John Garfield, as a boxer who refuses mob orders to throw a fight.
>
> *Body and Soul*, 1947.

Nebraska was locked in a struggle about embryonic stem cell research.

Anti-abortion groups are strong politically, in this reddest of Republican states. They can rally thousands of letters, phone calls, and emails, and woe betide the government official who crosses them. Vote "wrong" one time, and that can be the end of a political career.

For instance, in 2004, "then-Senator Curt Bromm … was denied a Nebraska Right to Life endorsement … after he didn't support a legislative ban on fetal cell tissue research." That was the same sort of research which led to the polio vaccine.

"[Bromm] had a pro-life voting record until that time [...], so we parted ways [...]," said Nebraska Right to Life Executive Director Julie Schmit-Albin, "It's a matter of holding them accountable. It's [strictly] pro-life business."[1]

But Nebraska University is a serious center of research. Embryonic stem cell projects were underway: to make new cells to fight liver failure, to relieve choking lung disease, and more.

And a patient advocate group called Nebraskans for Life-Saving Cures worked tirelessly to protect the freedom that science requires.

The scientists wanted to work without fear they might be put in jail or have their funding denied; the Right to Life folks viewed the research as abortion, and wanted to block it completely. It was all too easy to think of one side as God's side, and the other as Lucifer's — and who wants to back down in a fight against the devil? Meeting in the middle might seem like being halfway to Hell.

But Nebraskans are practical folks. In a state law, Legislative Bill 606, they did what seems impossible for Washington today — they compromised. The brainchild of people like Senators Mike Flood, Steve Lathrop, and Brad Ashford, LB 606 formed an acceptable middle ground.

Like all compromises, it had something for both sides to hate. The science side gave up a lot.

1. It would be illegal for the state to fund research involving Somatic Cell Nuclear Transfer (SCNT, or therapeutic cloning for cells).
2. No embryonic stem cell lines could be made with state funding.
3. Adult stem cell research would receive preferential treatment, $500,000 a year in state funds.

In exchange, the state's anti-abortion groups signed an agreement that there would be no new legislation concerning embryonic stem cell research.

LB 606 was voted unanimously into law — 48–0.

How did the public feel about it? In November 2008, Fairbank, Maslin, Maullin & Associates (FMM&A) completed a statewide survey of Nebraska voters, showing that a two-to-one majority of Nebraska voters supported LB 606.

The Right to Life folks seemed content — at first.

"We're just relieved it's finally over," said Ms. Schmit-Albin, "It's been a long haul. It's been since 2000 when the first bioethics issue came before the Legislature with the fetal tissue ban."

Greg Schleppenbach of the Nebraska Catholic Conference said, "It establishes a significant ethical boundary in the State of Nebraska that it's wrong to destroy human embryos."[2]

---

[1] Lee M. As pressure mounts, McClurg mum on stem cell vote. *Lincoln Journal Star* [Internet]. 2009 Nov 20 [cited 2015 Feb 17]. Available from: http://journalstar.com/news/local/as-pressure-mounts-mcclurg-mum-on-stem-cell-vote/article_c03df406-d579-11de-bcfa-001cc4c002e0.html

[2] Ross T. Lawmakers approve restrictions on stem cell research. *Lincoln Journal Star* [Internet]. 2008 Mar 24 [cited 2015 Feb 17]. Available from: http://journalstar.com/news/local/govt-and-politics/lawmakers-approve-restrictions-on-stem-cell-research/article_5466669e-c74a-5ba3-bd78-6ff0298d83e3.html

The Nebraska Coalition for Ethical Research, another group opposed to embryonic stem cell research, praised the agreement, saying that the "Nobel Prize people" should be aware of State Senator Steve Lathrop, who had negotiated a compromise bill on cloning and embryonic stem cell research in Nebraska.[3]

Victoria Kohout, Executive Director of the Nebraska Coalition for Lifesaving Cures, said, "Legislative Bill 606, approved 48–0 in March 2008, said human embryonic stem cell research with stem cell lines created elsewhere is ethically acceptable as ... state policy in Nebraska ...." — personal communication.

Ron Withem, director of governmental relations for the University of Nebraska system, said the bill "allows the core research using embryonic stem cells to continue, consistent with regents guidelines."

http://www.kearneyhub.com/news/local/article_3441b272-66dc-556b-85e2-99b213edd73d.html

However, Chip Maxwell, Executive Director of the Nebraska Coalition for Ethical Research, had this to say in the *Lincoln Journal Star*: "We were glad to get a ban on human cloning and destruction of human embryos at state facilities [...], *but we wanted more*" (italics mine).[4]

Technically, the Right to Life groups did not break their promise with what they did next. The agreement was that there would be no new laws; they did not try to make any.

But they turned their attentions to *the only place in Nebraska where embryonic stem cell research was done* — University of Nebraska — and found a way to shut it down. If embryonic stem cell research could not go forward at University of Nebraska, it was dead in the state.

Here is how the attack would work.

1. The governing board of University of Nebraska is made up of elected regents.
2. Candidates try for the endorsement of the Religious Right.
3. Five of the eight members of the Board of Regents had the backing of those groups.
4. The Regents were now told to sign a resolution banning all embryonic stem cell research except the little bit allowed by former President George W. Bush.

Here in one sentence is the guts of the resolution:

"No research with stem cells derived from human embryos may be conducted unless the stem cell lines used in the research were derived prior to [...] August 9, 2001."

Pre-2001 stem cell lines are old, fed on rat nutrients, and may be even unsafe to use.

Four of the eight regents immediately signed the resolution.

One more vote, and the Religious Right would impose their policy on Nebraska University.

---

[3] Maxwell C. Stem Cell Research: Nebraska Politics: Lathrop for Nobel Prize [Internet]. 2008 Feb 21 [cited 2015 Feb 17]. Available from: http://checkwithchip.blogspot.sg/2008/02/stem-cell-research-nebraska-politics.html

[4] Maxwell C. Find other stem cell options. *Lincoln Journal Star* [Internet]. 2009 Nov 4 [cited 2015 Feb 17]. Available from: http://journalstar.com/news/opinion/editorial/columnists/find-other-stem-cell-research-options/article_fd50bed2-c9a8-11de-8c99-001cc4c002e0.html

The future of Nebraska stem cell research came down to one man, Jim McClurg, a strong right-to-life Republican. But Jim McClurg also had a degree in biochemistry. He wanted time to think.

"There's a lot of good people on both sides of this issue who have heartfelt opinions on it. I'll decide once I've heard what everybody has to say," he said.

The Religious Right made his phone number public, and encouraged their membership to contact him. Some of the comments made were hard to believe. One board member stated:

"If our children have no defects or diseases, where will our Beethoven and Steven Hawking of the future derive intellectual nutrients for their endeavors when they have no physical impairment to focus their work?"[5]

He appeared to be saying that incurable disease was necessary so that geniuses had work to "focus" on!

But the research for cure message seemed to be getting through. Across the country, people of every walk of life, including the Religious Right, were reconsidering their positions.

Of embryonic stem cell research, Pro-life Senator Orrin Hatch said:, "I think it's the ultimate pro-life position, because … being pro-life is not just caring for the unborn but caring for those who are living."[6]

And when a crowd was called for to protest against stem cell research?

The anti-abortion rights group, Nebraska Right to Life, called for protesters to attend the regents meeting. A grand total of six (6) people rallied outside.[7] Not six thousand, not six hundred — just six: not exactly a throng.

Young people took note of the emerging new science. The Student Senate of Nebraska University voted 15–0 in favor of embryonic stem cell research.

And the President of University of Nebraska, J. B. Milliken, had this to say:

"When the LB 606 compromise was negotiated, I believe it was clearly understood by all parties that it [was] a resolution of the public policy considerations regarding embryonic stem cell research in Nebraska. […] Embryonic stem cell research holds enormous promise, and if the University of Nebraska is to be a leading research university it should be appropriately engaged in this research."

Why did Nebraska need embryonic stem cells?

Embryonic stem cells remain the standard against which all other cells are compared and evaluated. The deficiencies of these few Bush-approved stem cell lines for use in

---

[5] Ferlic R. Debunking some of the far-fetched arguments used against ESCR [Internet]. 2004 Jul 11 [cited 2015 Feb 17]. Available from: http://sci.rutgers.edu/forum/showthread.php?19859-Debunking-Some-Of-The-Far-Fetched-Arguments-Used-Against-ESCR

[6] Brand M. Sen. Hatch Backs Expansion of Stem-Cell Funding. *NPR News* [Internet]. 2007 Feb 16 [cited 2015 Feb 17]. Available from: http://www.npr.org/templates/story/story.php?storyId=7447914

[7] NU pres: Stem cell research policies are fine. *Victoria Advocate* [Internet]. 2009 Oct 23 [cited 2015 Feb 17]. Available from: https://www.victoriaadvocate.com/news/2009/oct/23/bc-ne-stem-cells-nebraska4th-ld-writethru/

research are widely documented. If the Board of Regents were to restrict research to only the Bush-approved lines, University of Nebraska researchers would be unable to contribute their skills and dedication to this fast-moving and promising research.

No one denied the value of adult stem cell research, using the body's natural supplies of stem cells. But to me the difference between adult and embryonic was crystal clear: adult stem cells were like gift certificates, redeemable only at certain stores; embryonic stem cells were like cash money, acceptable everywhere. We needed both.

Champion stem cell researchers, eminently busy folks, took time to state their positions.

James Thomson, considered the inventor of human embryonic stem cell research, and co-inventor of induced pluripotent stem cells, said, "The 'gold standard', embryonic stem cell research [...], has been recognized by all the primary world scientific bodies [...]. Research efforts on embryonic stem cells [and other forms] are not alternative pathways to discovery, but are, in fact, complementary and synergistic.... Universities must support freedom of inquiry .... The U.S. has been the world's center of biomedical innovation [...]. Millions of patients have seen their lives improved through therapies that resulted from these principles of freedom to pursue ideas and research directions without ideological bans. *This has clearly been the American way*" (italics mine).

The U.K.'s Sir Ian Wilmut, the scientist who cloned Dolly the sheep, but who was now working primarily with induced pluripotent stem cells, said, "Denying access to newly created human embryonic stem cell lines would be a significant impediment to medical research."

Nobel Prize winner Shinya Yamanaka, inventor of induced pluripotent stem cells, said, "Research on embryonic stem cells is essential to further advance induced pluripotent stem cell technology."

Irv Weissman of Stanford, often called the father of adult stem cell research, said, "No one stem cell type seems likely to provide a one-stop shop for all human ailments; each has advantages and disadvantages .... Research must proceed with all types of stem cells ...."

Dr. Wise Young, one of the most genuinely beloved figures in the paralysis community, said, "If this condition were imposed, no serious stem cell scientist would want to join the [faculty]. Such a decision would make the University of Nebraska a laughing stock among scientists, [and] would make recruitment of the best faculty and top students extremely difficult."

Lawrence Goldstein, author of *Stem Cells for Dummies*, said, "We need as many skilled scientists, engineers, and physicians as possible, working with the best possible and greatest variety of stem cells and approaches to solve problems in human disease. ... Stranding talented University of Nebraska researchers on the bench would be harmful .... We need "all the pieces" — both stem cells and researchers!"

And when it was time to push the button? Jim McClurg voted against the additional restrictions.

"Unfortunate," said Julie Schmit-Albin, Nebraska Right to Life's Executive Director. She noted that elected officials in the past had earned her group's endorsement but failed to vote accordingly — and subsequently lost that backing.

To which McClurg responded, "People didn't elect me to worry about getting elected again."[8]

Because of that act of conscience, embryonic stem cell research is still legal in Nebraska.

---

[8] Lee M. NU regents open door to expanded stem cell research. *Lincoln Journal Star* [Internet]. 2009 Nov 20 [cited 2015 Feb 17]. Available from: http://journalstar.com/news/local/education/nu-regents-open-door-to-expanded-stem-cell-research/article_e4ef1dc4-d606-11de-83c6-001cc4c002e0.html

53 STEM CELL TOURISM

If you had an incurable disease, would you go to another nation for an experimental treatment?

As you recall, when Roman was first paralyzed, a drug called Sygen was undergoing clinical trials in America. I was unable to get our son into the trials, but with the help of U.S. Representative Pete Stark, a permission slip from the FDA, and a prescription from a cooperative doctor, I was able to purchase the Sygen — from Switzerland.

Did that "stem cell tourism" help? To this day, I don't know. Roman got the function of his triceps back, but that could have been from the physical therapy. More than one doctor told me I was desperate then, which made me an easy target for anyone with a plausible-sounding promise of cure.

True enough: I would have done anything (and still would) to ease my son's suffering. Back then, if someone in a white coat had told me eating dirt cured paralysis, I would have leaped for a shovel. Today, I would ask that the mud be analyzed first.

Wanting cure is like needing air when you are drowning. With customers that eager, crooks may come with fake medicine, or "pseudomedicine" as Canada's Timothy Caulfield once called it.

On the other hand, what if there was a genuine cure, but another country found it first, and only politics was in the way? What if you lived in Mexico, for example, where embryonic stem cell research is illegal for religious reasons, and America developed a cure using embryonic stem cells for diabetes or blindness; would you not come to the U.S. for treatment?

"There is stem cell tourism, *and* stem cell tourism," said Brock Reeve, Christopher Reeve's brother, and Director of Harvard stem cell program.

http://news.harvard.edu/gazette/story/2012/11/the-rise-of-stem-cell-tourism/

The best fight I ever saw on stem cell tourism took place at the 2009 World Stem Cell Summit.

The first panelist to speak was Grant Albrecht, a patient advocate with a golden voice, an actor's stage presence, and Transverse Meyelitis, (TM) a nerve disease of progressive paralysis. It hurt to watch him slowly make his way to the microphone.

Grant had achieved the actor's dream, an important role in a Broadway play, but he had to give it up when his legs would no longer support him. He began a search around the world, hunting for a cure. Fifty thousand dollars went to one company, $25,000 to another. He would have considered almost any method, he said, feeling that he was on his "last legs". And his opinion on our own government's response to the need for cure research?

"We have a problem like a five-alarm fire, and the government is giving a drip–drip–drip response," he said.

And now, the central figure: Alex Moffett is Chief Executive Officer of Beike Holdings, one of the world's largest stem cell tourism destinations, which is headquartered in Thailand and China.

A distinction should be made between good and bad stem cell tourism, Mr. Moffett said, calling it bad when cures were promised but the physicians were not well trained, and good when there was professional oversight, top quality labs, and transparency.

According to Mr. Moffett, his company had 400 employees, including eight PhDs and 100 MDs. He claimed his company had treated more than six thousand patients with adult stem cells. He acknowledged that "mistakes had been made" in early years, but now he wanted to get things done right.

China, he said, just passed stringent new guidelines which included restrictions on research. The clear implication was that his company would not have been in business there if it was not reputable.

The next speaker, Doug Sipp, said China was "handsomely paid" by Beike. Mr. Sipp had worked between 2005 to 2009 for the International Society for Stem Cell Research (ISSCR), and was also Secretary-Treasurer of the Asia-Pacific Developmental Biotechnology Society.

Sipp asked: if Beike's work is legitimate, why not share their data and information gathered from their tests? He argued for "appropriately-designed experiments with proper peer review". Importantly, he suggested anyone in the audience considering stem cell tourism should first visit the ISSCR website and download the *ISSCR Patient Handbook on Stem Cell Therapies*, with points to consider before allowing yourself to become part of an experiment.[1]

Waiting to speak was Jeanne Fontana, MD, PhD, who serves on the California stem cell board as a substitute for Dr. John Reed of Burnham Institute. Dr. Fontana has a sunny disposition, a personality that engenders good feelings. But she was not smiling today.

Dr. Fontana lost her mother to Lou Gehrig's disease. When her mom was diagnosed in the late 1990s, Dr. Fontana did a worldwide literature search but found no cure. Today, there might be as many as 170 offshore clinics offering "stem cell treatments" — were they for real? She spoke about the need for independent testing for treatments; i.e., testing done by people with no financial connection to the companies.

She had two main questions for Beike:

1. Did patients improve after going for the treatments offered?
2. Was there data (records of testing and results), and if so, why not share it?

---

[1] International Society for Stem Cell Research. Patient handbook on stem cell therapies [Internet]. 2008 Dec 3 [cited 2015 Feb 17]. Available from: http://www.isscr.org/home/publications/patient-handbook

A long line of speakers waited for the microphone. Most expressed skepticism, a few gave support.

I wanted to hear both sides. But my teeth ground together when I heard that Beike treated patients with spinal cord injury paralysis — people like my son.

If there was a successful treatment for paralysis, I would know about it.

Some paralyzed friends had gone overseas for an operation known as the Olfactory Epithelial Glia (OEG) treatment, where a surgeon reaches a scalpel up through the patient's nose, scrapes off part of the brain, and spreads it like jelly on the injured spine — for about $40,000. They used the part of the central nervous system that regenerates, but in my opinion, it was useless. I talked to someone who had the operation. Some of the skin on his arms had become sensitized, so he could tell when his attendant put his sleeves on. But that was about it. He was still paralyzed.

One audience member rose in support of Beike. She stated that her loved one had been given a treatment from the company and did recover some function. (She did not provide details on what condition it was.) However, she added, there was no follow-up.

"Why is follow-up important?" I asked Dr. Fontana later.

"Because the treatments themselves may have negative consequences."

There must be follow-up to treatments to keep the patients safe and find out what happened.

A venture capitalist, Linda Powers of Toucan Industries, made an impassioned plea for courtesy for Beike's representative, saying she was "staggered by the savagery of the criticism" and the "sweeping generalities" of the comments from audience and panelists.

Dr. Jeanne Loring is always ready to step up to bat for the betterment of stem cell research and therapy.

Dr. Jeanne Loring stepped to the microphone. This top-notch scientist is always on the alert for anything which might help or hurt our endeavor.

She stated that she had analyzed hundreds of stem cell samples and had a standing offer to analyze cells for any patient who is considering an overseas procedure: she asked Mr. Moffett if *Beike would provide her with samples of the stem cells they use in patients.*

Mr. Moffett replied simply, "Yes".

Unfortunately, to the best of my knowledge, the stem cells never arrived.

I emailed Steven Marshank, a representative of Beike, and asked about their spinal cord injury treatments.

He responded, "For spinal cord injury we provide cord mesenchymal stem cells and sometimes we provide mesenchymal stem cells from the patient's own bone marrow. This is delivered by injections into the spinal cord fluid and at least one IV."

So, would I take my paralyzed son to Beike?

No, not at this point. In years to come, when a properly designed treatment is approved by the FDA, I will look around and find the best possible place to have a procedure done. I will offer suggestions to my son, after which Roman will make his decision. If he chooses Beike, so be it.

But I would never recommend a treatment that has not been systematically documented and tested. Things can always get worse, and sometimes much worse. A paralyzed person might suddenly have endless pain, like being on fire.

I put my faith in the grueling years of work put in by scientists like the aforementioned Hans Keirstead and Thomas Okarma of Geron, and Mike West and Jane Lebkowski of Asterias Biotherapeutics, battling their way toward human trials with embryonic stem cells. Their struggle is long and frustrating (22,500 pages of documentation — enough for 50 books!) but it will be worth it in the long run, because we will know exactly where we stand. If it succeeds, we know what worked, and we can do it again. If it fails, we can learn from our mistakes and not repeat them.

Should we still listen to what companies like Beike have to say? Of course. No one has a lock on progress. If Beike comes up with something great and can back it up with documentation, they should profit, and the world should know.

As Bernie Siegel put it when he introduced the speakers, "While the Summit does not endorse any stem cell treatment companies or organizations, *maintaining the conversation is vital*" (italics mine).

The lines of communication must remain open.

54 THE MAN WHO COULD FLY WITHOUT A PLANE

Charles Reid: The man who could fly, even with multiple sclerosis.

To celebrate my 69th birthday, Gloria bought me a ticket for an entertainment attraction called iFly, a vertical wind tunnel which hurls you up in the air using 165 mph chilly gusts.

"Oh, that looks like fun!" I had made the mistake of observing casually, not realizing she was taking notes. And suddenly there I was, wrapped in a blue flight suit, goggles on, and earplugs shoved "halfway into the brain" as the instructor cheerfully recommended.

His name was Charles Reid, a superbly athletic African–American. He showed us how flying was done — or rather, how he could do it.

Leaping, soaring, and bounding around in the enclosed blast of air, he made it look easy. He would disappear upwards into the darkness, only to reappear hurtling downward, changing direction, soaring, turning, flipping in the air, like Peter Pan without stage wires.

When it was my turn, I leaned forward into the roar of wind… and was suddenly weightless. Reid grabbed one of the numerous handles on my flight suit and we were off.

It was not quite the same as when he did it, feeling not so much like flying as it was like being luggage tossed by invisible giants. It might have been embarrassing, if it had not been such fun.

Afterward, he commented on my Proposition 71 T-shirt. I still have a couple, which I wear on special occasions, like when I plan on being scared out of my mind in wind tunnels.

"They should allow it," he said, meaning stem cell research.

"They do!" I exclaimed, leaping into my 60-second California stem cell program speech, which I can say awakened out of a sound sleep.

"I have multiple sclerosis," said Charles Reid.

Multiple sclerosis (MS) — "many scars" (on the brain), one of the ways the disease was once diagnosed after the patient had died — affects 160,000 Californians.

We swopped business cards, and I called him up next day.

"Three years ago I was having blurred vision," he said, "I went to the emergency room and found out I was 80% blind in my right eye. It was either cancer or multiple sclerosis, they said, and sent me over to Stanford where they kept me three days. I had multiple sessions of MRI, maybe seven hours in the machine. They put me on a steroid to reduce the eye's inner swelling…."

"And now?"

"Minor symptoms, like numbness in my fingertips. If I go to pick up a penny on the floor, I have to look at it, be sure I have it. And exhaustion is part of the MS life, of course."

Silence.

"What do the doctors say?"

"To avoid stress. That is pretty easy, my job is my passion, so I stay positive most of the time."

That was certainly true. He had made each of the fliers feel special and important.

"And about the… progressive nature of the disease?"

"Sooner or later I'll be in a wheelchair," said Charles Reid.

Naturally I told him there might be reasons to disagree.

Like our friends in Connecticut. A spinoff company from that state's outstanding stem cell program had identified a novel approach to treating MS using human embryonic stem cells which significantly reduced MS disease severity in animal models.[1]

That company is ImStem Biotechnology, Inc., and is Connecticut pride all the way.

"Connecticut's investment in stem cells, especially human embryonic stem cells, continues to position our state as a leader in biomedical research," said Governor Dannel P. Malloy recently.[2]

Champion scientists like Joel Pachter, Xiaofang Wang, Stephen Crocker, and Ren-He Xu used embryonic stem cells to make mesenchymal cells, the specific cells needed — and along the way showed the clear superiority of embryonic over adult stem cells for the job. They set up eight adult stem cell lines from the bone marrow and compared them with half that number of embryonic stem cells — and the embryonic stem cells beat adult stem cells all the way. Not only were human embryonic stem cells stronger and faster, but the adult stem cells actually made things worse, producing an inflammatory

---

[1] Embryonic stem cells offer treatment promise for multiple sclerosis. *UConn Today* [Internet]. 2014 Jun 11 [cited 2015 Feb 17]. Available from: http://today.uconn.edu/blog/2014/06/embryonic-stem-cells-offer-treatment-promise-for-multiple-sclerosis/

[2] ibid.

molecule called a cytokine. As part of the MS problem is nerve inflammation, the last thing you want is more of the same!

Closer to home was the Craig Walsh effort at the University of California at Irvine. With a California stem cell program grant, Primary Investigator (PI) Dr. Walsh was tackling the disease alongside co-PIs Jean Loring of Scripps Research Institute and Claude Bernard of Monash University in Australia.

Their effort focused on myelin, the natural insulation around the body's nerves. Important in the fight against paralysis, restoring myelin may also be key to defeating MS.

Think of a mouse chewing on an electrical cord. What happens? A short circuit goes off. The rodent gets fried, maybe the house burns down. At the very least, the electricity cannot do its proper job. Flick the switch all you want, but no current gets through and the lights won't go on.

Similarly, if the myelin insulation is destroyed, the nerve short-circuits. It cannot send the right messages between the brain, spine, and body, resulting in paralysis, numbness, or pain.

But if the myelin could be restored, reinsulating the nerves, it might cure the disease.

In the Walsh experiment, mice were paralyzed by a virus that works similarly to MS.

Then, neural precursor cells (NPC) were derived from the human embryonic stem cell line H9. According to the investigators, the paralyzed mice then received "intraspinal transplantation of these NPC cells, resulting in *significant clinical recovery*, beginning at 2–3 weeks following transplant"[3] (italics mine).

The investigators noted that there were problems of rejection. "Despite this striking recovery, these [...] NPCs were rapidly rejected." But the positive improvements remained. Paralyzed mice became unparalyzed.

In both experiments, the subjects got better.

How long would it take before people got well? There was no way to know right now.

But for the first time in the history of the world, we have a chance to defeat this terrible disease, to someday put its name — with a check-mark beside it — on the trophy walls of medical research.

"I feel a little bit more hopeful," said the man who could fly without an airplane.

---

[3] https://www.cirm.ca.gov/our-progress/awards/multiple-sclerosis-therapy-human-pluripotent-stem-cell-derived-neural-progenitor

55 THIEF OF LIVES

Jeff Sheehy, with husband Bill Berry and daughter Michelle.

Jeff Sheehy is a tall, broad-shouldered man, with more than a passing resemblance to a young John Wayne. A member of the California stem cell program's board of directors, Sheehy can be counted on to raise and deal with the most difficult issues, never shying away from a problem just because it is tough.

    He has a daughter, Michelle, aged nine; her eyes just shine when she looks at her Dad. So what is the problem?

    Jeff Sheehy is HIV (human immunodeficiency virus) — positive.[1]

    Pills, as part of anti-retroviral therapy (ART), keep Jeff alive. As long as he takes his medications every day, his HIV is survivable; if he stops, HIV advances to the death

---

[1]Spotlight on disease team awards — HIV/AIDS: Jeff Sheehy [Internet] [cited 2015 Feb 17]. Available from: https://www.cirm.ca.gov/our-progress/video/spotlight-disease-team-awards-hivaids-jeff-sheehy

sentence of AIDS (acquired immunodeficiency syndrome). HIV can be lived with; AIDS cannot.

But Jeff's disease is only managed, not cured.

People on long-term ART may face heart and lung disease, cancer, osteoporosis, and other life-threatening maladies.

To date, only one person in the world has been *cured* of HIV-AIDS, and that was Timothy Ray Brown, better known as the "Berlin patient".

Residing in Berlin, Germany, Timothy Brown had both leukemia and HIV, the latter of which he was managing by the ART pills. To save him from dying of leukemia, Brown was given a bone marrow transplant — *from a person immune to HIV*. Apparently, as many as ten percent of Northern Europeans are blessed with a mutation that prevents HIV from entering their cells and infecting them. (This might be an immunity passed down from survivors of the Middle Agee's Black Plague.)

Tim Brown overcame both leukemia *and* HIV-AIDS.[2]

But his experience would not work for everyone. A bone marrow transplant operation is not only painful and dangerous, but also depends on finding a suitable "match" for the donation.

Could there be a way to make a person immune to HIV?

Enter the California stem cell program. Two teams of researchers were selected to take on the challenge of HIV-AIDS: they would try to make changes in the stem cells of the blood, so what cured the Berlin patient might be done again across the world.

Team One is Calimmune, Inc., a company begun by David Baltimore, Nobel Prize winner and former board member of the California stem cell board of directors. That team is led by Ronald Mitsuyasu, MD, of the University of California at Los Angeles, and Jacob P. Lalezari, MD, of Quest Clinical Research in San Francisco.

Team Two is comprised of John Zaia, David DiGiusto, and Paula Cannon from the Beckman Research Institute of City of Hope, and the University of Southern California.

These teams face a battlefield of microscopic complexity: the gene chain of our lives.

Did anyone say complicated? Wow! Hearing them talk is like reading aloud from a bowl of alphabet soup.

Look at the following "word": CCR5.

CCR5 is a gene, which is part of our DNA structure. It is not always our friend. CCR5 sometimes lets HIV into our bodies, like a door that admits killers into a house.

But if the gene had a change, or a mutation — delta 32 ($\wedge$32) — added to it, the door locks and the HIV virus cannot enter.

In short, CCR5 — bad, CCR5$\wedge$32 — good.

Tim Brown, the Berlin patient, was injected with a bone marrow transplant from a person who had CCR5$\wedge$32, and from which Brown was cured.

---

[2]DeNoon DJ. Man cured of AIDS virus: 1st U.S. news conference [Internet]. 2012 Jul 24 [cited 2015 Feb 17]. Available from: http://www.webmd.com/hiv-aids/news/20120724/man-cured-of-aids-virus-news-conference

Would it be possible to put delta 32 into the gene structure permanently and lock the CCR5 "door" — or maybe snip out CCR5 from the DNA chain altogether? Hold that thought.

A question I like to ask when interviewing scientists is: why did he or she choose their career?

I interviewed Dr. Paula Cannon from the University of Southern California.

When I asked her "the question", she seemed embarrassed, saying she had chosen biology "almost at random" in college. She had done well, earning a degree in microbiology, but after graduation she "did a 180-degree career spin" and went into the music business in Liverpool, home of the Beatles. She did not like it, but it shaped her future.

This was a time when HIV-AIDS was ravaging the entertainment industry, and people were dying with no hope from this modern day plague.

Time passed; another 180-degree career spin happened and Paula Cannon came back to the science business, but this time with a goal — to fight the "thief of lives".

As Dr. Cannon can talk science in "people-talk", I asked for answers to three key questions:

1. What is a zinc finger nuclease (ZFN) and how do you use it?

"A zinc finger nuclease is a sort of genetic scissors. It can snip out bad stuff in the body's DNA chain. We hope to program ZFNs to snip out CCR5, so it won't let the AIDS virus in again. We will use a virus to insert the ZFN — or maybe just the information needed to make the ZFN."

2. What is ribonucleic acid (RNA) and why does it matter to the AIDS research project?

"DNA is like a thousand-page blueprint for a house; RNA might be considered one page — maybe for the upstairs bathroom. RNA may bring very specific and permanent changes in the body's structure, by transporting the ZFN where you want it to go."

3. How important was the California stem cell program to you as a scientist?

"The CIRM grant made it possible for me to fight against AIDS within a stem cell structure; without CIRM, this would not have happened."

The battle is joined; Calimmune's team is already in human trials, while John Zaia's team is following very closely behind. They are challenging the thief of lives: the virus called AIDS.

Both teams deserve the thanks of America (where more than one million individuals have HIV-AIDS) as well as the rest of the world (33.4 million infected).

These are not just numbers; these are friends and family, our neighbors and loved ones.[3]

---

[3] U.S. statistics [Internet] [updated 2014 Dec 2; cited 2015 Feb 17]. Available from: https://www.aids.gov/hiv-aids-basics/hiv-aids-101/statistics/

As Jeff Sheehy puts it, "The more the research moves forward, the more chance there is that when Michelle graduates from college, or when she gets married — then I'll be there."[4]

And I hope I'll be invited to the wedding, when Jeff Sheehy walks his daughter down the aisle.

---

[4] Progress and promise in HIV/AIDS [Internet] [cited 2015 Feb 17]. Available from: https://www.cirm.ca.gov/our-progress/video/progress-and-promise-hivaids

56 SINGAPORE, BIOPOLIS, AND THE POWER OF THE SMALL

Biopolis: Singapore's city of tomorrow.

In the year 2001, the most powerful nation on Earth was tying itself in knots trying to decide whether to allow stem cell research or not — and one of the smallest countries was taking a giant leap forward in biomedicine.

It is no insult to say that Singapore is a small country. Size-wise, it is not quite as large as Rhode Island. Combining one central island and 63 little ones, Singapore has a total population of roughly five million — and the most amazing can-do attitude toward life's problems.

Not enough land mass? Make more! Link some islands together, and build more territory![1]

But in 2001, Singapore needed an economic change. Although their previous concentration on the "three economic pillars" (electronics, chemicals and engineering) had

---

[1] Singapore. *Wikipedia* [Internet] [updated 2015 Feb 15; cited 2015 Feb 17]. Available from: http://en.wikipedia.org/wiki/Singapore#Geography

Sir Sydney Brenner, who famously advised Singapore to invest in biomedicine.

done well for them, the 1997 Asian economic crisis had made them consider the need for a fourth.

Could biomedicine (with an emphasis on stem cells) provide jobs for their people, growth for the economy, and health for their nation?

To be sure they were doing the right thing morally, they did what Britain had already done and California soon would: gathering scientists and ethicists, religious experts and economists, to discuss and debate the new science. Everyone had a chance to speak, and to learn.

Home to a diverse population of Chinese, Malays, and Indians,[2] the Singaporeans were somehow able to work together. Perhaps because religion was not allowed to be part of the government, Singapore's Buddhists, Christians, Muslims, Hindus, and non-believers could argue, debate, decide — and then get on with it and make things happen.

The government established boundaries — no reproductive cloning and no use of blastocysts past the 14th day; regulations in line with the biomedicine-supportive world — and then they went ahead.

Perhaps because they are an island nation used to interacting with travelers from around the world, they seemed to have no problem listening to the advice of outsiders. Nobel Laureate Sir Sydney Brenner is the man responsible for the scientific use of the tiny roundworm *C. elegans*. He is also a great friend of Singapore's. In 1983, he was invited by deputy Prime Minister Goh Keng Swee to make a speech on biotechnology

---

[2]Demographics of Singapore. *Wikipedia* [Internet] [updated 2015 Jan 16; cited 2015 Feb 17]. Available from: http://en.wikipedia.org/wiki/Demographics_of_Singapore#Religion.5B22.5D.5B23.5D

A*STAR Chairman Lim Chuan Poh defends Singaporean biomedicine with his every waking breath.

in industry, which speech was highly influential in that country's considerations of biomedical investment.

With the support of Singapore's first Prime Minister Lee Kuan Yew, the Institute of Molecular and Cell Biology was established. When Brenner suggested Singapore needed a *city built for biomedicine*, it was as if the whole country went, "Good idea!"

There were also four pillars of citizen support.

The previously mentioned Goh Keng Swee and a civil servant named Philip Yeo worked together amazingly well, even though the latter had what some might call an "attitude".

Mr. Yeo once said, "I may be a civil servant, but I am neither civil nor a servant!"

Goh Keng Swee recognized talent, and gave Philip Yeo the freedom he needed to bring change.

When someone complained about Yeo, Goh Keng Swee laughed and said, "That is just Philip, what can I do?"

His confidence paid off.

"Come, we are going to buy some land," said Philip Yeo one day.

"What for?" said the deputy Prime Minister.

"For Biopolis," said Philip Yeo.

Who would design the city of the future? Imagine how politicized such a decision could become; there might be pressure to give the job to a friend ("My son-in-law would just be perfect for this…") or whoever gave the most campaign contributions. But Singapore went for the best.

An Iraqi–British woman, Zaha Hadid, considered one of the world's greatest architects, designed an interconnected complex of 13 buildings, 9 of which would be for

Philip Yeo: Creative force behind BIOPOLIS, Yeo also arranged 5-year contracts for scientists, so they would not have to waste time seeking grants.

biomedicine. The idea was to make scientists' lives easier and more productive, providing for their needs in a close-knit city. A person could get up in the morning and be at work in half an hour. Jurong Town Corporation, famous for turning Jurong Island from a swamp into a bustling metropolis, took on the challenge — with spectacular results.

Huge (three and a half million square feet), Biopolis became 13 magnificent buildings, connected by skyways and underground tunnels, interspersed with lakes and lush greenery.

Completed in 2003, Biopolis was officially launched by Singapore's President, Dr. Tony Tan. Philip Yeo set up the National Biotechnology Committee — co-chaired by Lim Chuan Poh.

Soon to become Chairman of A*STAR (Agency for Science, Technology, and Research), Lim Chuan Poh had been head of education for the country, and a Lieutenant General in the Singaporean defense force. He not only had to make sure obstacles to progress were dealt with, but also to share Singapore's biomedical story with her people, and the world.

Behind the scenes working out "the seed ideas of Biopolis […] on napkins and white boards"[3] were Professor Tan Chorh Chuan, Dr. John Wong, and Dr. Kong Hwai Loong.

---

[3] Speech by Mr Lim Chuan Poh, Chairman of Agency for Science, Technology, and Research at the Biopolis 10th anniversary gala dinner [Internet]. 2013 Oct 16 [cited 2015 Feb 17]. Available from: http://www.a-star.edu.sg/Media/News/Speeches/ID/1894/SPEECH-BY-MR-LIM-CHUAN-POH-CHAIRMAN-OF-AGENCY-FOR-SCIENCE-TECHNOLOGY-AND-RESEARCH-AT-THE-BIOPOLIS-10TH-ANNIVERSARY-GALA-DINNER.aspx

As quickly as a building could go up, it was occupied: first by government-paid researchers and scientists (domestic and foreign), then by biomedical corporations like Merck and Glaxo Smith-Kline. Inviting the scientists was the personal mission of Dr. Christopher Tan, formerly of the University of Calgary, Canada.

When Severe Asiatic Respiratory Syndrome (SARS, or better known as the "Asian flu") struck, A*STAR scientists developed a SARS diagnosis kit so that, from a single drop of blood, the disease and its mutations could be quickly detected (less than two hours).

With stem cells, they concentrated first on basic research, and then moved on to translational medicine. To get things started, champions were brought in: experts like Edison Liu, Neal Copeland, George Radda, Jackie Ling, David and Birgid Lane, Edward Holmes, and Judith Swain. Then came students:

"We attracted over 1,000 international students, and provided scholarships to 650 of our own," said Lim Chuan Poh in a speech given during the 10th anniversary of Biopolis, "As Philip Yeo called it, we were kidnapping whales and nurturing guppies!"[4]

How is Singapore doing with biomedicine?

Between 2003 and 2013, manufacturing output by the biomedical sector grew five times, to over S$29 billion in 2012, contributing to five percent of Singapore's gross domestic product.

As said by Lim Chuan Poh, "Where there was one medical school before, we now have three [...].Where there was one [biomedical research center], we now have over 13 research institutes [...]. Over 30 of the world's leading biomedical sciences companies have set up their Regional HQs in Singapore; and biomedical employment] grew by 3.5 times to over 5,400, with [about half of these researchers] located in the Biopolis."[5]

According to Angeline Ng Chiu Yen of A*STAR, "Biopolis has a total of over 5,200 people: 4,300 are the scientists and engineers. It is international, with half of the scientists coming from 60 different countries."

And who is co-chair of the National Biotechnology Committee? Lee Hsien Loong, the current Prime Minister of Singapore, and son of Lee Kuan Yew, the first Prime Minister.

In American terms, it would be as if former Presidents George W. Bush and his father, George H. W. Bush were both strong supporters of biomedicine!

---

[4]Speech by President Tony Tan Keng Yam at the Biopolis 10th anniversary gala dinner [Internet] [cited 2015 Feb 17]. Available from: http://www.amc-singapore.net/news/presentation/10/
[5]Speech by Mr Lim Chuan Poh, Chairman of Agency for Science, Technology, and Research at the Biopolis 10th anniversary gala dinner.

57 SINGAPORE SCIENTISTS

Pioneering scientist Ariff Bongso (*centre*) has been called "Singapore's Father of Stem Cell Research."

Called Singapore's "Father of Stem Cell Research", Ariff Bongso is respected not only for his contributions to embryonic stem cell research, but also for his pioneering work on in vitro fertilization (IVF) techniques which help childless couples conceive.

In 1990, Professor Bongso's group identified several factors which are now used to develop the culture medium at almost all IVF clinics today. In 1994, his group isolated the first embryonic stem cells from blastocysts, which were an important step in the development of human embryonic stem cell lines by Dr. James Thomson four years later.[1]

Hoping to interview Dr. Bongso, I reached out to him systematically, but with no response. Were my emails not reaching him, or was he just saying no? Months went by, the deadline for this book approached, and I had to forego interviewing the Asian giant.

---

[1] Stojković M (2008). Celebrating 10 years of hESC lines: an interview with Ariff Bongso. *Stem Cells* **27**: 275–277.

Ng Huck Hui leads Biopolis scientists in bid to understand how stem cells regulate themselves, and how misregulation can cause disease.

Bing Lim: Serving his country at Genomic Institute of Singapore.

But if Singaporean stem cell research began with Ariff Bongso, it does not end with him, and the closer I looked, the more there was to see.

First, remember A*STAR, the Agency of Science, Technology and Research that administers Biopolis medical research? That is where the government-funded stem cell research is done.

Divide that into five jurisdictions — and here are some leaders who are making change happen.

1. "At the **Genomic Institute of Singapore** (GIS), Ng Huck Hui, Paul Robson, Larry Stanton, Tara Huber and Bing Lim lead a major effort to understand [how] stem cells regulate self-renewal and differentiation [...], and how misregulation can lead to disease."

Zhiyan Fu: Archiving and retrieving vast amounts of scientific data.

Ed Holmes: Commuting from San Diego to Singapore.

2. "At **Bioprocessing Technology Institute** (BTI), Andre Choo and Steve Oh are characterizing [antibodies which] can eliminate undifferentiated [and therefore dangerous] stem cells prior to stem cell therapy."
3. "At the **Institute of Bioengineering and Nanotechnology** (IBN), Charlotte Hauser and Andrew Wan are engineering novel biomaterials to support stem cell […] differentiation."

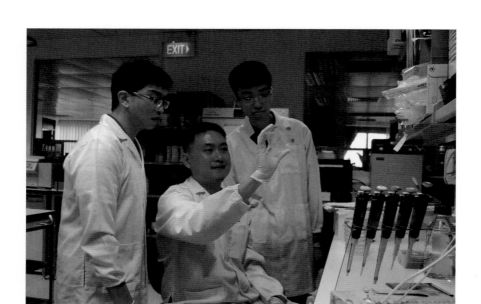

Jonathan Loh: Studying the changing genes of stem cell biology.

Paul Robson: How do stem cells regulate themselves?

4. "At the **Institute of Molecular and Cell Biology** (IMCB), Jonathan Loh conducts research into the epigenetic regulation [i.e., changing genes] of stem cell biology."
5. "The **Institute of Medical Biology** (IMB) has a major program on adult stem cells which supports skin biology. Basic research is carried out by Leah Vardy, Praba Sampath and Ray Dunn, while Sai Kiang Lim has been [working with] stem cells

Prabha Sampath: Trying to make wounds heal better.

in bone tissue. Nick Barker has identified the WNT target gene Lgr5 as a unique marker of adult stem cell populations in the intestine, skin, and stomach [...]. Simon Cool and Victor Nurcombe's [...] group has been developing heparin sugars [to help] wound healing."

One person, Jonathan Yuin Han Loh, caught my attention for a very unscientific reason: he has a great smile! A photo showed Dr. Loh with a beaming grin like Jackie Chan, whom I met once at a bookstore signing.

With Singapore having so many top scientists to choose from, I needed a place to start, so I looked up Dr. Loh to see what kind of success he was having and what his goals were.

As stated in *The Biopolis Story*, which was published to commemorate the tenth year of Biopolis, "Dr. Loh's research focuses on (what governs) the regenerative qualities of the embryonic stem cells."[2]

Loh and his colleagues had found a way to take one drop of blood and make stem cells.

He said, "It all began when we wondered if we could reduce the volume of blood used for reprogramming. Our fingerprick technique [...] used less than a drop of blood." Their reprogramming is a variation on the Yamanaka technique.

---

[2] Biomedical Research Council. *The Biopolis Story: Commemorating Ten Years of Excellence*. Singapore: A*STAR and Jurong Town Corporation; 2013.

Jackie Ying found a way to use green tea as a carrying device for stem cells.

Dr. Stuart Alexander Cook, Senior Consultant at the National Heart Centre Singapore and co-author of their study, said, "We were able to differentiate the [human induced pluripotent stem cells] reprogrammed from Jonathan's finger prick technique, into functional heart cells."[3]

There was also a picture of Loh and Dr. Huck Hui Ng, Executive Director of GIS, both with their arms full of trophies, awards they had just won.

Dr. Ng described his mission in seemingly simple terms: "What makes a stem cell? How [do we] make a non-stem cell a stem cell?" Of course, when he explained it, the simplicity went away.

"In general, transcriptional activation results from the tethering of sequence specific DNA binding transcription factors to their cognate sites…"

Sigh… If only scientists did not have to talk like… well, like scientists!

Looking at the two men side by side, something occurred to me. Singaporean scientists looked happy. There were one or two frowners, of course; that is inevitable. But most seemed glad to be working at their jobs, and proud of their country for helping them do it.

Some of their work was amazing, like using green tea to bring anti-cancer therapies to tumor cells. Dr. Jackie Yi-Ru Ying (another champion smiler!) was working on that,

---

[3] A*STAR scientists create stem cells from a drop of blood [Internet] 2014 Mar 20 [cited 2015 Feb 17]. Available from: http://www.a-star.edu.sg/Media/News/Press-Releases/ID/2582/ASTAR-Scientists-Create-Stem-Cells-From-a-Drop-of-Blood.aspx

when she was not busy being a Professor of the Department of Chemical Engineering at Massachusetts Institute of Technology, a job she has held since 2001.[4]

Baseball cap-wearing Zhiyan Fu looked like a computer programmer, which he was — taking vast amounts of scientific data, putting it into a cloud, and making it easily accessible.[5]

Researcher Wen Cai Zhang was attacking lung cancer. In some countries smoking may be viewed as a symbol of success, but lung cancer is the opposite of success.

Jingqiu Zhang had made a "model" of a premature aging disease, Hutchison's disease, as well as a heart condition, cardiomyopathy.[6]

Bing Lim was dividing his time between Singapore and Harvard, working to direct embryonic stem cell differentiation into cell types useful for therapy.[7]

Lawrence Stanton wanted to know how stem cells control cell fate decisions — and in so doing, trying to defend the brain against nerve-degenerative diseases like Alzheimer's.[8]

Wai Leong thinks big in the battle against cancer. As he puts it, "One of the grand challenges for cancer treatment is that current therapeutic strategies are ineffective due to [...] tumor recurrence caused by cancer stem cells (CSCs). The lab addresses this [...] by integrating the fields of cancer, CSCs, targeted therapy, and disease modeling, to translate biological findings [...] into innovative, targeted cancer therapies."[9]

Christopher Wong wants to make curing people profitable so it will attract investment. Calling himself a "technopreneur", he is trying to design a new business model by starting biomedical companies himself. One such company is making a pathogen chip to detect harmful viruses, while another develops therapies against various forms of cancer.[10]

Praba Sampath battles incomplete wound healing, and has come up with a molecular "on-off switch" to control migration of skin cells necessary for wounds to heal.[11]

---

[4] Ying Jackie Yi-Ru Internet] [cited 2015 Feb 17]. Available from: http://www.ibn.a-star.edu.sg/research_areas_7.php?id=1

[5] Dr. Zhiyan Fu [Internet] [cited 2015 Feb 17]. Available from: https://researchers.a-star.edu.sg/userprofile.aspx?userid=8319

[6] A human iPSC model of Hutchinson Gilford Progeria reveals vascular smooth muscle and mesenchymal stem cell defects [Internet] [cited 2015 Feb 17]. Available from: http://oar.a-star.edu.sg/jspui/handle/123456789/425

[7] Bing Lim, M.D., Ph.D. [Internet] [cited 2015 Feb 17]. Available from: http://www.gis.a-star.edu.sg/internet/site/investigator/investigator.php?user_id=30

[8] Lawrence Stanton, Ph.D. [Internet] [cited 2015 Feb 17]. Available from: http://www.gis.a-star.edu.sg/internet/site/investigator/investigator.php?user_id=27

[9] Wai Leong Tam, Ph.D. [Internet] [cited 2015 Feb 17]. Available from: http://www.gis.a-star.edu.sg/internet/site/investigator/investigator.php?user_id=172

[10] Christopher Wong, Ph.D., MBA [Internet] [cited 2015 Feb 17]. Available from: http://www.gis.a-star.edu.sg/internet/site/investigator/investigator.php?user_id=45

[11] A*STAR scientists discover "switch" critical to wound healing [Internet]. 2013 Mar 8 [cited 2015 Feb 17]. Available from: http://www.a-star.edu.sg/Media/News/Press-Releases/ID/1782/ASTAR-Scientists-Discover-Switch-Critical-to-Wound-Healing.aspx

Paul Robson, Senior Group Leader at the Genome Institute, has taken on a staggering research challenge: "To gain a full molecular understanding of the [...] mammalian blastocyst." How many diseases and disabilities would benefit from his success!

And then, just when I was wrapping up this chapter, I found an email from an unfamiliar source.

Dear Don,

Thank you for your e-mail. I was saddened to hear about your son.

I have been inundated with similar such requests for quotes, comments and interviews for books, magazines, monographs etc., but have declined such requests.

My entry into medical research was not for fame or fortune but for that great passion in searching for cures for the illnesses that plague mankind. I document my hard work in good journals and leave it entirely up to the reader to recognize its importance.

I therefore beg for your understanding in having to turn down your request for [...] an interview. However, below are two important publications of my early and current work for your reading.

Bongso A, Fong CY, Ng SC and Ratnam SS (1994). Isolation and culture of inner cell mass cells from human blastocysts. Human Reproduction, 9: 2110–2117.

Bongso A and Fong CY (2013). The therapeutic potential, challenges and future clinical directions of stem cells from the Wharton's jelly of the human umbilical cord. Stem Cell Reviews and Reports, 9: 226–240.

I wish you good luck with your book which I am sure will be useful and enjoyable reading material.

My warmest regards to you, your son and family.

Ariff

Professor Ariff Bongso PhD (Canada), MSc (Canada), FRCOG (United Kingdom), DSc (Singapore), DVM (Ceylon), DSc, FSLCOG (Sri Lanka)
Department of Obstetrics and Gynaecology
Yong Loo Lin School of Medicine
National University Health System (NUHS)
National University of Singapore (NUS)
1E Kent Ridge Road, NUHS Tower Block Level 12,
Singapore 119228

Singapore, for me, will always be a land of surprises — and smiles.

58 FIGHTERS AGAINST PARKINSON'S

Ann and Greg Wasson: Their marriage challenges Parkinson's disease, which has been described as "a slowly closing grave."

Is there an epidemic of Parkinson's disease (PD)? The world's population is growing older, and diseases associated with aging could therefore just be becoming more prevalent. All I know is that I can name too many people diagnosed with Parkinson's.

Muhammad Ali. Perhaps the greatest fight in boxing history was the "Rumble in the Jungle" match between Muhammad Ali and George Foreman.

Foreman today has a smiling image, like a favorite uncle. But back then he was like Frankenstein's monster. He ran up the sides of mountains, and lifted Joe Frazier off his feet with one punch.

Muhammad Ali? Denied competition three years for his refusal to support the Vietnam war, he looked soft, out of shape, and old. But he guaranteed victory.

In the first round, Ali was a whirlwind and was all over Foreman. But that was expected; he had amazing skills. The question was about endurance. Above all, the experts agreed, don't get caught on the ropes…

Ali ran *to* the ropes, leaning against them, and then he gestured to the giant, "Come on, big George!"

And Big George came, hardly believing his luck. Throwing punches with everything he had, he pounded Ali with tremendous blows that did internal damage; Ali urinated blood for days afterward. But he stayed right there and took it all. Head behind forearms, he absorbed or deflected every punch the giant had, trash-talking him all the way.

"Is that all you got? I thought you was tough! My grandma hits harder than that!"

In the 8th round, Foreman came out slower from his corner. The massive arms hung heavy. He seemed confused; indecisive.

This was the moment. Like a lion to the kill, Ali came off the ropes. Left-right-left, jolting snaps, setting him up — right fist collided with George Foreman's chin — and the giant went down.

Against all odds, Muhammad Ali was champion of the world once more. *Sports Illustrated* named him the Sportsman of the Century. Yet even this champion of champions was not immune to Parkinson's.

(The Muhammad Ali museum is an amazing and inspirational experience and is worth a visit: http://www.alicenter.org/.)

Robin Williams. When Gloria heard that the Robin Williams movie PATCH ADAMS was going to be filmed in the Bay Area, she hustled around and got us jobs as extras! We received $200 a day and free food for something we would have paid them to let us do.

I actually had a line (more accurately a grunt) in the picture. I played a butcher at a convention of meat-packers, and when Robin Williams' character shouted the question, "Do we like chicken?" we all had to answer, "NOOOOOOOO" like a cow, holding our fingers at the sides of our heads like horns. I repeated that noise about a hundred times, and it is actually in the movie.

But there was another reason we were there. Halfway through the shooting, Gloria grabbed my arm and dragged me through the crowd, saying, "Come on, come on!" I do not question my wife when she is in that mood, and stumbled after her until —

Robin Williams turned around.

Gloria had told him I wanted to talk about stem cells and he just said, "Sure!"

We had a half-hour chat about stem cell research. I knew he was a friend of Christopher Reeve, but had no idea of the depth of their friendship, nor that Christopher credited him with giving him a reason to live.

When the newly paralyzed Superman was seriously considering letting go of life, Robin Williams bounded into the hospital room. Acting like a crazy Russian proctologist, he snapped on a rubber glove and said, "Ve shall begin mit the rectal exam!" And

Christopher smiled. In that moment, he said later on, he knew somehow things would be all right.

I knew nothing of this, just that he was a friend of Christopher Reeve, and was portraying Patch Adams, a doctor who practiced unconventional medicine. Maybe we should talk. I told him about Roman and the state of paralysis research, as well as embryonic stem cell research; all the things you would expect me to say. He listened intently, asked intelligent questions, and never joked once.

"Whatever Christopher needs, I will make sure it happens," he said.

But the man who gave hope to a friend and laughter to the world would be stolen from us by emotional depression, undoubtedly made worse by his diagnosis with PD.

NOTE: The autopsy of Mr. Williams' brain showed indications of a disease called Lewy body dementia (LBD), very similar to Parkinson's (both involve a misfolding of a protein in the brain) but not identical. LBD is often "undiagnosed since it shares symptoms with better-known diseases like…Parkinson's…"

— "Autopsy" Robin Williams had Lewy Body Dementia," Joanna Rothkopf, Salon, November 13, 2014.

Greg and Ann Wasson. Pills can mask the symptoms of PD — the shaking and the sudden "freezing" where the body just stops — but the disease itself is vicious and incurable.

At the 2003 Congressional hearings on stem cell research, patient advocate Greg Wasson said:

"Currently, I take about 25 pills per day just for Parkinson's, and must re-dose every three hours. My medications […] allow me to sit here before you today, and speak, and be understood. Without them, I would be unable to walk, feed or clothe myself. But [they] do nothing to slow the progress of my disease. What you see […] is a 'chemical costume' I must put on every three hours."

"Parkinson's medications become less effective overtime […]. I now fluctuate 'off' my medications without warning several times each day. An 'off' fluctuation can leave me stranded at a mall, or in my living room, or at the movies."

"And for both my wife Ann and myself, the time will come when our medications fail us permanently. [We will] enter a twilight world of immobility, encased in our bodies as if in tombs, able to think but not speak, understand but not communicate. Eventually some complication of the disease will cause death, a death that by then maybe welcome."

I wish so much Greg and Ann Wasson could have had my 30 minutes time with Robin Williams. These are activists. Despite dealing with PD's symptoms every day, this married couple is a team. They fight, standing up for medical research at every chance.

Michael J. Fox. The world loves movie star Michael J. Fox, who raised awareness of PD to the world and answered the bullying anti-science tactics of Rush Limbaugh, while millions cheered. Not so many know the depth of his organization's financial contributions to research: in 2013, the Michael J. Fox Foundation contributed $67,429,325 in research grants.[1]

Joan Samuelson. Founder of Parkinson's Action Network and an original board member of the California stem cell program, Joan Samuelson reminds me of the late genius Buckminster Fuller's book, *I Seem to be a Verb*. To all who have seen her relentless struggle against the disease which shook her limbs but not her spirit, the name Joan Samuelson is a word for action.

Dianne Wyshack. For many years, she would hunt the internet for anything related to stem cell research, which she would then share with everyone. At first, I thought she was a foundation, because she did so much work! Dianne was always willing to do one thing more, or volunteer another sleepless hour. And now she is gone. We are diminished by losing her.

Because of fundraising efforts of people like these, PD is under scientific assault.

At the Buck Institute for Aging Research, Xianmin Zeng is leading the charge. With co-investigator Larry Couture of the City of Hope, and with collaborative funding from the state of Maryland, a way is being sought to defeat the disease.

According to Dr. Zeng, "PD is a devastating movement disorder caused by the death of dopaminergic neurons in the midbrain. These [make] dopamine a critical component of the motor circuit that ensures body movements are smooth and coordinated. [...] Human embryonic stem cells may offer a [...] source of the right kind of cell."[2]

He further adds, "These experiments may not only provide a blueprint for moving Parkinson's disease toward the clinic, but could also be a... blueprint for... stem cell therapy for (other) neurological disorders including motor neuron diseases and spinal cord injury…[3]

(To hear Dr. Zeng answer questions on stem cell research and PD, go to: http://www.cirm.ca.gov/our-progress/video/parkinsons-ask-stem-cell-expert-xianmin-zeng-buck-institute.)

Another approach might be to "rewrite the code" of a PD person's DNA. On a CIRM grant, Birgit Schuele of the Parkinson's Institute is working to develop a novel research tool to explore the molecular basis of PD.

She wants to develop a cell line from patients with PD. She will use the induced pluripotent stem cell technique, converting a skin cell to embryonic-like status and

---

[1] Eisner Amper Accountants & Advisors. Consolidated financial statements. The Michael J. Fox Foundation for Parkinson's Research; 2014 May 12. Available from: https://www.michaeljfox.org/files/2013-Audited-Financial-Statements-MJFF.pdf

[2] Banking transplant ready dopaminergic neurons using a scalable process [Internet] [cited 2015 Feb 18]. Available from: https://www.cirm.ca.gov/our-progress/awards/banking-transplant-ready-dopaminergic-neurons-using-scalable-process

[3] CIRM shared research laboratory for stem cells and aging [Internet] [cited 2015 Feb 18]. Available from: https://www.cirm.ca.gov/our-progress/awards/cirm-shared-research-laboratory-stem-cells-and-aging-0

then multiplying it. She will then remove the mutation which causes the disease. The Zinc Finger Nuclease technique will snip out the unwanted part of the DNA chain. The "healed" cells will then be injected into the brain.

Which approach will succeed? Both could win, and both could fail. The path to cure is crowded with roadblocks, and the greatest invention in the world can be stopped by an unforeseen detail.

Did you know the great writer Mark Twain went broke backing the invention of the typewriter? The version he spent all his money on was very good, almost the right one — but not quite. Someone else became rich off a slightly different version of the new typing machine, and Mark Twain had to make a world lecture tour to regain financial solvency.

But the California stem cell program raises the funding — so scientists can try.

News flash! As this is written, the *BBC News* is just reporting:

"Stem cells can be used to heal the damage in the brain caused by Parkinson's, according to scientists in Sweden [...]. The disease is caused by the loss of nerve cells in the brain that produce the chemical dopamine. [The scientists] converted human embryonic stem cells into neurons that produced dopamine. [These were injected into rats' brains, and] the damage was reversed."

These tests must be repeated by other scientists, of course, and eventually taken to human trials — but this is wonderful news nonetheless.

To our friends who are gone, we will fight on to honor your memory.

For those who work beside us still, let me share a note I sent to Rayilyn Lee Brown. PD has stolen her physical voice; she can no longer speak reliably. But her written words are strong and her fires remain undiminished. I wrote to the warrior-woman from Arizona, saying, in part:

"Please know that your personal efforts are making a difference. Generations yet unborn may never know your name; but there will one day be a path out of Parkinson's suffering, and you will have helped make it happen."

59 INTERNATIONAL FRIENDS

Brooke Ellison: Paralyzed in junior high, Brooke is now a college professor, and was the subject of a Christopher Reeve movie.

"You've got to see Niagara Falls while you're here," said my flight mate, as the jet drew closer to Toronto, Canada, "It is even more spectacular from the Canadian side."

"I couldn't without my wife Gloria," I said, "It is on our must-do-together list — plus I will be too busy at the stem cell conference!"

"Stem cells? We're going there too!" It felt like half the passengers in the airplane turned around.

Toronto made us welcome. Canada's magnificent newspaper, *The Globe and Mail*, did an in-depth coverage of the 2011 International Society for Stem Cell Research (ISSCR) convention.

The whole nation seemed to be paying attention, and well they should. In stem cell research, the country of the Maple Leaf is a major player. Maybe the rest of the nation

Brock Reeve is the Director of Harvard University stem cell program and brother of Christopher.

might be consumed with Stanley Cup hockey fever, but the Premier of Ontario, Dalton McGuinty, was attending a press conference on regenerative medicine.

Of course, he had to acknowledge the game:

"Tonight we will see if Canada's Canadian hockey players are better than America's Canadian hockey players," said the Premier at the press conference.

Premier McGuinty spoke of Canada's Stem Cell Network which has trained over 1,300 new stem cell scientists. In 2007, California Governor Arnold Schwarzenegger, Bob Klein of CIRM, and Premier McGuinty had met to create the international Cancer Stem Cell Consortium.

And in 2011, Dr. Mick Bhatia of McMaster University converted skin cells directly into blood cells. Think what that miracle could mean for the always short supply of usable blood.

The whole convention was like that, surprises built on the hard work of researchers.

There was also controversy.

New York had two substantive stem cell programs. The New York Stem Cell Foundation was privately funded and was led with elegance and dignity by Susan Solomon. Ms. Solomon mentioned the program's decision *to pay women who donate eggs for research.*

Bravo, New York! This is wonderful. When a woman donates eggs to help another couple have a child, she can be paid a significant amount of money, something like $5,000 or more. But if she wants to donate eggs to help research which might save lives and ease suffering — for that, she cannot be paid? That does not make sense to me. California should learn from New York on this important issue.

Joe Riggs is the founder of Students for Stem Cell Research, now in 19 countries.

But I wanted to hear more! had heard that a powerful advocate named Brooke Ellison was involved in that controversy. Christopher Reeve had made a movie about her, the *Brooke Ellison Story*, written by Camille Thomasson. Brooke was a Junior High School student who was suddenly paralyzed from the neck down — but when she regained consciousness from the accident, her first words to mother Jean were: "Please don't let me be left back in school!" (And she did indeed graduate on time!)

Today the determined Ms. Ellison is Assistant Professor at Stony Brook University, teaching courses on health policy, leadership, ethics, and stem cell research. She is a World Economic Forum Young Global Leader.

In a personal memo, Brooke told me, "I was a member of the Ethics Committee of the Empire State Stem Cell Board from its inception until 2014. [We implemented] progressive and scientifically-friendly policies, [like] providing women with compensation for donating eggs necessary in this research [...]. This is a basic justice issue. It is compensating women, in the very same way they are compensated for egg donation in fertility procedures, for their time and burden in facilitating potentially life-saving research [...]. New York is leading the way; other states should follow."

The second New York program was the Empire State Stem Cell Program, with the acronym NYSTEM, and it had a $600 million funding commitment over 11 years, which was spectacular.[1]

---

[1] NYSTEM. Empire State Stem Cell Board strategic plan. 2008 May. 44 p. Available from: http://stemcell.ny.gov/sites/default/files/documents/files/NYSTEM_Strategic_Plan.pdf

Sally Temple: One of New York's finest: Scientist, fund raiser, and co-founder of New York's Neural Stem Cell Institute.

One of their top scientists, Lorenz Studer, has been working for more than a decade to find a way to restore the dopamine cells taken away by Parkinson's disease. It is my unscientific opinion that he is is getting close. How wonderful it might be to see an end to this disease in our lifetime.

New York scientists also developed what might be the world's first patient-specific ALS cell line, a huge breakthrough in Lou Gehrig's disease, so the progress of the disease can be followed in a Petri dish instead of standing by helplessly as a terminally ill person leaves this earth too soon..

New York was investigating a stem cell model of type 1 diabetes as well — and there are 26 million diabetics in America! Financially, a young person with diabetes today faces "average annual medical expenses [of] $9,061, compared with $1,468 for youths without the disease."[2]

As usual, it was a struggle to understand what the scientists were saying. It was logical they would have a "secret language" for communication among themselves — but big words and Latin are hard for the rest of us, and they need to be able to talk "people talk" as well as "science speak". Or else, like auto-mechanics, they will be masters of a world the rest of us do not understand, and probably will not support. They need to listen to the National Public Radio Show, *Klick 'n Klack: The Tappet Brothers*, who make auto-problems clear, even fun!

My favorite presenter? Shosei Yoshida of Japan, who spoke about sperm stem cells in mouse testicles. This was not a subject to which I had previously given much thought.

---

[2] Herzog K. Yearly medical costly for kids with diabetes. *Journal Sentinel* [Internet]. 2011 Aug 28 [cited 2015 Feb 18]. Available from: http://www.jsonline.com/news/health/120933004.html

Professor Shinya Yamanaka, Nobel Laureate.

But if our field had an impersonator, someone who could imitate scientists, Shosei Yoshida would be high on his or her to-do list. The man was animated, enthusiastic, full of the joy of helping.

A question dogged my mind. "Drosophilia ovary"? Johns Hopkins' Daniella Drummond-Burbosa broached the subject, discussing the "control of stem cells by [...] systemic factors in the *drosophila ovary*". I turned to the scientist beside me, and asked:

"Why does that matter? I mean — drosophilia — those are flies, right?"

"A fly's insides are easier to understand," she said, and their life cycle is short "If we want to know how to reprogram cells, drosophilia, lets us go back to the beginning, figure out how each step works."

Information overload: One reporter said, "I wish I understood what I just learned." The confusion of undigested new knowledge was like a huge shipment of canned foods arrives at the back of a store; it takes a while to sort them onto their proper shelves. I did not know what I knew!

Sometimes it was just so amazing...

Could we cure paralysis by making a hollow spinal cord column — and stuffing it with stem cells? Robert S. Langer, a professor at the Massachusetts Institute of Technology with more than 1,100 published articles and 760 issued and pending patents, spoke of working with top embryonic stem cell scientists to do just that.

The result? Formerly paralyzed monkeys now galloped on a treadmill on the movie screen before us. There were gasps from the audience.

Greatness of the future? Robert Blelloch received the ISSCR Outstanding Young Investigator Award for his work on the signals regulating both embryonic and induced pluripotent stem cells.

A room full of junior researchers heard successful scientists talk about how to get grants.

I was able to chime in about the California stem cell program, suggesting new scientists should make the CIRM website their homepage, to find out what new grants were coming.

In the halls and on the elevator, you never knew who you might be talking to.

"Here for the stem cells?" I asked one distinguished-looking early-riser. He said yes, and we shook hands.

"I'm Rudy Jaenisch," he said.

"Whoa!" I said, and jumped back a little — then I shook his hand again and thanked the stem cell pioneer for doing so much to help the field. He was speaking on "Stem Cells, Pluripotency and Nuclear Reprogramming".

At the "meet the experts luncheon" you could sit with a favorite scientist, eat a boxed lunch, and ask anything. I got to sit with Margaret Goodell of Baylor College of Texas, and what a cheerful charmer she was. In addition to being an expert on hematopoietic stem cells, she was someone you could not help but like — I was glad she is in our field!

Even Shinya Yamanaka was there. The Nobel Prize winning Japanese scientist was ready to respond to several scientific articles that were critical of his new method. Everything boiled down to the comparison between embryonic and induced pluripotent stem cells, and he talked about it for half an hour

Because I always sit next to the microphone, I was ready for the announcer to call for questions. I also knew everyone else in the room had something to ask. Halfway through the magic word "Question", I bounced up as from a trampoline and was talking before the echo died.

"Is there a difference in the *immune properties* of induced pluripotent stem cells compared to embryonic?"

And he said… something. I thanked him, and sat down to think about what I thought I heard.

If I understood his remarks correctly, there were differences in the immune properties, both good ones as well as "bad". Many embryonic stem cells are like many induced pluripotent stem cells, he said, and vice versa. A good difference was that the immune response might be less severe because the cells were made from the patient's own body. More research needs to be done, he said.

He invited everyone to come to Japan, scheduled to host the ISSCR meeting the following year, despite the tidal wave and nuclear power plant disaster which had so riveted the world's attention.

"Join us in Japan! It is safe, beautiful, and you will like the food!" he said, as the audience burst into laughter and applause.

There was Freda Miller, who was fighting paralysis with skin cells, which may be turned into useful nerves. Elly Tanaka was studying how a salamander regenerates its severed spinal cord by first growing a living "tube" for the cord to grow inside. Amy Wagers of Harvard University spoke on regenerating muscle for the aged.

And speaking of Harvard University, Brock Reeve walked by. Naturally I had to jump up and run over and shake the hand of the director of Harvard University's stem cell program, the brother of Christopher Reeve.

It was a hall of fame for research for cure. Present were Fred "Rusty" Gage of Salk Institute and Elaine Fuchs, chair of the ISSCR itself — everybody in the hall deserved a book written about them.

The exhibit halls? Huge. Two phone book-sized volumes described thousands of posters.

I walked up to two Chinese scientists, and asked them where was the most stem cell research conducted in China, and they said, "Beijing, Shanghai, and Guangdong." One of them, Gang Li, is now in Mountain View, California, close to where I worked. We have a mutual friend, Deepak Srivastava of Gladstone, who worked on turning heart attack scars into useful tissue.

Hans Keirstead! The Canadian and California scientist was his usual beaming self, there to discuss another piece of groundbreaking research. He, Tanya Watt, and Gabriel Nistor were presenting work on stem cell therapy aimed at spinal muscular atrophy (SMA) trying to save the lives of children who might otherwise die before the age of two. I remembered my little friend Pranav, and Gwendolyn Strong, and their battles with SMA.

Passing by was George Daley of Boston's Children's Hospital, a great scientist and communicator. He could be a convention by himself! I asked to sit at his table during lunch, but the scheduler just laughed — Daley's table was booked solid before the conference began.

The hall of biomedicine was a continual surprise. For instance, Kawasaki, the famous Japanese motorcycle company, had gone into biomedicine.

Or Xyclone, with their laser technology, which may help somatic cell nuclear transfer (SCNT, sometimes called therapeutic cloning) become a practical reality. SCNT involves removing the nucleus of a woman's egg, and putting the patient's bit of skin inside. Set these things in a dish of salt water and provide an electrical shock, and then wait five days — the mixture can be taken apart for stem cells. But there was a problem. it seemed like every time the needle went into the egg, it would collapse. A method of making an ultra-small hole in the egg is needed, and that might be what the Xyclone laser can do.

Companies and conversations, theories and therapies, champions right now and those soon to be: international friends advancing the cause of regenerative medicine —

It was like touching tomorrow, today.

60 A TEXAS MIRACLE, OR THIRTY-TWO

Joe and Nina Brown: Texan fighters for research freedom.

During my hitch in the military, I spent a year at Fort Wolters, Texas, not far from Mineral Wells. It was swampland, poison snake country, and the locals told of a water-skier who was killed when his ski-tips hooked into a tangled mass of mating water moccasins. It was said he did not even scream, being bitten so many times. They dragged him out and he was dead on the spot.

But the people who lived there seemed unconcerned. It was just one more danger for them, something to deal with, like being careful crossing the street. I admired their courage, though I prefer my snake-less little portion of northern California.

As a medical research advocate too, it takes guts to work in the Lone Star State.

Consider the Cancer Prevention and Research Institute of Texas (CPRIT), the three billion dollar program inspired by Texan Lance Armstrong. Some folks trash talk Armstrong

Beckie McCleery and Judy Haley: Founders of Texans for the Advancement of Medical Research (TAMR).

now because he used performance-enhancing drugs in bike racing. But to me, that is like a traffic ticket compared to the lives he may have saved through advocating cancer research.

If I wore a hat, I would take it off, not only to Lance but to all the patient advocates of Texas.

Let me tell you a story about one of those groups and a woman called Nina Brown.

Nina (pronounced like the number) had Parkinson's disease. Some nights it got so bad she had to crawl to reach the bathroom.

But tough? There is an advocate saying, "If you want something done, hire a Parkinson's disease person." Some advocate groups shy from political confrontation, but not Parkinson's disease folks. They are ready for the fight. It might be partly because of the condition itself. Parkinson's disease is progressive, so every day it can get a little worse, and you never get used to it. People like Nina Brown may suffer the agonies of the damned — but they don't give up.

There was an anti-research bill proposed which would have criminalized embryonic stem cell research in the state of Texas: including CPRIT, the Lance Armstrong effort.

But Nina Brown and her husband Joe (ignoring a severe heart condition), with Beckie McCleery, Judie Haley, Ralph Dittman, Melinda and Wayne Rose, and other folks formed a group called Texans for the Advancement of Medical Research (TAMR). They made a video describing and supporting the research which the anti-science bill would ban.

They organized letter-writing campaigns, set up meetings with local disease groups, sought out reporters, held hearings with legislators' aides, on and on.

And when the fighting had to be done face to face, that was all right too. Our side would present the case against the bill even after a 160-mile drive to the State Capitol, a grueling trip for a person with Parkinson's.

The opposition was the Republican Religious Right — and they had a way to control the debate.

The bills were scheduled to be heard in a certain order. But when TAMR requested to speak, Republican leadership changed the order of the bills, inserting others before the stem cell bill, dragging out the waiting period.

One bill was on religion in the classroom, and the Republicans summoned school children, encouraging them to talk about their religious views. Some of the kids talked a half hour each.

This tactic had been used before.

On a separate bill about vaccinations, also opposed by the Religious Right, a woman with stage four cervical cancer wanted to speak. The Republicans made her wait an extra three hours, even though she was bending over in agony.

And Nina Brown? They made her wait *ten hours*. Nina was so exhausted that when her time came to speak — at two in the morning — she had to be carried to the microphone.

But she spoke — as did Judy Haley and Beckie McCleery, mothers of children with type 1 diabetes — and they prevailed. When they tell that story today, the TAMR folks always give credit to a courageous Republican, Bob Eisler, who stood with them that day.

TAMR had no budget. But they did have a super-lobbyist, Ellen Arnold. In one year, Ms. Arnold had dissected as many as 7,000 Texas bills, looking for those with an anti-research component. Of these, only the most dangerous could be fought. There was one year the Religious Right pushed 32 bills with anti-research components, and TAMR defeated every single one.

How could TAMR afford Ellen Arnold, a professional lawyer capable of charging several hundred dollars an hour? They couldn't. Ms. Arnold worked many hours, days, and weeks without pay. A shrewd businesswoman, she was not!

But in the history of medical research, Ellen Arnold's name stands tall.

Advocates help each other, too.

Once I was assisting a New Jersey stem cell research bill which would have provided $750 million for research. They needed some money for a television advertisement. I contacted Texan Beckie McCleery, asking her to make a call to somebody rich to help New Jersey — and she came through. (We lost the New Jersey bill by a tactical error, doing it during a non-Presidential year. In the off-years, Democrats tend to get lazy and their turnout may be down 20%.)

So when Beckie asked me for the names of scientists to provide written testimony in support of embryonic stem cell research, I got out the phone book, and sent her a dozen.

One of them, Dr. Larry Goldstein of the University of California at San Diego, was not content with providing written testimony. He hopped on a plane and testified in person.

And when the anti-research folks tried to insert a no-embryonic stem cell research clause in Lance Armstrong's Texas anti-cancer bill, TAMR was there. The clause was not inserted.

While writing this book, I reached out to Beckie McCleery, who co-founded TAMR. I found her at her and Judie Haley's unique place of business, the 50–50 Pharmacy. Half the profits they make go to medical research, which amounted to over $12 million in the past 20 years.

Joe Brown has passed on, and Nina is poorly. TAMR has gone quiet. Other groups like David Bales' Texans for Stem Cell Research are leading the charge now. They are cheerful, active and determined.[1]

But Beckie and Judie kept TAMR alive. They pay the fees, fill out the forms, and keep their records intact and updated. They are ready. If there is need, you can find them on Facebook: https://www.facebook.com/pages/Texans-for-Advancement-of-Medical-Research-TAMR/140246622663587.

And that is why, from that day to this, embryonic stem cell research is still legal in the Lone Star State.

---

[1] Executive team [Internet] [cited 2015 Feb 17]. Available from: http://www.txstemcell.org/content_5.html

61 THE STEM CELL MUSKETEERS OF BRAZIL

Stevens Rehen, advanced both practical and political aspects of stem cell research.

Mayana Zatz, patient advocate extraordinaire.

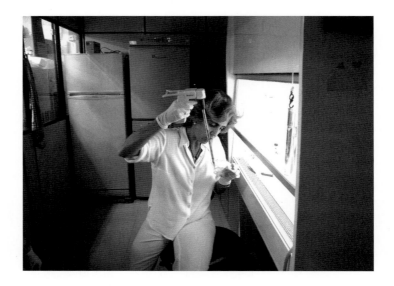

Lygia Pereira: Advancing stem cell research in Brazil.

When I first heard that Brazil had signed a letter of agreement with CIRM for research cooperation, I could hardly believe my ears. Brazil is such a religiously conservative country; how could they achieve the scientific freedom they would need?

I wrote to a friend, Dr. Lygia Pereira from São Paulo, who described "the road to legalizing embryonic stem cell research in Brazil" as "bumpy" — which was an understatement.

In 2005, her letter stated, the Biosafety Law approved in Congress allowed the use of in vitro fertilization embryos for research, but only if they had been frozen for more than three years. This was in itself was an improvement over the previous law, which flatly forbade embryonic stem cell research.

However, Brazilian law "guarantees to every Brazilian the right to life". If an embryo could be considered a life, with full personhood rights under law, embryonic stem cell research could be declared unconstitutional.

A lawsuit was filed, against the science.

Enter the three musketeers: Lygia Pereira, Stevens Rehen, and Mayana Zatz.

Like the fictional characters of Alexandre Dumas — Athos, Porthos, and Aramis — the three musketeers of Brazilian stem cell research had different abilities. All three were outstanding scientists first and foremost, but they also had individual strengths.

Dr. Pereira is a pioneer, the kind of scientist who takes giant early leaps of progress, on which others build. She is someone who will act rather than wait for others to go first. She wrote three books, five book chapters, and many papers. She began working with stem cells in 1992 and developed a mouse model for a condition called Marfan's syndrome, a potentially fatal disease.

Dr. Rehen is a visionary who looks at the situation now, and asks: how can we do it better? One improvement that may revolutionize the field is his method of increasing the number of stem cells. The old way was to put embryonic stem cells into a dish of feeder gel, and gently tap and shake the container. This "passaging" helped develop the stem cell colonies, but was awkward and time-consuming. Instead, Dr Rehen developed an automatic spinning method, which should be called the "Rehen Spin". It is reportedly seven hundred times as efficient as the older method — and for only half the cost.

And the third Musketeer? Mayana Zatz is the ultimate patient advocate. Having visited the shanty-towns of Brazil, she saw children with muscular dystrophy born to families who could not afford to give them *any* medical treatment, not even a wheelchair. She fought for them, organizing and raising funds for the Brazilian Association of Muscular Dystrophy. And when they requested to be in the court so the judges could see who was affected by a research ban, it was Dr. Zatz who made it happen.

The Catholic Church in Brazil has huge influence. "They used misinformation — 'adult stem cells are just as good as embryonic stem cells', 'embryonic stem cells cause tumors and thus should be prohibited', etc., instead of the truth, which was that their reasons were religious opinion, not scientific arguments," said Dr. Pereira.

The three scientists made hundreds of presentations. Everything was on the line. If they won, embryonic stem cell research could go forward. If they lost, it stopped.

When the Supreme Court argued the case, the courtroom was packed with patients and their families, invisible no longer.

The final vote? Six to five — in favor of allowing the research. Six to five... Without the efforts of Pereira, Zats, and Rehen, the stem cell musketeers, it is safe to assume there would be no embryonic stem cell research in Brazil today.

Instead, Brazil has taken its rightful place in world science. President Luz Ignacio Lula da Silva's previous approval was validated — and the scientists went back to their work.

Mayana Zatz is currently a member of the International Human Genome Project of the Academy of Sciences of the State of São Paulo, and of the Brazilian Academy of Science.

Her work has been cited 1,500 times in 102 publications, and she has had approximately 150 articles published in magazines such as *Nature Genetics* and *Human Molecular Genetics*. Mayana is also a columnist for *Brazilian Veja Magazine*.[1]

Steven Rehens is a full professor at the National Laboratory for Embryonic Stem Cell Research at the Federal University of Rio de Janeiro.

According to his profile on the Pew Fellows directory, Dr. Rehen's work is directed toward the generation of stem cells as disease models, with an emphasis on mental disorders. His research goals include the scale-up of stem cells in spinner flasks to facilitate high-throughput screening.

FLASH! As this is written, Brazil's Stevens Rehen has just taken "… an important step toward the implantation of stem cell-generated neurons as a treatment for Parkinson's disease. Using as FDA-approved substance for treating stomach cancer, Rehen and colleagues were able to grow dopamine-producing neurons derived from embryonic stem cells that remained healthy and functional (in) mice, for as long as fifteen months, restoring motor function (muscle control) without forming tumors."
— http://www.sciencedaily.com/releases/2015/04/150403104223.htm.

In 2008, Dr. Lygia Pereira established Brazil's first embryonic stem cell line, BR-1.

Her team has cultured the Harvard University stem cell lines, using those cells as a model for study. As the Brazilian government began investing in cell therapy, her team received a grant to establish lines of human embryonic stem cells — and in 2008, Dr. Pereira reported the generation of her first line, BR-1. Since then, she has established four additional lines, and has provided cells and training to several other groups in Brazil.[2]

Dr. Pereira said, "When we reported […] the first Brazilian line of [human embryonic stem cells], the public reaction was all positive."

Long live the three stem cell musketeers of Brazil! May they never need to take up their legislative swords again.

---

[1] Mayana Zatz. *Wikipedia* [Internet] [updated 2014 May 23; cited 2015 Feb 18]. Available from: http://en.wikipedia.org/wiki/Mayana_Zatz
[2] Member spotlight on Lygia V. Pereira, Ph.D. [Internet] [cited 2015 Feb 18]. Available from: http://www.isscr.org/visitor-types/members/member-spotlight/isscr-member-spotlight/2014/11/03/member-spotlight-on-lygia-v.-pereira-ph.d

62 ADVENTURES IN CHINA

Dr. Haidan Chen: Why is a returning student scientist called a sea turtle?

One of the joys of writing a book is the astonishing people you meet on the way.

Like Joy Yueyue Zhang, with whom I Skype-chatted with after reading her book, *The Cosmopolitanization of Science: Stem Cell Governance in China*, which includes interviews with Chinese stem cell scientists, some of whom I later contacted with her help.

Like Professor H. W. Ouyang, the Director of the School of Biomedicine at Zhejiang University, who may have a revolutionary stem cell answer to knee pain. When the meniscus, which is the cushion between the leg joints, wears down, the pain can be crippling; my wife Gloria plans her trips up the stairs in our house carefully, to spare her knees.

But what if that internal knee cushion could be restored? Professor Ouyang states that his group "initiated the clinical use of cartilage-derived progenitors for knee joint treatment."

Why did he get into stem cell research?

Wise Young: Co-inventor of methylprednisolone, and author of international clinical trials for paralysis, Dr. Young is genuinely beloved by the paralysis community.

Professor Ouyang explained, "I am an orthopedic surgeon [and] there are no satisfactory therapeutics for [cartilage-based] diseases. It is an uncomfortable experience when, as a doctor, you cannot provide hope to hopeless patients."

Dr. Haidan Chen explained to me what a "sea turtle" was, in terms of Chinese medical research.

Joy Yueyue Zhang, author of "Cosmopolitanization of Science: Stem Cell Governance in China".

The sea turtle as an animal lives in the wild ocean most of its life, but then returns to the place of its birth.

The Chinese character for the sea turtle, it seems, resembles that of a student who goes overseas and then returns. As Dr. Chen puts it, "Overseas education and training is crucial for the development of Chinese life sciences." When these students return to the motherland equipped with expertise in the chosen subject, they can teach the next generation.

Former Premier Deng Xiao Ping's open door policy has been called the "Spring of Science" because it encouraged Chinese scientists to go to other countries to learn and return. Sensibly, China makes the return attractive, offering strong salaries and laboratory set up inducements.

But what would it be like to go to China as an outsider, and try to set up human trials?

Sixty-ish, heavily-bearded, and with a smile that warms the room, Chinese–American Wise Young has fought to cure paralysis for four decades. The first book with even a glimmer of hope for paralysis cure was *Quest for Cure* by Sam Maddox — and Wise Young was featured in that.

When Roman was first paralyzed, he was given a steroid called *methylprednisolone*, which kept down the swelling of the spinal cord, preventing some of the additional internal damage that follows the accident. Wise Young was the co-inventor of methylprednisolone, to this day the only medication universally prescribed for newly paralyzed individuals.

Wise's website, CareCure.org, is not only a great source of tips on dealing with spinal cord injury from people who are living the experience, but there is also a paralysis-trained nurse to answer questions, and often Wise himself, who apparently does not find sleep necessary.

But lately, Dr. Young has been hard to find. He no longer attends all the dozens of paralysis research meetings up and down the East Coast.

Has he lost interest in spinal cord injury paralysis? Hardly.

First, his mother has been having memory lapses and requires care. So Wise commutes from New Jersey to China (!) to care for his Mom six weeks at a time, alternating with his brothers.

"She has taken up a new hobby — drinking beer!" he jokes, "But only two cans a day, and it helps her sleep. I am learning to cook; maybe I should write a cookbook, *Recipes for Mom*!"

But there is another reason Wise Young has racked up *two million* frequent flier miles.

In 1996, Chinese gymnast Sang Lan broke her neck and became paralyzed at the Goodwill Games. After the tragedy, she stayed in America for a year of rehabilitation. Her father flew in from Shanghai and reached out to Wise, whom *Time* magazine had named "America's Best" in spinal cord injury research.

"He asked me the question every good father would ask: is there a cure? I had to say no, but we are working on it," said Wise.

The gymnast and the scientist became friends. When Sang Lan took her first trip outside the walls of the hospital, it was to Dr. Young's research lab.

But when they sat together at a farewell dinner before Sang Lan went back to China...

"Suddenly she burst into tears," said Wise, "I asked her why, and she said, if the cure comes in America, how would it reach her in China?"

"I told Sang Lan, the only way there will be cures is after there are human trials. And I promised her I would bring human trials for paralysis to China."

The next ten years were filled with unrelenting struggle. China, of course, has no intention of allowing her people to be experimented on for the benefit of outsiders!

But Wise Young spoke on the benefits to the Middle Kingdom: "No country needs spinal cord injury cure more than China. Roughly one-third of the world's paralyzed people are Chinese."

Speaking both Mandarin and Cantonese, Wise looked for ways to meet Chinese scientists to build support for nationwide clinical trials.

He found an unexpected ally. The giant pharmaceutical corporation Pfizer, wanting China to purchase methylprednisolone, paid for Wise's travel expenses to talk to Chinese scientists — and paid the Chinese scientists themselves to come and listen!

"Over a four year period," Wise estimates, "I met minimally 2,000 spinal cord injury doctors, essentially every spinal cord injury surgeon in China. In 2002, I met Suzanne Poon, the great philanthropist from Hong Kong. She has a paralyzed son, and asked me if there was a cure. I said no, not yet, but could we work together?"

Since the only way the Chinese government would allow clinical trials would be an absolute guarantee that no patients would be hurt, a fully safe therapy had to be found.

So what approach did Wise Young choose to try and benefit paralyzed people?

Stem cells derived from cord blood, plus orally administered lithium (a drug used for psychiatric purposes, intended to ease manic depression) — and a special operation, intradural decompression surgery, which cleans out dead matter from between a person's vertebrae.

Dr. Achim Rosemann, an anthropologist who studies scientific progress across the globe, quotes Dr. Young saying: "… the discovery of the intradural decompression is a Chinese story, and there is great pride in this. […] And lithium was discovered in Hong Kong University. It is not […] something discovered in the U.S., but [was] discovered by one of our centres."[1]

But just as the Wise Young multi-site study's planning was almost complete, a major scandal occurred.

The head of the Chinese Food and Drug Administration (CFDA) Zheng Xioyu was accused of corruption, taking money to approve drugs. He was arrested, put on trial, convicted — and shot. His own family had to buy the bullet that executed him.

Even though this had nothing to do with spinal cord injury, it put a damper on the advancement of medical science.

For Wise Young, it meant years of delay, shuttling back and forth between various government agencies.

He said, "We […] met senior officials of the Ministry of Health. One department head told us that our (clinical) trial was not the responsibility of his department and referred us to another. That department in turn said that it wasn't their responsibility and referred us to the CFDA. We wrote to the CFDA — [who referred us to] the Ministry of Health!"

But the project did not end. The Chinese military took them under their wing — one of their hospitals is the legendary Hospital #328, where Mao Tse-Tung, the founder of Modern China, was cared for as he lay dying. What happened? What follows is Wise Young's interpretation.

Years ago, Deng Xiao Ping, former Premier of China, had fallen into disfavor, and had been exiled to the island of Xian. His son, Deng Fu Puan, fell (or was pushed?) off a third-story building — and broke his back.

All during his exile of 1969, Deng Xiao Ping was caregiver for his paralyzed son. The experience made a deep impression. When Deng Xiao Ping returned to power, the top levels of government seemed to feel an obligation to fight paralysis. Today, Chinese spinal cord surgeons are among the world's most expert.

The next hurdle was to make the clinical trials themselves in line with the best in the world. To accomplish this, the doctors at every participating hospital were trained in an identical manner, so there would be no difference from one hospital to the next.

One test was of the intradural operation, cleaning out dead tissue from between the vertebrae. Of 20 people who were treated, Wise said, 60% are walking, some with assistance, some without. About the same number achieved independence in bowel and bladder control." Lithium turned out to be a pain-reliever, reducing paralysis nerve pain on a permanent basis.

---

[1]Rosemann A (2013). Scientific multipolarization: its impact on international clinical research collaborations and theoretical implications. *Sci Technol Soc* **18** (3): 339–359.

Today, Dr. Young is raising money for a giant set of clinical trials, involving not only China, but also India, Norway, and the U.S. The tests will decide the efficacy of the combination of lithium, umbilical cord stem cells, and the intradural decompression spinal operation — plus strenuous rehabilitation.

Will this combination therapy be the full answer to paralysis? Is umbilical cord blood a sufficient source of stem cells? Will lithium increase the power of the cord blood cells? Only time will tell.

The idea of multi-hospital participation, where there is no additional expense because it would be considered part of the hospital's normal work, is potentially game-changing. Dr. Young estimates that doing trials in China can be done for *ten percent* of what it would cost in the States. Add to that the pain relief from lithium, and the intradural spinal cord injury operation —

And these are gifts from China to the world.

63 PROFESSOR FOREVER AND THE GIANT SQUID

Tom Piccot and the giant squid, 1873: Artist unknown.

October 26, 1873. The fog was heavy on Portugal Cove, Newfoundland, Canada.

Twelve-year-old fisherman Tom Piccot, his father Theophilous, and their neighbor Daniel Squires were rowing out to check on their codlines when the mist parted.

Something huge and unknown floated half-submerged before them. It might be a dead whale: tow it ashore and it could be boiled down for oil, meat for the dogs, and bones for fertilizer; nothing is wasted in Newfoundland.

The elder Piccot sank a boat hook into the body… of the sleeping giant squid, which woke.

Huge black eyes blazed green. Lashing arms clutched the wooden boat — and tipped it down. Water flooded over the stern, and the two older men shook hands, accepting their time to die.

But the boy, Tom Piccot, had other ideas. He snatched up the bait axe and attacked the giant squid. With two blows he chopped off two tentacles where they wrapped around the boat.

As a spider yanks its limbs back from the fire, the creature recoiled — and shot away, backwards, down into darkness, gone.

Tom Piccot sold one of the tentacles to a local minister, who donated it to the British museum. It resides there to this day. You can see the actual dried arm coiled in a circle and numbered #28 on Wikipedia's list of specimen images at http://en.wikipedia.org/wiki/List_of_giant_squid_specimens_and_sightings (note the descriptive phrase "hacked off").

Researching for a book I wrote about this real life Kraken,[1] I visited Portugal Cove.

At nearby St. John's museum, scientists showed me the preserved carcass of a giant squid, *architeuthis dux*, king of the cephalopods. Even in death, the black beak could still emerge by reflex, twisting and snapping, like the monster in the movie *Alien*.

I looked up the Piccot families. There were about a dozen in the phone book, but at last I heard a cheerful voice say, "Oh, sure, tha's me great-grand-dad!" Gloria and I stayed a week in the village, interviewing anyone who would sit still.

But it surprised me how unimpressed Canadians were about the amazing story that had taken place in their own backyard. (I'd have put a life-size statue of the boy and the squid in the town square!) Remarks were very casual: "Oh, I heard summat about it, years ago…"

And so it is with Canadian stem cell research. Judged by accomplishments, the Maple Leaf country should have stem cell research as its second national sport (nothing can replace ice hockey), or at least have it as part of the educational syllabus (for a good text, I recommend *Dreams and Due Diligence: Till and McCulloch's Stem Cell Discovery and Legacy* by Canada's Joe Sornberger). But as the Canadian Stem Cell Foundation noted, "For 50 years, Canadians have been at the forefront of stem cell science, yet received shockingly little support."[2]

Whenever Bob Klein spoke before a Canadian audience, he always thanked the nation which purified insulin, which keeps his diabetic son Jordan alive. If time permitted, he would mention Ray Rajotte and James Shapiro who established the Edmonton Protocol, retrieving pancreases from corpses to fight diabetes.

But there are just so many top Canadian researchers!

Even the briefest mention of champions would have to include Alan Bernstein, Allen Eaves, Mick Bhatia, Derek Van der Kooy, Freda Miller, Guy Sauvageu, Keith Humphreys, Tim Kieffer, Peter Dirks, Michael Rudnicki, Charles LeBlond, Samuel Weiss, John Hassell, Mark Freedman, Andy Becker, Henry Friesen, Andras Nagy, and on and on.

If the gathering was about cancer, Bob could show respect to Canadians like John Dick, credited with being first to detect cancer stem cells in human leukemia, or Connie Eaves, best-known for her work in blood stem cell biology but whose career also furthers breast cancer cure.

But medical history can be foggy as an October morning on Portugal Cove.

---

[1] Reed DC. *The Kraken*. Honesdale: Boyds Mills Press; 1997.
[2] About — Canadian Stem Cell Foundation [Internet] [cited 2015 Feb 19]. Available from: http://stemcellfoundation.ca/en/the-foundation/about/

Even the Canadian insulin victory by F. G. Banting and Charles Best was controversial in its time. Half of the 1923 Nobel Prize money was given to the man whose laboratory it was, J. J. R. Macleod, with not even a mention of Charles Best. This infuriated Dr. Banting, who donated half his award to Best, whereupon Macleod divided his share with a fourth member of the team, J. B. Collip.[3]

And the beginnings of stem cell research? Every breakthrough depends on the contributions of others; we all stand on the shoulders of the past. If you ever visit the Chicago Air Museum, in the section on the Wright Brothers, notice the wall behind the hanging planes. You will see letters from other aeronautic pioneers, offering advice to Wilber and Orville Wright.

In 1960, James E. Till and Ernest A. McCulloch, both of the University of Toronto, were studying the effects of radiation.

As recollected by James, while dissecting an irradiated mouse, McCulloch said, "Hmm. Um — this is interesting…"[4]

On the mouse's spleen, he had found little bumps, nodules, *apparently started by just one cell.*

That single cell changed the world.

In 2005, McCulloch and Till won the Lasker Award,[5] often called the American Nobel Prize, crediting the scientists for "ingenious experiments that first identified a stem cell — the blood-forming stem cell — which set the stage for all current research on adult and embryonic stem cells."

Had the two Canadians been the first to identify a stem cell?

The Canadian Stem Cell Foundation thinks so. On their website, it is stated that "Dr. James Till and Dr. Ernest McCulloch [stunned] the scientific world when they [discovered] transplantable stem cells at the Ontario Cancer Institute in Toronto. Stem cell science in Canada, and around the world [was] born."[6]

The *New York Times* called Dr. McCulloch "a father of the stem cell research that scientists say holds promise for the treatment of many ailments."[7]

A differing opinion is offered by University of California Davis stem cell researcher and author Paul Knoepfler, who described the invention of stem cell research as a "group effort".[8]

---

[3] Bliss M. *The Discovery of Insulin*. Chicago: University of Chicago Press; 1982.
[4] Till JE, McCulloch EA (1961). A direct measurement of the radiation sensitivity of normal mouse bone marrow cells. Radiat Res 14: 213–222.
[5] Strauss E. Albert Lasker Basic Medical Research Award [Internet] [cited 2015 Feb 19]. Available from: http://www.laskerfoundation.org/awards/2005_b_description.htm
[6] Canada's contribution [Internet] [cited 2015 Feb 19]. Available from: http://stemcellfoundation.ca/en/about-stem-cells/canadas-contribution/
[7] Altman LK. Ernest McCulloch, crucial figure in stem cell research, dies at 84. *The New York Times* [internet]. 2011 Feb 1 [cited 2015 Feb 19]. Available from: http://www.nytimes.com/2011/02/01/health/research/01mcculloch.html?_r=0
[8] Who really discovered stem cells? The history you need to know [Internet] 2012 Apr 11 [cited 2015 Feb 19]. Available from: http://www.ipscell.com/2012/04/who-really-discovered-stem-cells-the-history-you-need-to-know/

Dr. Knoepfler noted there had been talk about "stamzelles" as early as 1868, and mentioned the scientist and categorizer Ernst Haeckel,[9] who coined such terms as "ecology" and "phylum", as well as a phrase many high school students have memorized: "ontogeny recapitulates phylogeny".

Other scientists, like Theodore Boveri, speculated about the existence and function of stem cells, "describing a capacity of stem cells for self-renewal as well as differentiation".[10]

The effect of stem cells in bone marrows had been noticed in 1942, during radiation experiments in the development of the Atomic bomb.[11]

(For more information on the term "stem cell", see Miguel Ramalho-Santos and Holger Willenbring's excellent article, On the Origin of the Term "Stem Cell", in *Cell Stem Cell*.[12])

Here is Dr. James Till, as quoted in the Canada Science and Technology Museum:

"We are known as the 'fathers of stem cell research' because we proved the existence of stem cells and characterized them. This significantly altered the landscape of cell biology, especially cancer research. In 1961, we published a paper which proved the existence of stem cells. In 1963, we defined the two key properties of stem cells, [namely] the capacity for self-renewal [and differentiation]. We could not have done this without the contributions of Dr. Andrew Becker and Dr. Louis Siminovitch [...]. What made the biggest difference in our research was that we switched from focusing on what stem cells looked like to what they can do. Our contribution was to explore what they can do — which we did." (Dr. Becker and Dr. Siminovitch are still researching at age 90!)

The work of previous scientists cannot be denied. But it is one thing to point out a miracle, and another to revolutionize its use. Scientists believed the world was round long before Columbus, but it was the shipmaster's voyage to America that unflattened the perception of the earth.

McCulloch and Till, giants of the North, brought the research to the world.

But the best is yet to come. On June 18, 2008, there was a meeting between California Governor Arnold Schwarzenegger, Canadian Minister of Health Tony Clement, CIRM President Alan Trounson, and CIRM founder and board chair Bob Klein.

"This is an historic agreement. Canadian researchers have long been at the forefront on stem cell research. Now we are working across borders, bringing together the best minds

---

[9] Ernst H. *Wikipedia* [Internet] [updated 2015 Feb 18; cited 2015 Feb 19]. Available from: http://en.wikipedia.org/wiki/Ernst_Haeckel

[10] Maehle A-H (2011) Ambiguous cells: the emergence of the stem cell concept in the nineteenth and twentieth centuries. *Notes Rec* **65** (4): 359–378.

Also c.f. Lancaster C (2009). The conceptualisation of the stem cell: the influence of the spleen colony assay by JE Till and EA McCulloch. Unpublished M.A. dissertation. Durham University.

[11] Kraft A (2009) Manhattan transfer: lethal radiation, bone marrow transplantation, and the birth of stem cell biology, ca. 1942–1961. *Hist Stud Nat Sci* **39**: 171–218.

[12] Ramalho-Santos M, Willenbring H (2007). On the origin of the term "stem cell." *Cell Stem Cell* **1** (1): 35–38.

from both countries [...] to prevent and treat cancer for the benefit of all our citizens," said Tony Clement.

"Collaborations such this [...] have great potential in improving the lives of not only Californians, but people around the world," said Governor Schwarzenegger.

CIRM's press release stated: "The Canadian partners [will] make an initial investment of more than CAD 100 million in the collaboration, with Genome Canada, the Canadian Institutes of Health Research, and [...] the Ontario Institute for Cancer Research having already confirmed commitments..."

I picked up the phone and called Canada. Every interview pointed me in new directions.

One familiar name popped up repeatedly, and him not even a Canadian — Bob Klein. In 2007, Bob was appointed to the Board of Directors of Genome Canada. In 2008, he joined the board of the Ontario Institute for Cancer Research.

And who nominated him for both these positions? Calvin R. Stiller, an amazing man, has dedicated his life to the fight against chronic disease. In fundraising alone, he may have gathered as much as 250 million Canadian research dollars.

Tragedy influenced him early. As a boy, Calvin Stiller witnessed his father's slow and painful death by kidney failure. As a man, Stiller began and presided over the first clinical trials of kidney transplants, so that others could avoid the pain and suffering his Dad had experienced.

A scientist himself, Dr. Stiller was the principal investigator on a Canadian multi-site study on Cyclosporine, important in preventing the body from rejecting transplanted cells and organs. He published 265 scientific papers, co-edited five book chapters, and authored a book, *Lifegifts*, about organ donation. Stiller is considered the voice of Canada's drive for organ donation.

He began and led four successful biomedical companies and four venture funding companies, employing thousands and advancing research. He started the Canadian Medical Hall of Fame, the Ontario Institute for Cancer Research, the Multi-Organ Transplant Service, and much more.

The University of Western Ontario named him Professor Emeritus, an honor which means literally "professor forever".

Scientist Henry Friesen was once asked what Stiller's contribution to Canadian research was, and his answer was almost poetic.

"Go outside and look around; what you see will be his gift."

I contacted Dr. Stiller in September, 2014 about the beginnings of the Canada–California Cancer Stem Cell Coalition. You will not be shocked by whom he brought up

"My friend Bob Dynes, former President of the University of California, and a Canadian himself, had told me about Bob Klein, a man who does not recognize obstacles, only opportunities. I sat next to Bob at a meeting of Genome Canada, and at break we went outside and just started talking about how something should be done, and could be done. California and Canada have about the same population, and they share a problem-solving attitude. When the conversation was through, we knew something would happen — it felt like building a train track between California and Canada, and the first train car was going to be cancer."

Today, Canadian champions Tak Wah Mak and John Dick are working on major cancer projects with California leaders Catriona Jamieson and Dennis Slamon. Two teams, four projects!

But the last word should come from one of the greatest Maple Leaf scientists of all.

"Many of the most important discoveries [on cancer stem cell research] have been made by Canadian and Californian researchers. An international collaboration involving Canada and California […] can be expected to raise research in this field to a much higher level."

— James Till, co-recipient of the Lasker Award with Ernest McCulloch

Canadian champions James Till and Ernest McCulloch receive the Lasker Award for their stem cell discoveries.

64 A Double Baker's Dozen of Disease Team Grants?

Want to stay on track toward clinical trials? Former Senior Vice President for Research and Development Ellen Feigal gives valuable tips in a video: https://www.cirm.ca.gov/our-progress/video/stem-cell-clinical-trials-staying-critical-path-workshop-intro-ellen-feigal-cirm

As much as $20 million each, disease team grants are the most spectacular research projects offered by the California stem cell program. How do they work? First, the principal investigator (PI) applies for a small grant ($50,000 or so) saying roughly what he or she wants to do. If approved, that grant is used by the PI to choose a dream team of top scientists. The hopeful team figures out (and the PI proposes) a multimillion dollar attack on a chronic disease — *a therapy they must bring to human clinical trials within just four years.*[1]

At every step of the way, milestones of progress must be met or the money stops. Some of the projects are listed below, and I am grateful to Dr. Ellen Feigal, formerly Senior Vice-President of Research and Development of CIRM, for her comments on the status of these projects:

---

[1] Search all CIRM grants [Internet] [cited 2015 Feb 19]. Available from: http://www.cirm.ca.gov/grants

Alzheimer's disease

A top scientist wanted to transplant neural stem cells directly into the hippocampus portion of the brain, hoping to markedly restore memory for patients.[2]

Status: It was a great experiment, and may well succeed another time, possibly with a different source of cells. But it did not meet a milestone, and lost its funding.

Blindness

Mark Humayun, from the University of Southern California, estimated that by 2020, over 450,000 Californians will suffer from vision loss or blindness due to age-related macular degeneration, which is the loss of a layer of cells at the back of the eye. This can be overcome by transplanting new cells derived from embryonic stem cells.[3]

Status: This therapy is advancing to the clinic.

Blindness

Henry Klassen, from the University of California at Irvine, hopes to use progenitor cells to treat retinitis pigmentosa, a severe disease of the eye. In so doing, Klassen hopes to save the light-sensing cells of the eye. An application will be made to the FDA, and following approval, a small number of patients with severe retinitis pigmentosa will be injected with cells in their worse-seeing eye.[4]

Status: This treatment is advancing to the clinic.

Brain tumors

Karen Aboody, from the City of Hope, has built a stem cell bank to make her prospective attempt at destroying brain tumors widely available. Her techniques may also remove solid tumors throughout the body, which is applicable to other deadly growths.[5]

Status: This technique is advancing to investigational new drug filing and requesting FDA for clinical trials.

---

[2] Restoration of memory in Alzheimer's disease: a new paradigm using neural stem cell therapy [Internet] [cited 2015 Feb 19]. Available from: https://www.cirm.ca.gov/Grant/restoration-memory-alzheimers-disease-a-new-paradigm-using-neural-stem-cell-therapy

[3] Stem cell based treatment strategy for age-related macular degeneration (AMD) [Internet] [cited 2015 Feb 19]. Available from: https://www.cirm.ca.gov/our-progress/awards/stem-cell-based-treatment-strategy-age-related-macular-degeneration-amd

[4] Use of retinal progenitor cells for the treatment of retinitis pigmentosa [Internet] [cited 2015 Feb 19]. Available from: https://www.cirm.ca.gov/Grant/retinal-progenitor-cells-treatment-retinitis-pigmentosa

[5] Stem cell-mediated therapy for high-grade glioma: toward Phase I-II clinical trials [Internet] [cited 2015 Feb 19]. Available from: http://www.cirm.ca.gov/content/stem-cell-mediated-therapy-high-grade-glioma-toward-phase-i-ii-clinical-trineurological-diso

Cancer

Dennis Slamon, from the University of California at Los Angeles, points out that a minor population of cancer stem cells drives the growth of an entire tumor. Even if a cancer appears gone, these trigger-like cells can bring back the cancer. Slaymon believes his lead drug can inhibit the growth of cancer stem cells.[6]

Status: This treatment is advancing to the clinic.

Cancer

Judith Shizuru, from Stanford University, hopes to enable chemotherapy-free transplants with the use of an antibody. The technique might also save the lives of children with an immune-system defect, known as severe combined immune deficiency (SCID), patients who often die before the age of two.[7]

Status: This treatment is advancing to the clinic.

Cancer

Antoni Ribas, from the University of California at Los Angeles, seeks to reprogram stem cells in hopes of redirecting the patient's immune response to specifically attack the cancer.[8]

Status: This treatment is advancing to the clinic.

Critical limb ischemia

John Laird, from the University of California at Davis, hopes to benefit two million Americans at risk of leg amputation or death. A key problem associated with ischemia is low oxygen level in the limbs. Laird states that stem cells can migrate into the areas having the lowest oxygen levels, wrap around the damaged blood vessels, and secrete helpful factors where needed most — essentially a stem cell hug of life![9]

Status: This treatment is advancing to the clinic.

---

[6] Therapeutic opportunities to target tumor initiating cells in solid tumors [Internet] [cited 2015 Feb 19]. Available from: http://www.cirm.ca.gov/content/therapeutic-opportunities-target-tumor-initiating-cells-solid-tumors

[7] ibid.

[8] Genetic re-programming of stem cells to fight cancer [Internet] [cited 2015 Feb 19]. Available from: http://www.cirm.ca.gov/Grant/genetic-re-programming-stem-cells-fight-cancer-0

[9] Phase I study of IM injection of VEGF-producing MSC for the treatment of critical limb ischemia [Internet] [cited 2015 Feb 19]. Available from: https://www.cirm.ca.gov/Grant/phase-i-study-im-injection-vegf-producing-msc-treatment-critical-limb-ischemia

Diabetes

ViaCyte, Inc., is challenging diabetes using embryonic stem cells to build beta cells, which have been shown to cure diabetes in mice and rats. The effort could become the most significant stem cell-based medical treatment of the coming decade.[10]

Status: This treatment is advancing to the clinic.

Heart attack

Rachel Smith of Capricor, Inc., wants to use stem cells to turn heart attack scars, which make a second heart attack more likely, back into healthy tissue.[11]

Status: This treatment is advancing to the clinic.

Heart attack

Eduardo Marban, of Cedars-Sinai Medical Center, states that the adult human heart contains cardiac stem cells that are able to partially repair the heart following a heart attack. His team has developed a way to isolate these cells and grow them into large numbers. Clusters of these cells can be injected into the heart, even in the middle of a heart attack.[12]

Status: This treatment is advancing as an allogeneic to the clinic.

Heart failure

Robert Robbins, from Stanford University, intends to use embryonic stem cell-derived heart cells for patients with end stage heart failure, which enables the transplanting of cells as an alternative to transplanting the whole heart.[13]

Status: This treatment is advancing to the clinic.

---

[10] Cell therapy for diabetes [Internet] [cited 2015 Feb 19]. Available from: https://www.cirm.ca.gov/our-progress/awards/cell-therapy-diabetes

[11] Allogeneic cardiac-derived stem cells for patients following a myocardial infarction [Internet] [cited 2015 Feb 19]. Available from: https://www.cirm.ca.gov/Grant/allogeneic-cardiac-derived-stem-cells-patients-following-a-myocardial-infarction

[12] Autologous cardiac-derived cells for advanced ischemic cardiomyopathy [Internet] [cited 2015 Feb 19]. Available from: https://www.cirm.ca.gov/our-progress/awards/autologous-cardiac-derived-cells-advanced-ischemic-cardiomyopathy

[13] Human embryonic stem cell-derived cardiomyocytes for patients with end-stage heart failure [Internet] [cited 2015 Feb 19]. Available from: https://www.cirm.ca.gov/Grant/human-embryonic-stem-cell-derived-cardiomyocytes-patients-with-end-stage-heart-failure-0

HIV/AIDS

Irving Chen, from the University of California, Los Angeles, believes that the process which can change the color of a flower (through RNA interference) may also block the spread of HIV infection. Right now, surviving with AIDS requires several dozen pills a day. But Chen's method may lead to therapies that will require only a single treatment.[14]

Status: Chen is focused on early translational work. His co-investigator at Calimmune is being funded by CIRM for the clinical trial.

HIV/AIDS

John Zaia, from City of Hope, may have found a way to duplicate the "cure" of AIDS patient, Timothy Brown, who received a bone marrow transplant with a delta 32 mutation on the CCR5 gene. This mutation apparently blocks entry of the AIDS virus. It may be possible to alter a patient's cells so that they can block the virus.[15]

Status: This treatment is advancing to the clinic.

Huntington's disease

Vicki Wheelock, from the University of California at Davis, hopes to fight the fatal disease with stem cells as a delivery vehicle: bringing a nerve-growth factor called brain-derived neurotrophic factor to the at-risk nerve cells.[16]

Status: This treatment is advancing to the clinic.

Leukemia

Dennis Carson and Catriona Jamieson of the University of California at San Diego are fighting an after-treatment problem with the deadly disease; it often leaves small amounts of leukemia stem cells still alive which come back stronger later. The Carson team hopes to make anti-leukemia drugs available soon.[17]

Status: This treatment is advancing to the clinic.

---

[14] HPSC based therapy for HIV disease using RNAi to CCR5 [Internet] [cited 2015 Feb 19]. Available from: https://www.cirm.ca.gov/our-progress/awards/hpsc-based-therapy-hiv-disease-using-rnai-ccr5

[15] Therapeutic opportunities to target tumor initiating cells in solid tumors.

[16] MSC engineered to produce BDNF for the treatment of Huntington's disease [Internet] [cited 2015 Feb 19]. Available from: https://www.cirm.ca.gov/Grant/msc-engineered-produce-bdnf-treatment-huntingtons-disease-0

[17] Development of highly active anti-leukemia stem cell therapy (HALT) [Internet] [cited 2015 Feb 19]. Available from: https://www.cirm.ca.gov/our-progress/awards/development-highly-active-anti-leukemia-stem-cell-therapy-halt

Leukemia

According to Irv Weissman of Stanford University, the body's protective cells, macrophages, should "eat" the leukemia invaders. Unfortunately a marker cell called CD-47 on the leukemia cells acts as a "don't eat me" sticker. His team strives to remove the marker, which would allow the body's natural immune system to do its job.[18]

Status: This treatment is advancing to the clinic.

Lou Gehrig's disease

Larry Goldstein, from the University of California at San Diego, hopes to prevent the progression of amyotrophic lateral sclerosis by using astrocytes, which are nerve support cells, developed from embryonic stem cells.[19]

Status: The team is working on early translational components.

Lou Gehrig's disease

Clive Svendsen, from Cedars-Sinai Medical Center, wants to use nerve "fertilizers", which are more technically known as glial-derived neurotrophic factors (GDNF), to treat the disease. Neural progenitor cells can serve as "Trojan horses" to carry the GDNF to the nerve cells.[20]

Status: This treatment is advancing to the clinic.

Muscular dystrophy

Stanley Nelson, from the University of California at Los Angeles, is tackling the muscle-wasting disease. Nelson may bring repairs directly inside each muscle cell to lessen the disease severity.[21]

Status: The team is working on early translational components.

---

[18] Development of therapeutic antibodies targeting human acute myeloid leukemia stem cells [Internet] [cited 2015 Feb 19]. Available from: https://www.cirm.ca.gov/our-progress/awards/development-therapeutic-antibodies-targeting-human-acute-myeloid-leukemia-stem

[19] Stem cell-derived astrocyte precursor transplants in amyotrophic lateral sclerosis [Internet] [cited 2015 Feb 19]. Available from: https://www.cirm.ca.gov/our-progress/awards/stem-cell-derived-astrocyte-precursor-transplants-amyotrophic-lateral-sclerosis

[20] Progenitor cells secreting GDNF for the treatment of ALS [Internet] [cited 2015 Feb 19]. Available from: https://www.cirm.ca.gov/Grant/progenitor-cells-secreting-gdnf-treatment-als

[21] Combination therapy to enhance antisense mediated exon skipping for duchenne muscular dystrophy [Internet] [cited 2015 Feb 19]. Available from: https://www.cirm.ca.gov/Grant/combination-therapy-enhance-antisense-mediated-exon-skipping-duchenne-muscular-dystrophy-0

Osteoporosis

Nancy Lane, from the University of California at Davis, wants to strengthen the bones of the elderly, and thereby lessen frequency and severity of bone fractures. In osteoporosis, bones can grow almost transparently thin and are all too easily broken. Lane has developed a small molecule that directs stem cells to grow the skeleton stronger.[22]

Status: This treatment is advancing to the clinic.

Sickle-cell disease

Donald Kohn, from the University of California at Los Angeles, offers a stem cell therapy for the agonizing and often fatal disease. Kohn's approach will transplant adult cells that are genetically corrected by the gene that blocks the sickling of red blood cells.[23]

Status: This treatment is advancing to the clinic.

Skin disease

Alfred Lane, from Stanford University, hopes to use induced pluripotent stem cells to combat a hideous disease, dystrophic epidermolysis bullosa, in which the skin literally rips apart. His method may benefit other skin disorders as well.[24]

Status: The team is working on early translational components.

Spinal cord injury

Jane Lebkowski, now of Asterias Biotherapeutics, is continuing the paralysis work she previously led at Geron, Inc. This is the famous embryonic stem cell research begun by Hans Keirstead with a grant from the Roman Reed Spinal Cord Injury Research Act.

Status: Clinical trials were initiated in 2010 after testing that was conducted in more than 20 non-clinical studies. Initial clinical safety testing was conducted in five human subjects, with no safety concerns observed.

---

[22] Treatment of osteoporosis with endogenous mesenchymal stem cells [Internet] [cited 2015 Feb 19]. Available from: https://www.cirm.ca.gov/Grant/treatment-osteoporosis-with-endogenous-mesenchymal-stem-cells

[23] Stem cell gene therapy for sickle cell disease [Internet] [cited 2015 Feb 19]. Available from: https://www.cirm.ca.gov/our-progress/awards/stem-cell-gene-therapy-sickle-cell-disease

[24] iPS cell-based treatment of dominant dystrophic epidermolysis bullosa [Internet] [cited 2015 Feb 19]. Available from: https://www.cirm.ca.gov/our-progress/awards/ips-cell-based-treatment-dominant-dystrophic-epidermolysis-bullosa

Stroke

Gary Steinberg, of Stanford University, identifies stroke as the number one cause of disability, the second leading cause of dementia, and the third leading cause of death in adults. His team will use embryonic stem cells to make neural stem cells in order to augment the body's repair process after stroke.[25]

Status: This treatment is advancing to the clinic.

Twenty-five teams of California's top scientists, almost a double baker's dozen, working to defeat some of the world's most terrible diseases and disabilities — what can one say, except:

Good luck — and Godspeed!

---

[25] Embryonic-derived neural stem cells for treatment of motor sequelae following sub-cortical stroke [Internet] [cited 2015 Feb 19]. Available from: https://www.cirm.ca.gov/our-progress/awards/embryonic-derived-neural-stem-cells-treatment-motor-sequelae-following-sub

65 THE GREATEST SPEECH YOU NEVER HEARD

Bob Klein: "Seize the moment!"

It is said that Lincoln's greatest speech was never written down; that the reporters present were so caught up in it, they depended on somebody else to take notes — and nobody did. (Others have said the reason the speech was "lost" was that it was so powerful and incendiary that Lincoln himself agreed that it should not be reprinted during his campaign.) We are reasonably certain the speech was delivered on May 29, 1856, in Bloomington, Illinois, and that it contained the first use of the famous words: "A house divided against itself cannot stand."[1]

The greatest speech I ever heard was on the last day of the 2011 World Stem Cell Summit. I am sure its scheduling was intended to be the climactic finish of the gathering, but what happened instead was that many scientists were already in taxis, rushing for the airport.

---

[1] Lincoln's Lost Speech. *Wikipedia* [Internet] [updated 2014 Dec 31; cited 2015 Feb 19]. Available from: http://en.wikipedia.org/wiki/Lincoln's_Lost_Speech

Bob Klein's greatest speech — a message about bringing cures faster to a suffering world — was given to a half-empty audience.

But I have heard it nine times — because it was videotaped. Bernie Siegel posted virtually the entire convention online, available free. CIRM recorder Beth Drain very kindly transcribed it.

It has a somewhat forbidding title: "A New Governmental and Philanthropic Paradigm for Funding Stem Cell Research." ("Paradigm" means a new model, a new way of thinking.)

I have shortened it substantially; editing errors (and parenthetical additions) are mine.

Bob Klein

"[…] The current financial crisis, […] I expect, will crush the traditional funding model for medical research. At very best we may be able to maintain current levels of funding, but these will inevitably decline with inflation.

"The fight for resources has never been more difficult.

"But in California we have Proposition 71, which of its original $3 billion has committed $1.3 billion approximately, as well as attracting a billion in matching funds. It will have enough funding to take it actively through mid-2017, and meet funding commitments till 2020.

"In 1955, it was estimated that by 2005 it would cost one hundred billion dollars a year just to keep victims of polio alive in iron lungs, in hotels designed only for that purpose.

"[That cost] has been avoided. There was an investment in intellectual capital, [the Salk vaccine], to cure those patients or avoid the disease, and that has been a long-term benefit.

"We should be issuing bonds that spread the cost of such intellectual capital development over a long benefit period, [and] not trying to fund it on the backs of the current taxpayers alone.

"As we all know, it's a decade before you get the beginnings of the benefit of the investment, […] which is why I set up the first five years of California's Proposition 71, so there were no general fund payments [required].

"[…] We really need [to] spread the cost over the period that receives the benefits.

"[…] In 2006, Gordon Brown and Bill Gates […] launched a program called the International Facility for Financing Immunizations (IFFI). They were looking at (how much it would cost to defeat) tuberculosis in the underdeveloped world. They issued $5 billion in bonds through the World Bank.

"[…] The World Bank issued the bonds up front. [The nations helped pay for it, but the] cost was spread over 20 plus years.

"The value to […] nations of spreading their cost over a large number of years is immense.

"[…] By looking outside our current funding models, we might be able to find […] a way to expand by billions of dollars or tens of billions of dollars, a new funding resource.

"The capital markets [...] are hugely risk-averse.

"[...] One of our problems in fully engaging biotech or Big Pharma is that the capital model for a stem cell therapy that [cures] a disease breaks the current model of a business plan.

"Look at long-term cancer therapies. [...] They are a long-term income stream.

"[...] If you come up with a new approach to kidney disease that cures the kidney disease, [how can you make that financially attractive to investors? Also] how are you going to be able to pay all of that (research and testing) cost up front?

"How many diseases can [any nation's budget] make massive front-end payments on?

"[If you wanted to develop] stem cell transplants for 40 diseases, what's that going to do if it has to be a front-end payment?

"[...] There won't be enough [financial resources]. You are going to get severe rationing or limitations on what conditions can be addressed. That is not our goal, [which is] to treat diseases with the best science that humanity can bring to bear.

"[...] We can look to the Department of Defense, which in the United States has a cost savings formula they did experiments with. They essentially said to defense contractors, if you save a percentage of this new military system, we will share with you in those savings.

"Now, think about the implications of that. Ignore for a moment the fact that they limited the savings to a very short period of years and they capped it at such a small amount of money, that the transaction costs for most military contractors weren't worth dealing with it.

"But redesign the program so it works. If you say to biotech, the federal government will approve contingent compensation contracts, you decide what disease you're going to treat, and how you're going to treat it.

"If you [...] cure the disease, we will give you a percentage of the savings for 30 or 40 years.

"What happens? You've just unleashed hundreds of billions of dollars. Why can't the Congress do that right now? Why can't they just appropriate the money? Because we have a pay-go congress.

"Essentially a pay-go congress means you have to cut a cost to make an appropriation, or you have to raise taxes to make an appropriation — which I think is not too popular.

"[Instead, we would be] signing federal contracts that are subject to the performance of the company on reducing medical costs. In kidney disease, for example, is the stem cell therapy providing a repair, so the individual doesn't have to be on dialysis? [We need a measureable way to prove a therapy is lowering costs.]

"Can we create a compensation system where companies can take these contracts and, as soon as they start to see positive results, [...] they can start borrowing against them? And as the population using it increases, they can borrow more. What does that do for the company?

"[...] The company could borrow long-term at a very low rate, [instead of] extraordinarily expensive venture capital up front.

"What does that mean? [...] Now they can take a risk on a broader number of diseases. Now maybe they can take a risk on ALS (Lou Gehrig's disease) which is a small population. Now maybe they can risk a bigger portfolio *because they have a business model for stem cell research.*

"Right now they don't have that business model, and there're a lot of skeptics out there [who ask], What are you going to do? Charge a million dollars [for one person's] therapy? Who's going to pay that up front? How can any medical system afford that for a broad range of diseases?

"[We have to] create a business model that can really incentivize business with hundreds of billions of dollars of potential payments spread over 30–40 years, spread over the benefit period [...]. Those generations that are getting the benefits of improved quality of life and/or lower cost are paying that out of a portion of the savings.

"I suggest to you that getting the current Congress of the United States to raise taxes to fund more research is not probable. But [they may be receptive to] a program where you can take a percentage of the savings and pay it to a company under a long-term contract [to] create a therapy that will reduce human suffering and save the government money.

"Currently the business model for biotech [is] long-term therapy; [i.e., sick but alive, paying for medicines for many years.]

"There are a lot of great people in biotech who really want to cure people, but we put them into a quandary, because their economic interests are rewarded if they create a long-term therapy, not if they create [a cure].

"So we need a business model that really motivates them toward cure, on a large scale. Under this model we've re-aligned the interest of biotech to be the same as the patients and scientists, the same as the government. The ultimate goal is to cure [...] or at least to substantially mitigate.

"[...] Expanding the bond model, [like] California's Proposition 71, or Texas' $3 billion bond finance program for cancer research, or the Third Frontier program in Ohio, which included biotech funding; that is an important goal. But we have to go the extra mile and look at how we're going to get therapies to commercialization.

"Philanthropists and donors: as we create the new business model, and expand the bond financing to spread the cost of research over the generations that benefit, we will find that we've leveraged an entire new group of financial donors. [Why? Because this] leverages their money [...].

"(If) they can really leverage their money — then the donors can put up major dollars: [...] quantum levels larger donations than they're doing now, and with the confidence that they are driving a legacy investment that can improve the human condition.

"In today's economy, donors are skeptical. Some philanthropists don't believe the downstream dollars will be there. Donors are cutting their commitments.

"So(how did we attract major dollars) in that challenging climate; where did the *billion dollars in matching funds* come from for CIRM's facilities? (CIRM provided $272 million for new labs; outside sources donated enough to increase that amount to $1.15 billion.)

"[That money] came because donors saw a model that would take them through the stage of human trials. [Companies could] pick up these therapies and carry them forward to patients.

"To the extent that we create [a new business model] to really cure or substantially mitigate disease, we also changed the game for donors — and for the seed money so critical up front for high risk experiments, for brilliant new ideas.

"So how are we going to get the message out there? How do we mobilize the public to support biotech, and new medical research models?

"We have a remarkable task in front of us.

"But, with Proposition 71, we handled a very tough subject. We put scientists on the television. We mobilized the patient advocates.

"Clearly, it was impossible to pass six billion dollars of funding authorization: three billion dollars for the research and three billion dollars for the interest on the bonds over 35 years. Clearly it was impossible, but the patient advocates and the scientists got together and it happened.

"Scientists and advocates [...] are leaders in the stem cell revolution. [...] We must get the scientific communities in every media market, in every state, in *every country* to reach out.

("For you as a scientist, or advocate): don't wait for someone to approach you and ask for an interview. You've got to go to the media yourself, and educate them on science.

"We have to aggressively engage the media [...] so there is a broad public understanding of the value of stem cell research to every family and every child in this country — or we're going to get run over by this financial crisis.

"And that engagement needs to start — yesterday. Because we all have 'promises to keep, and miles to go before we sleep.'

"We are the hope of an entire generation. We are the hope that in this narrow window of opportunity, a revolution in medical care will not be crushed by an economic cycle.

"Because we, you and I, have children and families and people to whom we would give our every breath to rescue from suffering — *suffering that may within a decade prove to be largely [...] unnecessary.*"

"Thank you."[2]

---

[2]Robert N. Klein videos [Internet] [cited 2015 Feb 19]. Available from: http://bobkleinpublic-policyprofile.com/videos/

“ A STEM CELL MYSTERY: THE RESIGNATION OF MAHENDRA RAO

Mahendra Rao...One of the most brilliant scientific minds of our age... why did he resign from his position as Director of the National Institutes of Health's Center for Regenerative Medicine?

Here is a stem cell mystery. Something very strange happened which I do not fully understand to this day. See if you can make heads or tails out of it.

On March 28, 2014, Mahendra Rao resigned from his position as Director of the National Institute of Health's Center for Regenerative Medicine (CRM). (Note: CRM is *not* CIRM, the California Institute for Regenerative Medicine, and has no connection to the California stem cell program.)

(Funded at $52 million over 7 years, the CRM was a federal program, apparently designed to fund grants for induced pluripotent stem cell (iPSC) research. I say apparently, because it only funded one major project, and then shut its doors.[1]

The highly respected Dr. Rao had no trouble finding employment after the Center's closure, and is now on the boards of no less than four biomedical efforts: Cesca Therapeutics, the New York Stem Cell Foundation, Stemedica, and Q Labs.

I scoured the internet, but all I found was confirmation of confusion; i.e., nobody else really seemed clear as to what was going on either.

---

[1] Reardon S. NIH stem-cell programme closes. *Nature* [Internet]. 2014 Apr 8 [cited 2015 Feb 19]. Available from: http://www.nature.com/news/nih-stem-cell-programme-closes-1.15004

For instance, there was a report on April 11, 2014, by Zachary Brennan for *BioPharma-Reporter.com*. Here is an excerpt of that report:

CRM was established in 2010 to centralize the NIH's stem cell program. [...] Its goal was to develop useful therapies from [iPSCs and] shepherd them toward clinical trials.

Rao took the helm in 2011. Relations [soured] due to NIH's decision to fund only one project [...] to move iPS cells to clinical trials. Rao said he resigned after this became clear. He [hoped for at least] five trials.

James Anderson, who administered the CRM, [explained that only one application] received a high enough score [to justify funding (!)]. Anderson says the CRM will not continue in its current form. "The field is moving so fast we need to rethink [...]."

Anderson [says: "It is just smart science.] If something's not on track you don't keep spending money on it. This is a fast-evolving area of science. NIH decided to step back and reassess what the field needs..."

In a news release quoted in the *California Stem Cell Report*, Bernie Siegel said:

"The dismantling of the program appears to be a retreat by the United States [which is] a setback for the field at large. [...] Other countries are pouring in resources and moving full steam ahead. The NIH's failure to continue the program represents more than just a case of bureaucratic bungling. What we see here is [...] years of valuable work and planning just tossed away."[2]

A workshop was held on May 5, 2014.: http://commonfund.nih.gov/sites/default/files/CRM_May_6_2014_Summary_finalv2.pdf.

A key sentence reads: "The overarching goal of the CRM is to resolve translational challenges associated with the use of induced pluripotent stem cells." (The word "translational" means turning the research into actual usable therapies or medicines.)

"Currently, the CRM is at a transition point," the report said. Transition means change, but changing to what?

Only one scientist was receiving significant funding — Dr. Kapil Bharti, who was working on iPSCs for advanced macular degeneration (a form of blindness). Before him, several little grants were offered for technological challenges, but these did not sound like much.

One thing was clear: it would take a lot of effort to bring iPSCs into general use.

Despite all the fuss, iPSC research is not going to be a cure-all. It has substantive value, no question, but it is just one of several methods. Each problem and project has different needs; the scientists will need to decide which stem cell methodology to use: adult, embryonic, induced pluripotent, direct reprogramming, somatic cell nuclear transfer, or something else completely different.

But there was something deeper at work here, and I was missing it.

I contacted Jeanne Loring, who had advised the committee. She had a different perspective.

"They are making tools — specific techniques and technologies," she said.

---

[2]Jensen D. NIH action on stem cells: more than 'bureaucratic bungling' [Internet]. 2014 Apr 17 [cited 2015 Feb 19]. Available from: http://californiastemcellreport.blogspot.sg/2014/04/nih-action-on-stem-cells-more-than.html

Okay, fair enough. The report had listed a bunch of problems with iPSCs, and the solving of those problems might help advance that method.

But I did not have a sense of closure, like oh, okay, that is what that means... !

During the week, Tuesday through Thursday, I work for the Americans for Cures Foundation, which is led by Bob Klein. This advocacy group sprang from the original campaign to pass Proposition 71. I worked as a volunteer from 2003 to 2011, and now receive a small stipend. I have a little office there, walls full of binders, and all day long I study and type.

So, I asked Bob Klein about CRM.

"The NIH funds a program called NCATS, National Center for Advancing Translational Sciences, which is funding the CRM," he said, then frowning slightly, "But with the best will in the world, the NIH is stretched too thin financially."

A-ha! Not enough money — that was something I could understand!

As reported by Emily Mullin for *FierceBiotech Research* in October, 2013, "Cuts to federal science funding [have hit a historic high]. On top of several years of government budget reductions, most scientists are spending more time than ever writing federal grants today — yet are receiving less funding then they were three years ago [...]. Federal investment in scientific research has fallen off in recent years, suffering a 20% hit adjusted with inflation."[3]

But what did this mean to scientists actually doing the research? I Again, Loring:

"Over the last 10 years, the NIH has lost 23% of its budget's purchasing power. It funds far fewer grants, and *the odds of receiving a research grant have fallen by half* over that time. The funding rate ten years ago was around 30 to 35%; now it's about 15%. For the type of work I do in neurobiology and stem cell research, the odds are worse — I have about a one in ten chance for having a grant funded. I spend half my time writing and submitting (and resubmitting) grants, and it's getting worse. I've submitted seven grant applications in the last four months."

When it had sufficient funding, the NIH was the crown jewel of research. For example, "Roman's Law" by itself funded about $17 million in research funds, over the life of the program. But our scientists gathered an additional $87 million from other sources, which is mainly the NIH. So a $17 million program became a $104 million project.

For decades the NIH was the bedrock foundation of American medical research. But now?

Formless worries gnawed at me. Was I misunderstanding something?

I wrote to Mahendra Rao, the man who resigned, and asked him. He responded with the following:

"This is accurate... The NIH was supposed to respond to a rapidly changing field. [However] the response seems to have been 'let us think' [...] as opposed to 'let's do something'."

The NIH is the base of American research. If the base crumbles, science will fall.

---

[3] Mullin E. Survey details impact of federal cuts, sequestration on scientific research [Internet]. 2013 Oct 13 [cited 2015 Feb 19]. Available from: http://www.fiercebiotechresearch.com/node/10863/print

67 ARTHRITIS AND THE FIFTY STATES

Denis Evseenko is fighting to ease the agony of arthritis. In which states would his use of stem cells be rewarded — or against the law?

I woke in the middle of the night with surprising pain, like my knee was on fire.

"Arthritis," the doctor said, and prescribed Tylenol and physical therapy.

The prescribed exercises, such as sliding one's buttocks up and down a wall, allowed a therapist to make sarcastic comments. "You are weak in the knees," the therapist said, "Weak!" I am not weak, I told her; I used to do deep knee bends with 455 for a triple — I had been a competitive weightlifter! She was not as impressed as you might think.

"Weak, weak, weak!" she carried on.

The pain encouraged new ways to move. I get up from the floor like a camel rising, one awkward section at a time. Sometimes, to cross my legs, I use my hand to lift my knee. Getting out of a chair is not as exciting as it once was. I have even begun to consider using the elevator, instead of the stairs' free exercise.

The knee joint consists of big leg bones meeting on a cushion of cartilage. If the cushion is thick and healthy, you are comfortable. If not, not.

My aunt had a knee operation to replace the cartilage with something artificial. But during the operation, a blood clot went up to the brain, blocking circulation, causing an aneurysm and taking her life.

So I won't have an operation any time soon. Perhaps cortisone shots in the knee, which Gloria has done. They help for about six months, I am told, but cannot be endlessly repeated.

Visiting the CIRM website (www.cirm.ca.gov), I found several stem cell attempts to ease arthritis.

One was Dr. Dennis Evseenko's project, RB5-07230, at the University of California, Los Angeles. His grant was described as active, and the stem cell of choice was embryonic.

"Degenerative joint disease, also known as osteoarthritis, currently affects more than 20 million people in the U.S. alone, making [...] cartilage restoration one of the major priorities in medicine. The [cells needed] are likely to be present only early in development, which explains why [...] using adult stem cells have been unsuccessful."

"[...] This proposal explores the question of what [...] molecular signals control the survival and maintenance of cartilage cells, [and what might] provide new transplantation opportunities for patients with cartilage injury or arthritis [...] (These might delay or even prevent patients from needing joint replacement procedures [...]."

"All the scientific findings and technical tools developed in this proposal will be made available to researchers throughout California [...]."

As Dr. Evseenko put it in a phone interview, "Cartilage is formed early during human development, and adult cartilage has little to no regenerative potential."

He has studied how cartilage is generated, and applied these principles to generate similar cells from human embryonic stem cells (hESCs). He states that "tests have shown these ESC-derived cartilage cells to be proliferative and functional." (Lots of them, and they work.)

His lab is now focused on generating these cells on a large scale and testing their ability to repair cartilage injuries in large animal models.

Obviously, I wish him all the success in the world — and that means he must not work alone.

Every state in the union needs a regenerative medicine program. How close are we to that?

I made a project to determine where our 50 states stand on regenerative medicine programs in 2011. Some of this information is out of date. Still, it is the only list of its kind I know of, and you can get an idea of how far we have to go. Here is the status code:

1. ESC research is being funded.
2. Nothing is happening — there is no personhood threat, but also no good ESC legislation.
3. Personhood threats were attempted but struck down.
4. ESC research is legal but might be under threat.
5. No ESC research and a Personhood threat.
6. Other threats to research exist, such as bills prohibiting ESC research or somatic cell nuclear transfer.
7. The stem cell position is unknown.

Alabama — 6
Alaska — 7
Arizona — 6
Arkansas — 5
California — 1, 3, 4
Colorado — 5
Connecticut — 1
Delaware — 5, 6
Florida — 2, 6
Georgia — 5, 6
Hawaii — 2
Idaho — 2
Illinois — 1, 6
Indiana — 5, 6
Iowa — 2, 6 (and also possibly 1 and 4)
Kansas — 3, 5, 6
Kentucky — 5
Louisiana — 5, 6
Maine — 2, 5, 6
Maryland — 1, 6
Massachusetts — 1, 6
Michigan — 3, 5, 6
Minnesota — 5, 6
Mississippi — 3, 5, 6
Missouri — 3, 6
Montana — 3, 5, 6
Nebraska — 2, 3
Nevada — 7
New Hampshire — 2
New Jersey — 1, 6
New Mexico — 2
New York — 1, 6
North Carolina — 1
North Dakota — 5, 6
Ohio — 5, 6
Oklahoma — 5, 6
Oregon — 5, 6
Pennsylvania — 5, 6
Rhode Island — 2, 6
South Carolina — 5, 6
South Dakota — 5, 6

Tennessee — 2
Texas — 2
Utah — 2, 6
Vermont — 7
Virginia — 5, 6
Washington — 7
West Virginia — 7
Wisconsin — 1, 6
Wyoming — 6

68 WOULD YOU DRINK FROM A FOUNTAIN OF YOUTH?

Helen Blau: Uses stem cells to strengthen muscles of the aging hand and eye.

Gloria asked: would I want to live forever?

No. I take joy in the days that are given me. But when my chores are done, and the aches and pains pile up too high, I'm sure there will come a time when it will be a relief to shut my eyes and let the world roll on without me.

That time, I hasten to add, is hopefully decades away. My father, Dr. Charles H. Reed, is 92, and plays tennis three times a week. He can speak 12 languages, wrote a Vietnamese–English dictionary, and plans on going back to some of the wilder parts of Africa again next year. I like the idea of that century of productive life; it is my goal.

A better question might be: would I want to live the years I have — in good shape?

Absolutely! I would love to run fast enough to catch up with my grandkids, and feel again that strength which once invigorated my limbs. But those days are gone forever, are they not? Surely it is impossible to rejuvenate organs, limbs, mind — to replace old cells with new, to become… younger?

Irina Conboy: Fighting aging as if it were a disease.

When I first heard about Dr. Irina Conboy and her attempts to deal with aging as a disease which might be healed, I thought this was over the edge, impossible, snake oil, fakery.

But then I looked closer. I typed out her ten-page CIRM grant document, which is what I do when there is something too complicated for my small brain to understand.

I was surprised.

There was substance here. Naturally, she might not succeed. Science proceeds at its own pace, and it could take generations. Plus, anybody can be wrong. But one thing was certain. She was going at it scientifically, step by documented step.

Below is my non-scientist's interpretation of her work, grant number RN1-00532 of the California stem cell program. Before I put my interpretation of it in the book, I offered it to her for corrections. She said it "accurately summed up" her work.

Like most scientific papers, it began with big words: "Identification of hESC-mediated molecular mechanism that positively regulates the regenerative capacity of post-natal tissues".[1]

Could embryonic stem cells regulate aspects of aging?

"The tissue regenerative capacity deteriorates with age […] leading to loss of organ function."

Translation: As we get old, the body's natural repair kit does not work as well, and the parts wear out.

---

[1] Identification of hESC-mediated molecular mechanism that positively regulates the regenerative capacity of post-natal tissues [Internet] [cited 2015 Feb 20]. Available from: https://www.cirm.ca.gov/our-progress/awards/identification-hesc-mediated-molecular-mechanism-positively-regulates

"Our recent work [shows] that factors [in] human embryonic stem cells [can] enhance the regenerative responses of organ stem cells. [Identifying these] factors will help counter the loss of tissue maintenance and repair in the [elderly and] boost the performance of [...] cells struggling to repair deterioration, thus countering degeneration and improving organ function."

Translation: Something in stem cells can make an old person's body fix itself better.

"The body's capacity to regenerate new tissue can no longer keep up with tissue death. This lack of tissue repair [...] leads to the loss of organ function [and] causes many degenerative disorders [such as] Parkinson's, Alzheimer's and muscle atrophy. [Practical therapies will] only emerge when the balance between the regenerative and degenerative processes [is] properly understood."

Translation: The body breaks down, and tries to build itself back up. When the breakdown is more than the build-up, bad things happen. Therapies require understanding of this process.

"This proposal describes steps to rejuvenate stem cell responses in the old and [also] to rescue tissue repair in people suffering from debilitating degenerative diseases."

Translation: There may be a way to make an old person's body processes work like young again... exciting stuff!

Was she able to identify exactly what was in the human embryonic stem cells (hESCs) that had the "youthifying" effect?

"We found that hESCs produce both positive growth factors and negative regulators of the TGF-beta family [...]. This fits with our recently published work showing that young muscle regenerates well from strong growth factor signaling and low TGF-beta signals — while old muscle regenerated poorly due to weak growth factor signaling and high TGF-beta signaling."

What I gathered from that is:

Bad — Weak growth factor and *high* TGF-beta means the body does not fix itself very well. (I looked up "TGF-beta" hoping there was an easy way to understand it. If there is, I couldn't find it; I just think of it as an undesirable protein, "beta bad" — Sorry!)

Good — Accentuate the growth factor and inhibit TGF-beta, and the body's repair kit works better.

"Our [...] hypothesis is that positive growth factors triggered by hESCs [will] initiate and maintain regeneration, [and] the TGF-beta inhibitors produced by hESCs [will] reduce the TGF-beta signaling. [This] assures the robust regeneration of muscle."

Materials provided by embryonic stem cells might repair old muscles... like mine!

Do other scientists take Dr. Conboy's efforts seriously? She has had three papers and four book chapters published on it, not to mention a patent application filed with her university.

How was it going?

"Preliminary data indicates the effects on regeneration of old muscle look very promising. What was surprising is that administering these inhibitors [...] appears to reduce TGF-beta in the whole animal, suggesting [...] effects on other tissues as well as muscle."

"Our work is at the stage of understanding the molecular mechanisms by which the aging of the regenerative potential of organ stem cells *can be reversed* by particular human embryonic factors […], restoring high regenerative capacity to old muscle (emphasis added)."

"Additionally, our data […] suggest that muscles and the brain age by similar molecular mechanisms […]. Thus, therapeutic strategies for rejuvenating muscle repair might be applicable to the restoration of neurogenesis (growth of nerves) in the aged brain."

"The use of hESC-produced pro-regenerative factors for boosting the regenerative capacity of organ stem cells is likely to yield healthy, young tissue."

Instead of accepting as inevitable the decline of our body and brain, we might be able to reverse that process.

But however reasonable Dr. Conboy's work might sound on paper, was she alone on this?

Were others taking on the challenge of aging?

Dr. Helen Blau of Stanford University was also interested in aging, particularly as it related to muscle — even tiny muscles, like those which let us swallow, smile, roll an eyeball, or flick a marble with a finger.

Her work pointed out the vicious cycle of muscle weakness, aging, and injury.

If you are weak, you are more likely to incur an injury. Immobility while recovering causes the loss of muscle mass — resulting in more weakness, and increasing the likelihood of worse injuries. The second broken leg is more likely to occur than the first, and at a certain age, it does not take many fractures to lay you up forever.

However, in Dr. Blau's grant (TR3-05501) description, she states:

"We propose to develop a human stem cell therapy to prevent and/or reverse localized skeletal muscle atrophy in muscles of the aged […] to isolate muscle stem cells (MuSCs) from human muscle biopsies, [rejuvenate and increase their number], and demonstrate that transplantation of human MuSCs results in [increased] strength in atrophied muscles of aged recipients."[2]

She pointed out the need for such research:

"Since California is projected to be the fastest growing state in the U.S. in terms of population, with an elderly population that is projected to grow twice as fast as the total population of the state, it is more important than ever to develop strategies that positively impact the health of this demographic."

What areas of the body would be targets for strengthening?

Dr. Blau suggests small muscles of the eye critical to vision, the pharynx (essential to swallowing), and the muscles of the hand. A major effort will also entail the treatment of muscle atrophy in aged knee or hip rehabilitation patients.

She is working with mice right now, but her goal is to "isolate, characterize, and transplant" adult human cells. So she would take a biopsy (tissue sample) from a person's

---

[2] Local delivery of rejuvenated old muscle stem cells to increase strength in aged patients [Internet] [cited 2015 Feb 20]. Available from: https://www.cirm.ca.gov/our-progress/awards/local-delivery-rejuvenated-old-muscle-stem-cells-increase-strength-aged-patients

back, for example, isolate the functioning stem cells from it, and expand their numbers in a culture.

This expansion could be achieved by treating the biopsied cells with a drug, and then putting them back into the patient. Indeed, this has a proven potential to work, as Dr. Blau and her group have discovered a drug that can increase the numbers and rejuvenate the aged muscle stem cell population so that it has improved function in regeneration.

In mice, these studies showed that transplanting the rejuvenated cells increased strength of injured tissue in aged mice, back to that of young mice.[3] The group's goal is to use a person's own cells to treat muscle weakness in another part of the body. Importantly, this strategy would avoid immune rejection of cells.

Natalie DeWitt, Director of Research Development at the Baxter Lab at Stanford University is supportive, saying that "using muscle stem cells is a unique therapeutic approach. Since they can only form muscle, they do not form tumors, and they can uniquely improve muscle strength upon transplantation, as Blau and her team have shown."

A third scientist, Amy Wagers of Harvard University, states on her homepage that she has "identified a subset of satellite cells [...] that act as muscle stem cells [which] can restore muscle function when transplanted into injured or diseased muscle."

She refers to stem cell dysfunction that typically arises with advancing age, and says that "exciting new data from our studies suggests that stem cell aging is controlled at least in part by blood-borne mediators, which can change with age and can be manipulated to reverse age-associated dysfunction. [...] Future work will use the new insights gained from these studies to enable novel interventions to delay or reverse the onset of age-related diseases and extend the healthful life of aging individuals."[4]

And in a multi-author paper about the aging mouse brain, Dr. Wagers stated:

"Age-related decline of neurogenesis and cognitive function is associated with reduced blood flow and decreased numbers of neural stem cells."

"[However], we found that factors found in young blood [...] can improve cerebral vasculature (i.e., blood flow in the brain) and enhance neurogenesis, [which] may constitute the basis for new methods of treating age-related neurodegenerative and neurovascular diseases."[5]

And who raises funds for Harvard University's magnificent stem cell program?

The indefatigable Brock Reeve, Christopher's brother.

So would I drink from a fountain of youth?

Maybe just a sip.

---

[3] Andersen RE, Lim DA (2014). An ingredient for the elixir of youth. *Cell Res* **24**: 1381–1382. doi:10.1038/cr.2014.107; published online 2014 Aug 12.

[4] Wagers laboratory home [Internet] [cited 2015 Feb 20]. Available from: http://www.scrb.harvard.edu/lab/57/home

[5] Katsimpardi L, Litterman NK, Schein PA (2014). Vascular and neurogenic rejuvenation of the aging mouse brain by young systemic factors. *Science* **344** (6): 630–634.

Jill Helms.

"…Research over the past decade has conclusively demonstrated that aging is reversible … old cells can become young again… we may not be able to stop the march of time but our growing knowledge of stem cell biology and aging offers new avenues to treat old diseases. My group at Stanford has identified and characterized a potent stem cell activator, Wnt, and shown that it activates both embryonic and adult stem cells. With CIRM funding we developed a strategy for activating a patient's own bone stem cells (for) rapid healing… our goal is to discover new ways to remain fit and vigorous, well into old age. Supporting these discovery efforts ensures that "aging" is no longer synonymous with "disease."

— Jill Helms, CIRM grant recipient, Stanford University.

69 WHEN THINGS GO RIGHT

Roman reaches for tomorrow, using the power of his voice and mind to fight for cure research for everyone.

"With these words, and in this spectacular venue, I hereby announce my candidacy for State Senator of the great state of California," said the muscular man in the wheelchair.

It was not a political meeting, of course. Bernie Siegel, founder of Genetics Policy Institute, goes to great lengths to make everyone well-fed and welcome, regardless of their political persuasion.

In that cheerful audience were giants of the stem cell world, including donors of unparalleled generosity like Malin Burnham and Denny Sanford, whose financial gifts had given life to the Sanford-Burnham Consortium. Denny Sanford's recent gift of $100 million to establish the Stanford Stem Cell Clinical Center had stunned the stem cell world.[1]

In that special room that night, the Awards Banquet of the 2013 World Stem Cell Summit was held, commemorating everyone who had brought the world closer to cure.

---

[1] Robbins G. Sanford donates $100 million to UCSD. *UT San Diego* [Internet]. 2013 Nov 3 [cited 2015 Feb 19]. Available from: http://www.utsandiego.com/news/2013/nov/03/ucsd-dennysanford-stemcells/

When my son received his advocacy award, I knew he would tell the story of the law he inspired, the Roman Reed Spinal Cord Injury Research Act, and the Alabama law patterned after "Roman's Law," the T. J. Atchison Spinal Cord Injury Research Act, and the building of the Alabama Institute of Medicine (AIM).

http://aim4cure.org/

He would credit Dr. Jane Lebkowski, now of Asterias Biotherapeutics, for continuing the paralysis work she previously led at Geron, Inc. This was the famous embryonic stem cell research begun so long ago by Hans Keirstead, with grants from the Roman Reed Spinal Cord Injury Research Act. It was back on track with new funding, both from Asterias and the California stem cell program.

Roman might talk about his efforts with Stanford University's paralysis research group, or how Steve Young of the San Francisco 49'ers and other champions had helped put on fundraisers for the Roman Reed Foundation. He could mention his years of service in local government as City Council Commissioner. Driving a wheelchair did not keep him from getting things done.

But I had no idea he would take the opportunity to announce his candidacy for California State Senator — some moments are beyond pride.*

Paul Knoepfler, America's premier stem cell scientist blogger, told about how surviving cancer helped him realize the importance of taking on challenges — like his forte, sharing science issues in language everyone can understand.

Jeanne Loring was there, which was hardly a surprise — she was everywhere, speaking at or moderating on no less than seven panels. Her latest goal was the Stem Cell Matrix, to catalog and characterize all the working stem cell lines in the world.

Bob Klein spoke on the need for a patient advocate network ready to speak as stem cell research *begins human trials*. Things may well go wrong with early stem cell treatments, as they have with virtually every new medical advance. We should learn from the HIV/AIDS example. When their clinical trials began, the patients made clear immediately they knew some of the participants would actually die along the way — they were dying already, and research was their only hope.

Alex Richmond of Children's Neurobiologic Solutions reminded us of the youngest sufferers of chronic disease.

An entire new industry, biomedicine, was growing stronger, championed by leaders like Stephen Minger of General Electric, Devin Smith of Pfizer, Alain G. Vertez of NxR Biotechnology, and Mike West of BioTime, Inc., who essentially began biomedicine by founding Geron, Inc., in 1990.

Careers spanned the globe, like Edward Holmes, Chief Executive Officer of the Sanford Consortium for Regenerative Medicine AND Deputy Chair of a biomedical government agency in Singapore.

---

*With almost no budget, Roman's Senatorial campaign was a learning experience; he came in third. But he spread the message; no one who heard him can forget Roman Reed, and his impact on the world has just begun.

With 40 nations represented at the summit, almost every continent was heard from. Dr. Fanyi Zeng, Associate Director of the Shanghai Institute of Medical Genetics, gave a concise but wide-reaching summary of China's involvement and accomplishments in stem cell research. Fabiano Arzuaga of the University of Buenos Aires is undertaking research on the regulation of stem cell research in Latin America. David Fransen, Canada's Consul General, spoke all too briefly on some of Canada's accomplishments in our field. Japan was massively represented by at least a dozen scientists, including the magnificently bearded Norio Nakatsuji of Kyoto University. The booming research of Qatar was highlighted by Abdelali Haodi, Executive Director of Qatar's Biomedical Research Institute. Rosalia Mendez-Otero, recipient of the Medal of the National Order of Scientific Merit from the President of Brazil, was present, as was Avi Treves of Israel, who co-authored more than 90 scientific publications and held 10 patents.

Brock Reeve, Executive Director of the Harvard Stem Cell Institute, shared some of the mountain-sized problems (and possibilities) of fundraising.

Mary Ann Liebert was there! Anyone who loves science knows the Liebert publishers — more than 70 scientific magazines are published by them today. And there she was, honored by the Genetics Policy Institute and a roomful of appreciative readers.

The magnificent California stem cell program was well-represented: cheerful Chairperson Jonathan Thomas, always making new friends for the cause; former CIRM President Alan Trounson who spoke with passion on the urgency of ending HIV/AIDS — "We must eradicate it!" he said;

Amy Adams, veteran science writer who works to make sense of the issues surrounding the research; Kevin McCormack, another great wordsmith, the unofficial "translator" for the CIRM; Elona Baum, Vice President of Business Development, crucial as biomedicine turns theories into therapies; Ellen Feigal, Vice President for Research and Development, hard worker for patient advocate outreach; Don Gibbons, invaluable as Communications Officer; Uta Grieshammer, Science Officer, working on the development of a stem cell bank; Ian Sweedler and Michael Yaffe, men to talk to about CIRM's international cooperation; Geoffrey Lomax, Senior Officer for Medical and Ethical Standards; Board member Dr. Oswald Steward, whose expertise in spinal cord injury research allows him to oversee both the Reeve/Irvine Research Center at UC Irvine, and the Roman Reed Core Laboratory.

Does it seem sometimes like only bad things happen in the world?

Not in that room.

Not in that field.

70 WHERE DID THE MONEY GO? (AND A NEW YEAR'S DELIGHT* AT THE END...)

FACES OF THE FUTURE: Bob Klein with Mary Bass, David Bluestone, and Robert Klein (Bob's son). "As a young person, it is amazing to be part of this revolution of medical advances — history that's being made daily, hourly.... Mary Bass."

If you visit the CIRM website, www.cirm.ca.gov, you will find more information than most folks can read in an average year! It is always growing and jam-packed with interviews, news releases, updates, videos, announcements, availability of grants, access to scientific papers, and much more. But for me, everything boils down to one question:

Where did the money go?

At the end of the program's first ten years, we still have about a billion dollars left to allocate, so we are nowhere near being done. But it is the end of the year (and this small book),

---

*And a New Year's delight for me personally — my cancer test. Half a year after surgery, radiation, and hormone therapy, I had the blood test. The cancer result was "undetectable." I will monitor it of course, like any man of my age — but I am back on track to live to be a hundred!

and time for taking stock. It is my hope there will be another round of funding, a Proposition 71, Part Two — but for that, the program must prove worthy of the people's trust.

I believe all reasonable expectations of success have been met or exeeded.

How many diseases are we bringing to human trials? CIRM modestly claims "only" ten.

This is amazing progress — "ten potential treatments expected to be approved for clinical trials by the end of the year"? These include: HIV/AIDS (2), spinal cord injury, heart disease, solid tumors (2), leukemia, sickle-cell anemia, diabetes, and blindness.[1]

Adding up all those in various stages of development, we may have as many as 29 (see Chapter 64)!

But all that research must be paid for — by research grants.

Visualize a scientist filling out long and detailed grant request forms, then sending them off, one by one, and after that waiting... and waiting... for without these grants, there is no research.

New CIRM President Randy Mills is trying to make the process faster through a system of efficiencies he calls CIRM 2.0.

In an interview towards the end of 2014, President Mills said, "In the past it could take up to two years to go from us requesting applications for funding to getting the money out the door as part of an approved contract. CIRM 2.0 simplifies and accelerates the process, cutting that two years down to just four months. And if you miss that deadline, you only have to wait one month for the next application headline to come around."[2]

For researchers planning their futures, that is an early Christmas present.

And now, here is where the money went in terms of how much California's stem cell program spent on the various forms of chronic disease and disability, complete as of September 2, 2014.

Disease	Dollar Amount
Aging	$25,441,784
ALS (Lou Gehrig's disease)	$53,957,528
Alzheimer's disease	$61,989,729
Anemia	$16,222,341
Arterial limb disease	$15,406,649
Arthritis	$15,630,500
Autism	$37,714,473
Blood disorder	$50,908,649
Blindness	$107,344,609

(Continued)

---

[1] Holden A. 10 years/10 therapies: 10 years after its founding CIRM will have 10 therapies approved for clinical trials [internet]. 2014 Nov 20 [cited 2015 Feb 20]. Available from: http://blog.cirm.ca.gov/2014/11/20/10-years10-therapies-10-years-after-its-founding-cirm-will-have-10-therapies-approved-for-clinical-trials/

[2] Knoepfler P. Radical, supercharged vision for future of CIRM: interview with new Prez Mills [Internet]. 2014 Dec 5 [cited 2015 Feb 20]. Available from: http://www.ipscell.com/tag/c-randal-randy-mills/

(Continued)

Disease	Dollar Amount
Bone or cartilage disease	$55,775,038
Cancer	$184,009,974
Cancer: Brain tumor	$67,786,253
Cancer: Leukemia	$72,685,060
Cancer: Melanoma	$25,003,361
Cancer: Solid tumor	$33,835,221
Developmental disorders	$26,690,975
Deafness	$5,880,225
Diabetes	$50,526,400
Epidermolysis bullosa	$25,990,910
Fertility	$13,641,460
Genetic disorder	$36,689,452
Hearing loss	$5,880,225
Heart disease	$157,575,288
HIV/AIDS	$47,903,129
Huntington's disease	$30,309,173
Immune disease	$95,144,995
Incontinence	$5,025,428
Infectious disease	$28,129,811
Intestinal disease	$11,778,708
Kidney disease	$7,838,963
Liver disease	$19,944,202
Multiple sclerosis	$10,040,777
Muscular dystrophy	$17,802,533
Neurological disorders (includes Alzheimer's disease, ALS, autism, dementia, epilepsy, Huntington's disease, multiple sclerosis, neuropathy, Parkinson's disease, Rett's syndrome, spinal cord injury, spinal muscular atrophy, and stroke)	$308,195,026
Osteoporosis	$55,775,038
Parkinson's disease	$43,055,969
Pediatrics	$133,844,715
Respiratory disorders	$38,436,529
Severe combined immunodeficiency	$95,144,995
Sickle-cell disease	$50,908,649
Skeletal muscle	$133,844,715
Skin disease	$25,990,910
Spinal cord injury	$24,054,594
Stroke	$44,681,457
Thalassemia	$50,908,649
Toxicity	$2,680,118
Trauma	$13,709,972
Vascular disease	$15,406,649
Vision loss	$108,044,592

The California stem cell program is a gift of hope to every ill or injured person in the world.

From my family to yours, may you have all the joys of health through cure research.

71 THE END?

Bob Klein and Don Reed, embracing a moment of success.

As you might recall, one of my first stem cell fund raisers was a play I wrote, *A Night for No Mexican Tears*, about the great Mexican–American revolutionary Juan Cortina.

Opening night, I was onstage, introducing my grandson Roman "Little Man" Reed. I was explaining to the audience that he was our miracle child because the doctors had told us there was only a two percent chance that his father Roman would ever have children and —

Roman Reed Part Two, three years old, leaned into the microphone and said:

"The end, the end! Come down and play, Grampa!"

"Little Man" is 17 now, and we hope he will go to University of California, Berkeley next year. His little brother Jason, aged ten, hits home runs seemingly at will, and Katherine, five, just started learning taekwondo. "Big Roman" has so many stem cell projects I can't keep track of them all. My daughter Desiree is now Senior Deputy Athletic Director for Virginia Tech, and Desiree and Josh's son Jackson (also 10) knows every college football mascot in the land.

Roman Reed's daughter, Katherine Reed, age 4, with Grampa.

Gloria is still the spice in our relationship — and I turn 70 this year. Seventy… or what the Bible refers to as "three score and ten", a natural lifespan for a man.

So is this the end? Is it time to quit doing stem cell advocacy and "come down and play"?

If you visit my old website, www.stemcellbattles.com, you will see a picture of a black wheelchair. It belongs to Karen Miner. Below it are the words: "For temporary occupancy only". That goal has not been achieved: not for Karen nor my son, not for anyone.

But the California stem cell program has changed the game of chronic disease. Remember the grim phrase doctors used to say, "There is no cure"? Now they can say, "There is no cure *yet*".

And when cures do come? Imagine the impact of just one effective therapy… one actual cure.

For example, if it works and is brought to commercial availability, the CIRM-supported ViaCyte diabetes therapy will be astonishing. Not only would it eliminate the pain of all those daily bloodlettings, but also the risk of coma, blindness, and limb amputation.

And financially? What a blessing cure will be!

Diabetes cost America $245 billion in 2012.[1]

Lessening that by just two percent would repay California's entire $3 billion investment in Proposition 71 — in one year.

---

[1] The cost of diabetes [Internet] [cited 2015 Feb 20]. Available from: http://www.diabetes.org/advocacy/news-events/cost-of-diabetes.html

People who encourage miracles: the staff and board of directors of the California stem cell program: author on far right.

It will not happen all at once, of course. Nothing great comes fast; but when it does arrive, cure will change everything.

The science-deniers will have their fun, like those who yelled "Get a horse!" when they saw Henry Ford's new cars, but they cannot block us forever.

In here-and-now reality, the California stem cell program has helped build 12 new stem cell research centers, complete with equipment, up and down the state.

How important is this? Scientists at one of these centers recently helped children fight back against a terrible condition. Remember the "Bubble Boy"? David Vetter had to stay locked up in a plastic isolation unit because his body's immune system could not protect him from germs like the common cold. Tragically, he died.

But 18 little babies who had the same "Bubble Boy" condition are now living normal happy, playful lives. Their parents can hug and kiss them, take them outside, and let them run — because their once-faulty immune systems have been replaced.[2]

Those children's parents have Dr. Donald B. Kohn to thank — and the California stem cell program, without which that shining new Eli and Edith Broad Center of Regenerative Medicine at University of California, Los Angeles would not exist. And for families whose children are afflicted with sickle-cell disease, a very similar treatment is being developed at the same CIRM center right now.[3]

Cure grows: The founding Director of that Center, Dr. Owen Witte, was recently appointed to the President's Cancer Panel, a three-member panel reporting directly to President Obama.

Cure grows: Almost three thousand CIRM scientific breakthroughs have been written-up in peer-reviewed journals, each publication a piece of the puzzle of cure. And the knowledge gained is not secret, but (with intellectual property exceptions) made available free to all.

Cure grows: More than a billion extra dollars has poured in, matching grants and donations, new money, and jobs for Californians — careers that are richly satisfying. Biomedical people go home tired at night, but they are making the world a better place: easing suffering, saving lives, and improving the economy.

Should the California stem cell program continue?

Bob Klein has said that when the time is ripe, he will poll California and see how the state feels about continuing the medical revolution — with another five billion dollars.

It will be the greatest stem cell battle of all time. The campaign itself will probably cost $50 million. But by far the greater expense will be in energy, dedication, and time.

I remember talking to a friend once at a public meeting. She was an advocate for research, an effective fighter, but she was in a wheelchair now; her multiple sclerosis had grown worse.

---

[2] Jackson H. New "bubble baby" treatment means kisses for 18 kids. *NBC News* [Internet]. 2014 Nov 20 [cited 2015 Feb 20]. Available from: http://www.today.com/health/new-bubble-baby-treatment-means-kisses-18-kids-1D80305637

[3] Stem cell researchers use gene therapy to restore immune systems in "Bubble Boy" disease. *Science Daily* [Internet]. 2012 Sep 11 [cited 2015 Feb 20]. Available from: http://www.sciencedaily.com/releases/2012/09/120911111626.htm

She "would really like to meet Bob Klein," she said. Naturally I ran to hunt him down.

I found him behind an auditorium pillar, leaning back against the cold concrete. His eyes were shut, his face was gray. He looked exhausted; I hated to disturb his rest. But I remembered my friend in her wheelchair, and blurted out the situation.

Without opening his eyes, Bob listened, not wasting a scrap of energy.

"Yes," was all he said. I rushed to find my friend; we hurried back.

As we came toward the pillar, a different Bob Klein stepped out to greet us. Beaming, fairly radiating energy, he shook her hand, thanking her for the gift of her hard work.

"Proposition 71 could never have happened without you," he said, "Thank you for your advocacy."

There was a sunbeam framing my friend. Her face was up and in the light; her eyes were shining...

A trivial incident, perhaps. But it shows the spirit of the man, and of his cause: our cause. Because if Bob Klein will work to rouse the friends of cure, fighting on through exhaustion toward a world of health for all — he will not fight alone.

Remember the words of Christopher Reeve?

"Go forward," he said, "Go forward."

Our champion has fallen, but the flame of his faith still lights our way.

We will go forward — and we will prevail — because scientists and patient advocates across the world have taken up the torch.

CURE GROWS!
Don C. Reed
March 3, 2015

APPENDIX 1
INTERVIEW WITH LIM CHUAN POH, CHAIRMAN OF A*STAR AND BIOPOLIS

How did your interest in science begin?

I've always had an interest in science — not only from university but also from the military, where technology is a very big part of the game. Whenever there was any interesting announcement in the biomedical sciences, notwithstanding I had zero background, I would attempt to at least understand the big topics that were discussed, in particular the sequencing of the human genome.

When I was asked to run this organization, the book, *Cell of Cells: The Global Race to Capture and Control the Stem Cell*, written by Cynthia Fox in 2007, was one of the first things I acquired. Singaporeans were very important in opening up this field of science. Ariff Bongso's name is repeated many times in this book and to read the chapter about his work that was brought to Singapore alongside Alan Trounson and Ben Reubinoff, competing almost neck-and-neck with groups in Israel and the US — that was exciting.

Who would you credit for the dream of Biopolis?

There is both institutional and individual effort. In Singapore, the government must be given credit because they are prepared to invest in research and they stay committed. Both the present Prime Minister Lee Hsien Loong and President Tony Tan in Singapore have been instrumental in leading the government support for research. In fact the present Prime Minister was the one who started the structured investment in research in 1991 when he was Deputy Prime Minister. So there is continuity and a steady view of how we should invest in research. He also chairs the Research Innovation Enterprise Council, comprising both cabinet ministers involved in research, and also international members, people like Sir Richard Sykes, former Chief Executive of GlaxoSmithKline, Professor Bob Brown, President of Boston University; and senior scientists, academic leaders from around the world, as well as industry captains involved. Likewise, the President of Singapore, Dr. Tony Tan, has always been overseeing higher education and research. One of the key people that translated strong governmental support to realize Biopolis is Mr. Philip Yeo. I'm sure you've heard of that name before! He is just such a bundle of energy and he's passionate about science. So combined with strong government support, he made Biopolis happen, alongside some very exciting young scientific leaders from the universities and also from the clinical community. He traveled around the world to attract some of the best scientists, including from your country, to Singapore, even as we were just developing our plans.

Next year, the Prime Minister himself is coming to this location to open one more set of buildings, and I'm going to tell him that over the span of that 14 years, we started from no one doing research in this location to 9,000 people involved in biomedical sciences, physical sciences, and electronics research, coming from both the public and private sectors. Likewise, we started with zero space allocated for research, to around eight million square feet of space by next year, and all these on a tiny footprint of just 40 hectares. Singapore is land poor. We intensify the usage so we are one of the few locations where research is vertical and not sprawling. But the key thing is that you need good people and opportunities for them to interact. In this small compact location, this is what the scientists are able to do, both with their colleagues from the other research institutions and also with colleagues from the universities and, very importantly, also from the industry.

Is it correct that Singapore is ranked number five in stem cell research papers produced?

I think you may be referring to Nature Asia Pacific's high-impact journal ranking of Singapore. If you are referring to that, indeed Singapore is now ranked number five in the Asia-Pacific region, with number one of course being Japan, followed by China; China is growing very fast. In third place is Australia, number four is South Korea, and number five is Singapore. But of course, these are all much larger countries compared to Singapore, so on a per scientist basis, in terms of high-impact productivity, Singapore is actually ranked first out of the five countries — we have just over five million people.

I read that approximately 1,000 scientists and engineers work in your labs?

That figure is correct, so I'll just give you a broader context. A*STAR today has almost 5,400 people and we have about 4,400 involved in research, comprising research scientists, engineers, and technical staff. In the Biopolis campus, we have about 2,000 people, and in the Fusionopolis campus by this time next year, we will have over 3,000 people. We have one complex waiting to be completed, after which we will bring some of the other institutes that are now in the universities and in the Science Park into this common location.

I should send you a copy of this publication called *A*STAR Research*. We do this in collaboration with *Nature* so there is an online version that comes out every fortnight and a hard copy version that comes out every six months. I took the latest issue, which is from April to September 2014. When I turned to the section on cell biology and immunology, guess what, every single article on cell biology is on stem cells. Every single article! I should add further to that: some of the names that you are familiar with are still doing stem cell research in Singapore, like Edison Liu and Bing Lim. But what is very exciting for me is the talent development program we put in place, beginning in 2001. That has resulted in many young scientists doing stem cell research in Singapore. One of them is Dr. Jonathan Loh. He did his post-doctorate for three years with George Daley at Harvard University and in fact continues to have a collaboration there. He recently used a blood cell to derive induced pluripotent stem cells, making this very accessible. This may have a huge impact on drug discovery efforts.

We now have a lot of younger generation stem cell scientists in Singapore, in addition to those that we recruited in the initial wave and this is really how this effort, I think, is viable over the long-term. We have managed to create and plug Singapore into the international network.

You mentioned Ariff Bongso. Can you talk about him a little bit more?

Professor Bongso really excited many scientists in Singapore because of his contribution in discovering the human stem cell back in 1994. Younger scientists look to him as having a can-do attitude and that it is possible to do path-breaking science. In Singapore, we pay a lot of attention to make sure young Singaporeans in the schools are exposed to interesting science and given opportunities to know what happens in the labs so that we will continue to have many generations of young Singaporeans wanting to pursue research, and remain engaged with scientific discoveries. Ariff Bongso has an important role to play in that space too. Also, because of his very impactful work, he attracted international collaborators to do their discovery work in Singapore, which is another reason why Singapore was propelled to an international hub within the Asian context.

You mentioned a program to cultivate young people's interest in science?

That program is continuing, both in terms of outreach during the time when they are in the schools and also as we reach out to secondary schools, to make opportunities for them to send students to spend meaningful periods in our laboratories and enthuse them about scientific discovery. Every year, we bring in about 200 students from the schools to spend a month in our laboratories to do actual research. Many of them eventually apply to us for scholarships to pursue research or training in the best laboratories in the world for both undergraduate and postgraduate training. So if you look at our outreach footprint, we are reaching tens of thousands, if not hundreds of thousands of students. When I met the President of Stanford University, Professor John Hennessy, some years back when he came to build a collaboration with A*STAR, he told me that Singapore has the highest per capita PhD students in Stanford University. The highest! Just to give you a sense of what that number is, as I speak, I have nearly 60 PhD students in Stanford University just from A*STAR alone. And in Massachusetts Institute of Technology and Harvard University, just these two universities, you're looking at about 60 PhD students as well. And if you're looking at London — Oxford University and University of Cambridge — you're easily looking at another 70 to 80 PhD students.

How do you reach out to the general public about what you are doing?

We have a lot of programs to enthuse and inform the public about research and discovery in Singapore. It is almost the case that you will see a publication on scientific discovery practically everyday in our widely circulated newspapers. Of course, we now have *A*STAR Research* fortnightly; we provide this as a link to all the schools in Singapore so that students can read about the research taking place in A*STAR as well as the universities. We also have a scientist-resident in school program. This is the alumni of the schools and they are linked back to the schools so they can go back to the schools to give talks at their *alma mater*. In addition, A*STAR itself and our scientists work with the Science Centre of Singapore so that we can reach out to the general public and also the young school students. Because the *Science Centre is very popular with primary school students, when they visit the Science Centre, their parents have to come in too, so parents are also informed* about science and discovery both in the world and also in Singapore. So we do that a lot. We are starting a new exercise to reach out to the community to bring science to the community level just to speak about certain topics, again to inform people about the latest issues and discoveries and their impact.

How do you balance funding between basic science to advance the field and translational research to bring therapies commercially?

When we started Biopolis, many people that came to advise Singapore pointed to this as an issue to grapple with. Our initial investment was getting the basic science in place, recruiting the scientists and getting started. If we just loosely allude to the period from 2001 to 2005, which was Phase 1 of the biomedical science effort, that was when the foundation in basic science was laid. From 2006 to 2010, we started looking at translational and clinical research, both in terms of the mechanism but also supporting and recruiting clinician-scientists to help in this endeavor. These are people who are already doing clinical research. At the same time, there were the scholarships that I spoke about. We also support the MBBS PhD as well as the MD PhD program to ensure we have a pipeline of young clinician-scientists who will go into research. We are now in the third five-year plan for the biomedical science effort. In this particular phase, we add on to the initial effort of building basic science, as well as translational and clinical research. We have funding to incentivize collaboration between the scientific and clinical research community and industry to make sure that the research is able to reach the public. Arising from this, Singapore in biomedical science research and in particular in stem cell research has got one of the highest collaborations with industry.

What is the annual research budget for Singapore stem cell research?

I cannot give you the precise number because the stem cell research is now infused into the whole system, but I can tell you the overall budget for the agency and how much is given to biomedical science just to give you a sense of the order of magnitude. Firstly, the Singapore government is now committed to giving one percent of GDP to public sector research. Out of that one percent, the agency itself has a share of about 40%. The one percent is about S$16 billion so our share is about S$6 billion, which is about US$5 billion. So our annual investment is about US$1 billion, and about half of that amount goes to biomedical sciences research, once you take away about 10% to

15% for the overhead costs, so the rest actually goes straight into the research. So, this gives you a sense.

A*STAR is collaborating with the California stem cell program. Would you comment on the importance of international cooperation?

Absolutely. Firstly, science is international. Whether it is in stem cell research or biomedical sciences research, a lot of interesting work is happening around the world. It is very important that as we pursue the research, we can bring new knowledge and complementarity to what is already happening. We should avoid overlaps because the world has limited resources. We should build synergies through collaborative effort. In the space of personal or stratified medicine as we look at disease biology and therapy, the basis for international collaboration becomes even more important. In the case of Singapore, we have the major Asian phenotypes, whether it is the Chinese, Indian, or Malay population. Other research communities around the world bring different populations to the collaboration, and we learn a lot more through the collaborative efforts than if we were to do it by ourselves. So we believe in creating an environment conducive for this. We started scholarships for Singaporeans ten years ago to go to the best institutions around the world to do research, and then bring the international networks back to Singapore. We have also, since about seven years ago, started a complement to that effort. We now have a very big international scholarship program to bring students to Singapore to pursue research in A*STAR or the universities and after that, when the students go back to their own country, they then bring the network of Singapore back to their own country. This is what we call sending Singaporeans to the world and bringing the world to Singapore, as part of this whole open talent circulation strategy. Just to give you a sense of the level of interest and commitment to this effort, since we started seven years ago, we've had six academic years and in each academic year, we have two cycles. On average, we receive 1,000 applications per cycle and typically they come from over a hundred different countries. So you can find in Singapore now researchers from places like Africa and South America — countries that previously do not have students doing research in Singapore. Of course, they also come from the U.S., Canada, and the whole of Europe, including the former Eastern European countries. And very importantly, we also have a lot of researchers coming from the Middle East so we are now very connected to the wider world, beyond our traditional strong connection to Asia.

With Singapore increasingly viewed as the world hub for Asian international biomedical research, could you share with us the opportunities and obstacles for advancement?

In terms of opportunities, we always go where we can add complementarity or bring a unique strength to the contribution. For instance, if you recall the so-called swine flu a few years ago that started in Mexico, our Bioinformatics Institute very quickly put up a protein structure on the website that allows us to know if the mutation is such that it will render the drug Tamiflu ineffective. In fact, that has now become the reference site for World Health Organization when it comes to effectiveness of the drug for the different types of flu.

Recently we supported a very big effort across the whole of Asia for heart disease. This is the "Attract" program linking all the different communities in Asia together to study a common disease challenge. Much coordination and research work is done in Singapore. And of course, by virtue of Singapore studying Asian diseases, any important discovery will then flow out to the wider Asian community. For instance, we invested significantly in looking at gastric cancer. This is very much an East Asian disease. Unfortunately, by the time a patient discovers the disease, typically it is in the late stage so it is almost impossible to treat them. So we have a very high mortality rate: almost two-thirds mortality. Looking at this particular disease, we have found biomarkers to screen patients from their family history and begin treatments earlier.

But of course there are also a lot of obstacles, including the industry's different regulatory authorities. If only we can have a more harmonized regulatory approach, I think it will help both the industry and also the research community, so you don't have to replicate some of the efforts. That is certainly one set of issues. Obviously, we also need to get the different ethical regulations to be, in some sense, more convergent. I don't think it is possible or easy to bring about a single kind of ethical advisory system but to have greater convergence I think would be very facilitative. Likewise, some of the facilitation for movement of tissues and so on also may in itself pose some challenge to our research collaboration. We also need to think about that.

Chairman Lim, you have been very generous with your time. Thank you!

APPENDIX 2
INTERVIEW WITH HANS KEIRSTEAD

Note: Hans Keirstead has always seemed to me to be the face of modern science. He embodies the possibilities of stem cell translation, turning theories into therapies, and bringing cures into people's lives. He can think the most gigantic thoughts and express them in a way that seems natural, inevitable, and even easy — like, I could have thought of that myself! But it is also like watching films of Babe Ruth hitting homeruns out of the park — bang, there it goes, I could do that — except maybe not.

How significant were the two grants you received from the Roman Reed Spinal Cord Injury Research Act?

The Roman Reed grants enabled two core projects that I am known for.

One was the embryonic stem cell research application that led to the Geron trials. Being the world's first such stem cell-based clinical trial required us working with the FDA to create the pathway for stem cell-based trials. This had the effect of paving a highway through the jungle of FDA approval.

The second was an IP-10 program to stop the patient's immune system from attacking the patient: auto-inflammation diseases such as arthritis, dermatitis and osteomyelitis. Our team created an inhibitor of IP-10, which otherwise starts an immune attack that is directed at the self. The IP-10 technology was bought by Bristol Meyer Squibb, and is currently in Phase 3 trials.

Both of these were started because of funding from Roman's law.

Will you continue working on spinal cord injury?

As you know, I resigned from the university to work full-time at my company. My time at the University of California Irvine was precious to me, and it is a great sorrow to leave it.

Frankly, I never thought that I would be able to step up to a better biomedical development platform than the one I made at University of California Irvine, with the help of the institute and individuals there. But good fortune and hard work has presented me with a more efficient way to develop treatments; my sandbox is bigger now. And it is my intent to continue with spinal cord injury research, after getting this platform solid enough to support it. I feel strongly that the spinal cord field needs an injury leader, but it is my opinion that a spinal cord injury research-dedicated company will never be a viable one. Rather, spinal cord research and development in industry will be viable if it is the focus of a division of a strong company with multiple platform technologies addressing larger indications.

I recently merged my company, California Stem Cell, with NeoStem, with this vision in mind. When it is big enough, and with the right spinal cord technologies, I will propose the formation of a spinal cord injury division.

What first excited you about science?

At the age of 11, I decided to become a neuroscientist. I checked with my mother recently, asking if my memories had become romanticized, and she said no, I had never wanted to be anything else. It was a calling to help people. At first, I was pulled to the medical field to become a doctor. But then I switched over to the PhD track and the biomedical industry because these sectors invent and develop, whereas medicine applies. I felt my strengths were in inventing biomedical technology.

How did you realize that re-myelination (re-insulating damaged nerves) with embryonic stem cells might become a strategy for fighting paralysis?

When I first took a position as assistant professor at the University of California Irvine, I spent about six months staring at spinal cord injuries under the microscope. I wanted a new therapeutic target, because the others were not yielding results fast enough for my liking. After half a year of staring through the microscope, I came up with a new therapeutic target. I saw demyelination, where the natural insulation is stripped away from the nerve during the secondary degeneration, the wave of additional damage the spine incurs after the injury.

You chose human embryonic stem cells (hESCs) as your method to approach this problem — why?

I researched all the ways I could find to make new human oligodendrocytes, which re-insulate the damaged nerves. I wanted a source of renewable human tissue that would be inexhaustible. In the study of stem cells, I decided hESCs had a high likelihood of being both commercially and clinically viable. This was important. If a treatment is not commercially viable, it will not be clinically viable. If it can't make a profit, it won't reach the people.

At present, would you still stay with hESCs, or go to induced pluripotent stem cells (iPSCs) or direct programming?

Human embryonic stem cell technology is scalable, meaning you can make lots of it, as much as needed. It is a limitless source, with very few safety concerns if manufactured properly. For instance, you should never put an embryonic stem cell directly into the body; it must be differentiated into the cell you want first. If manufactured properly, you can virtually eliminate any negative risk.

Some politicians say that since we have iPSCs, which includes both direct and indirect reprogramming of cells, it is no longer necessary to work with human embryonic stem cells. Why is it important (or is it?) to support hESCs?

Stem cell types differ dramatically in behavior, scalability, and risk profiles. We need to protect the scientific freedom to find out which kind of stem cell is best for these various conditions. Only with scientific freedom will we be able to turn the potential for cure into reality.

Do you have any thoughts on therapeutic cloning, also known as somatic cell nuclear transfer (SCNT)?

There is tremendous potential, as the recent breakthroughs by Shoukhrat Mitalipov and others make clear, but there is still quite some way to go before it becomes a routine procedure.

What about the argument that SCNT should be banned because it might be used as a stepping stone to human reproductive cloning?

I can choose to pick up a hammer and hit my neighbor on the head with it. But I prefer to build a house instead. Should we criminalize hammers because they might be used wrongly?

Would you like to see Proposition 71 be renewed?

Absolutely! The value of Proposition 71 cannot be overstated. The California stem cell program built an unshakable foundation for stem cell research. Already it has taken applications into translation, raised awareness tremendously — and truly enhanced international collaboration, adding efficiency to both research and translation. Part Two will clearly advance treatments into the clinic; instead of testing rats, we will be treating humans.

Preparing for this interview, I read an article that said your original thesis on a new way to alleviate paralysis was selected as the best PhD thesis in Canada. Is this true?

Yes.

If you had to choose a few key moments in your life story of how that thesis led you to the recent $128 million sale of California Stem Cell, what would they be?

The first application of embryonic stem cells to a human being when the paralyzed T. J. Atchison was treated — that was a poignant event to me. Prior to that, some great moments were being admitted to Cambridge University in England, and receiving the grants from the Roman Reed Act; there have been a lot of special moments.

How about discouraging moments?

I am a positive person and try not to dwell on negatives. When you come to pitfalls and challenges, either blast through or go around. We are going up against spinal cord injury and cancer — if my staff and I were intimidated by big challenges, we would not be in this field.

What do you like best about science?

It's an innocent intrigue — the forum, the sandbox, the playground… science breeds and nurtures, and encourages one to have one's eyes wide open, to be naïve, surprised, and to discover — again, innocent intrigue.

Do you have a personal schedule, like a time to wake up?

I try to follow the dictates of my body. I wake up when the body wants to wake up, and go to sleep when it wants to sleep; I do my reading in the morning, when I feel hungry and excited and I can devour literature. I exercise in the middle of the day when I feel like jumping around. Deep thinking is for the evening, when the brain is calm and you can sense the *gestalt* — the whole picture. The brain and body are accustomed to certain routines; I try to pair the state of the body with tasks that need to be performed.

How important is patient advocacy in modern science?

This is impossible to overstate: people trust patients and patient advocates because their motivation is clear. Patient advocates can speak eloquently for research funding and freedom, which sometimes scientists cannot. Going back to Proposition 71, that was begun by a patient advocate, Bob Klein. Scientists tend to live and work in silos and are not always communicative; too busy perhaps. In general we do a terrible job of lobbying on our own behalf. Patient advocates are irreplaceable.

Without invading on your privacy too much, may I ask about your family?

Niki, my wife of eight years, is a PhD doing important work on Alzheimer's disease. Our son Connor aged four and a half years loves to experiment. Yesterday, he and I were on the kitchen floor mixing borox with glue, and we came up with a slimy goo like Silly Putty — he loved it.

Was there a key individual or mentor influence in your early years?

Professor Sam David of McGill University in Canada was a very big influence. He stressed the importance of honest research; he is a very solid person. Santiago Ramon y Cajal, considered the founder of neuroscience and a great artist whose drawings of nerves are still used today, opens the mind to newness.

I know you like to fly your own helicopter. Do you have any other hobbies?

I like to play the guitar. Also, I work out regularly with taekwondo, which I have been practicing for 20 years.

Do you have your black belt?

Yes.

Tell me more about the African project you embarked on with your father to build hospitals.

He sets up and assists sustainable aid projects, and I help. Hospitals and clinics should not be a flash in a pan; it is no good to just dump some money and leave. Our projects were structured around the idea that it should be self-sustaining and run by the people who live there.

How about your parents?

Sandra and Ken are my mom and dad. From my dad I learnt that anything is possible. My mom taught values like humility, kindness, compassion. She was an absolute angel, and I grew up surrounded by her sense of goodness, plus the energy and enthusiasm from my dad.

What advice would you give to young scientists?

Respect your ambitions. Don't let them be attenuated by the system you are in — change the system to accommodate your ambitions. *There is no path*. Make one.

Do you have any thoughts on sequestration when research funding is cut?

Infuriating. Diminishing research funds has immediate and long-term consequences. It culls future scientists, and it kills the treatments they might have generated. The amount we spend on medical research is so little when compared to other congressional priorities. Funding is fundamental to developing treatments. For officials to cutoff or diminish funding today? They are killing advances not yet made and killing the recruitment of new researchers. These are losses that cannot be quantified.

What is the importance of NIH funding?

Tremendous. The NIH is the largest source of research funding in the world. For most scientists, NIH funding is their primary source of funding, where they must go if they want the research to go forward. The NIH directly affects research and translational medicine, and it is getting harder and harder to get. What does that mean to a young scientist just starting off? If he or she cannot get funding, they must leave the field.

What is the future of stem cell research?

The pipeline is broad and deep. We are going to see an increase of applications to more and more diseases, more success in drug discovery and testing, and personalized medicines made from the patient's own body. We will continue translation of treatments for multiple conditions. We must generate treatments. There is so much more to do, and we don't intend on stopping.

Speaking of personalized medicine, how does your cancer therapy work?

During the surgical resection, you take out a piece of the cancer tumor.

You mean like when they removed my prostate gland, and cut it apart to find the cancer?

Yes. A small piece of the cancer is pulled out. We then isolate cancer cells from it. The cancer cells are the engine of cancer. We isolate and multiply the cells in a dish. Then we take blood from the patient, and from that we generate dendritic cells. Dendritic cells are the frontline soldiers of the immune system. We mix these with the cancer cells. The dendritic cells eat the cancer cells in the dish. Now the dendritic cells are changed, expressing the markers of the cancer cells, priming them to attack cancer cells. We take these primed cells and reintroduce them into the patient where they orchestrate an immune attack against the patient's cancer cells.

Wow. I can feel the hair on the back of my neck standing up.

It has been approved for Phase 3 trials. It is used against melanoma, which begins as skin cancer, then metastasizes, spreading internally. Melanoma is deadly.

Does the therapy work?

Fifty-two percent of those who underwent the procedure are alive five years later, more than double the current rate. We have been working on this since 2002. The five-year data are just being presented this week at the American Society of Clinical Oncology.

Will your technique work on other kinds of cancer?

We are working on ovarian cancer and liver cancer, attempting to apply the same technology. I am also attempting retinal regeneration with another stem cell technology, but that is in early stages right now.

What are your plans for the future?

To devote myself to the development of NeoStem, with the intent that it will become the indisputable global leader in stem cell application.

We wish you long life, health, and every success.

Note from Hans's father, Kenneth Keirstead[1]:

As Hans Keirstead's Dad, I take delight not only in his stem cell accomplishments, but in an adventure we have shared for the past several years. The ancestry both of us share goes back to Africa for the last 100 years where my grandparents served the indigenous people of South Africa as missionaries. They emigrated there in the early 1900's, and that is where I was born. As a family, we have kept a strong connection that resulted in Hans becoming engaged in Africa as well.

---

[1] Keirstead, Kenneth E. Message to: Don Reed. 2014 Jun 3.

Over ten years ago, I started a humanitarian program to improve healthcare to a very rural part of Guinea in West Africa. In 2006, Hans and his wife Niki visited Guinea to help launch a program called Sante et Espoir, or Health and Hope in English. Hans's help associated with the University of California Irvine started the flow of donated medications and wheelchairs that have made a significant contribution to disadvantaged and poor people there.

Hans's mantra has always been to live life with a capital "L". This symbolizes his attitude as a caring human being who helps those without hope to live better lives.

APPENDIX 3
INTERVIEW WITH BOB KLEIN

Can you remember when you realized what the CIRM could be?

I was in the car with Paul Berg in the summer of 2004. Dr. Berg won the Nobel Prize for recombinant DNA, which was the precursor for artificial insulin that keeps my son, who has type1 diabetes, alive. That first product of recombinant DNA has been followed in the next ten years by hundreds of therapies that have saved millions of lives of heart and cancer patients. Recombinant DNA has, in the most recent decade, provided the tools needed to unlock the human genome. I said to Paul, "Proposition 71's funding for stem cell therapies and research and the possible medical advancements that could result have been compared to the impact of recombinant DNA. Is that really possible, that it could have that profound of an impact on a broad spectrum of disease and injury?" I was not prepared for his answer. Dr. Berg said, "Bob, you don't understand. In all of human history, we've never been able to regenerate cells for any part of the human body. Stem cell research will give us that power. It will change the future of medicine and humanity. In 20 years, we will look back to today and wonder how medicine could ever have been practiced as it is now." I didn't say anything for the rest of the drive — I just thought about the magnitude of the promise and the responsibility that we must pass this initiative.

When it was time to actually write Proposition 71, who were the people in the room?

When I was writing Proposition 71, I was usually either alone or working with Amy DuRoss. There were seven different lawyers who reviewed drafts and who worked on different sections of the initiative with me to convert the narrative draft into statutory language that would pass intensive reviews of state laws and regulations. The financial sections involved intensive reviews and joint drafting discussions with Chad Cardall of Orick Harrington, a bond and tax expert. Of the Merksamer Law Firm in Sacramento, Richard Martland did the regulatory language that interfaced with existing regulations and modified it to work with the scientific rules of the agency. There are a number of other notable attorneys that have had notable roles like Mueller and Merksamer. For a comprehensive overview of all sections other than finance, James Harrison was the key individual who worked on incorporating the concepts into the state governmental structure.

Can you recall the first major donor beside yourself? How much was your total contribution, and how much was the total cost?

The major individual who assumed a leadership role in raising funds was John Doerr, who served as finance co-chair. Michael Goldberg is another individual who, early on, assumed a leadership role in fundraising. I gave $4.5 million during the campaign. The total campaign cost was $34.5 million.

Do you have a favorite memory of the campaign to pass Proposition 71?

When my children were on the stage with me on election night.

Aside from the outcome, was there a particularly satisfying moment of the lawsuit process?

The quality and thoroughness of Superior Court Judge Bonnie Sabraw's court decision was inspiring. My faith in the judicial system protecting important parts of the democratic process in California was absolutely confirmed and elevated by the excellence of this decision, which covered a vast area of law and an extraordinarily large number of complaints and documents that reached from areas of science through election law into sophisticated questions of finance.

What made you choose a large board of directors representing certain constituencies?

California's universities and non-profit research institutions represent a vast treasury of expertise in advancing medicine. A large board permitted me to capture the breadth and spectrum of that expertise, along with the diversity of backgrounds and experience with chronic disease and injury among patient advocates, and a few key seats for biotechnology leaders who were willing to exclude stem cell therapies from their personal financial portfolios.

Can you mentally reconstruct a typical work day for you during the seven years you worked as Chair? Is my memory correct that you worked as a volunteer for the first six years?

I can't recall ever having a typical day. The day would generally run from seven in the morning until ten at night. Much of the agency worked late into the evenings. I worked as a volunteer for the first five years. In the last two years, I received enough compensation to pay salary for my staff who worked on the extensive requirements of the chairman's position.

Do you have a favorite memory of your almost seven years' leadership of the ICOC board of directors?

In October of 2009, when the Board approved the first round of disease team grants that involved collaboration of California's scientists with grants with the U.K., Canada, and Australia. We knew we were on the path to translating a therapeutic application that would actually reach patients. There was a sense that we would change the quality of life and the human condition.

The opposition says there is no need for embryonic stem cell research. What are your thoughts?

The most recent research restates the need for embryonic stem cell research. Human embryonic stem cells (hESCs) represent the gold standard. Cells derived from induced pluripotent stem cells (iPSCs) from skin fragments must be validated against cells derived from hESCs to even know if the derived cell is an accurate replica. Critically, recent studies demonstrate that there are thousands of mutations from iPSCs that are not present in cell types derived from hESCs. Scientists will not know for several more years whether those mutations will change the interactions of these derived cells with the human body. The agency must remain committed to finding the best cell type for each disease. For some diseases, progenitor cells may provide the best option, but for blindness from age-related macular degeneration, affecting one in five Americans between 65–85 years old, or Type one diabetes impacting children throughout their lives, the leading candidate appears to be hESCs.

Some argue that Proposition 71 is a failure because there are no cures. How do you respond?

First, there are human trials of therapies that resulted from Proposition 71 research that have already saved some patients from some forms of deadly blood cancers. These are important milestones of progress that are backed by FDA approved science. This will be a challenging road, but we can see the path to success, which in 2004 was only a dream, and today has stepping stones in place. We must remember that behind these human clinical trials, there are more than 2,470 published scientific discoveries by research funding through the agency that creates an intellectual framework for supporting a major pipeline of therapy candidates headed toward more human trials. Some patients have benefited already, but their therapies are only interim products that must be refined and tested further. Twelve nations have decided to join our teams because they believe that Proposition 71 is one of the world's best bets on reducing the future of human suffering.

Should there be a Part Two of Proposition 71? If so, why?

At the beginning of 2016, the people of California need to be polled to understand their view on results of the funding that they gave California scientists and patients. We must listen to the people. I hope that they will re-up their commitment. We need that final tactical mile, a very challenging mile, from the initial human trials to broad availability.

What are your feelings on the difficulties of getting grants from the NIH?

All areas of the biological sciences owe a tremendous gratitude to the NIH funding program in advancing new therapies for individuals with chronic disease and injury. If young scientists cannot find funding, this generation is losing its best and brightest who can change the future of medical therapies. If we do not have health as a priority for the people of our country, what do we really have? Our health care system is collapsing under increasing costs of chronic disease as the population ages. NIH funding, in particular for early-intervention therapies like stem cell research, has to become a priority.

What do you see as threats to research for cure today and tomorrow?

The personhood movement by evangelicals in the U.S. led to a bill initiated by Vice Presidential candidate Paul Ryan. That bill would effectively outlaw human embryonic stem cell therapies and/or therapies derived from other types of cells like fetal stem cells, which was involved in the production of the original polio vaccine. Unless we uphold this constitutionality of research in Congress and at the local school board, science and medicine will be forced to abandon extraordinarily promising paths to mitigate human suffering.

Funding for Proposition 71 is a long-term, 30 years payment, after a five-year grace period. Would you explain why you structured it that way?

The taxpayers and their families on the first day of funding should not pay for a benefit that may start 14 years in the future. The taxpayers and their families who start receiving the benefits should share the costs with the next 20 or so years of families who benefit. This bond structure is the best structure for lining up the families and their children who should benefit with the responsibility of those who pay for it.

Your work as a patient advocate was inspired by your son's struggle with diabetes and your mother's Alzheimer's disease condition. Is there a message you would like to share with other patient advocates?

Every American is a phone call away from learning that their son or daughter, their husband or wife, or their parents are in a battle with chronic disease or injury for any quality of life or life itself. I pray that other parents don't face the tragedy of their son or daughter suffering through Type one diabetes, or childhood leukemia, or hear a message that their son or daughter is paralyzed in a car accident or sporting accident. Yet, we must all face the truth, that even if it is not today, our sons and daughters, our husbands and wives, will all eventually face chronic disease and injury in their lifetime — maybe even tomorrow. Anyone who is not part of a patient advocacy organization or who doesn't do a cancer walk for a friend or personally reach out to find political candidates who will fight for their family's future is leaving their family vulnerable to that tragic call announcing years of suffering without a therapy. *Each of us is the answer.* Each of us has a strong voice to reach families and friends to ask them to wake up, defend science, protect new therapies, and support those in medicine and science who are willing to give their lives to reduce human suffering. Today, we have a historical privilege of unimaginable dimensions to change the future of human suffering. Seize the day!

NAME INDEX
(PARTIAL)

Albrecht, Grant, 295
Alexander the Great, 197
Ali, Muhammad, 321, 322
Allickson, Julie, 265
Ames, John, Genevieve, David, 30
Anderson, Aileen, 174
Ashford, Brad, 289

Banting, Frederick G., 229, 351
Berg, Paul, 57
Bernard, Claude, 301
Bernstein, Alan, 197, 198
Best, Charles, 229, 351
Bharti, Khapil, 370
Bhatia, Mick, 328
Bing, Lim, 398, 402
Blau, Helen, 377, 380
Bluestone, Jeff and David, 228, 230
Blurton, Matthew, 138
Bongso, Ariff, 313, 315, 320
Bromm, Curt, 289
Brownback, Senator Sam, 25
Brown, Gordon, 364
Brown, Governor Jerry, 257, 258
Brown, Nina and Joe, 335, 336, 337, 338
Brown, Timothy Ray, 304
Burnham, Malin, 383
Burton-Brown, Kristy, 126
Bush, former President George W., 2

Campos, Nora, 256, 258
Cannon, Paula, 304, 305
Carson, Dennis, 359
Chen, Bertha, 252, 253

Chen, Haidan, 343, 345
Chen, Irving, 359
Choo, Andre, 315
Cibelli, Jose, 119
Clegg, Dennis, 164
Clinton, President Bill, 241
Cody, Dana, 42–44
Conboy, Irina, 378
Cuenda, Natividad, 119

Daly, Amy, 12, 13
D'Amour, Kevin, 231
Darnovski, Marci, 90
Davis, Susan, 264
Deisher, Theresa, 239
Deng, Xiao Ping, 345, 347
Detroit Zoo, 102
DeWitt, Natalie, 381
Dickey-Wicker Amendment, 241
Dick, John, 131
Dillman, Robert, 173
Doerr, John, 412
Domingo, Carmen, 154
Doyle, Jim, 193
Dracula, 65
Drown, Stuart, 277
DuRoss, Amy, 12

Ellison, Brooke, 327, 329
Evseenko, Dennis, 374

Feigal, Ellen, 355
Feinstein, Senator Dianne, 105, 106, 111
Flood, Mike, 289

Name Index (partial)

Florez, Dean, 276
Fogel, Susan, 76
Fontana, Jeannie, 296, 297
Foreman, George, 321, 322
Fox, Michael J., 323
Fuchs, Elaine, 150
Fu, Zhiyan, 319

Gage, Rusty, 149
Gallegos, Leroy, 227
Gates, Bill, 364
Gibbons, Leeza, 137, 138
Gladstone Institute, 50, 51, 52
Goldberg, Michael, 412
Gold, Joe, 158, 159
Goldstein, Larry, 30

Hadid, Zaha, 309
Haley, Judie, 336, 338
Hall, Zach, 57
Hans Keirstead, 171
Havton, Leif, 145, 146
Heimlich, Henry, 168
Heller, Stefan, 163
Helm, Jill, 382
Hoffman, Bob, 75
Huckabee, Mike, 140, 141
Hui, Lijian, 203
Humayun, Mark, 164

Ivey, Kathy, 50

Jamieson, Catriona, 129, 131, 357
Jarvik, Robert, 168
Jensen, David, 89, 93
Jintao, Hu, 286
Joan of Arc, 72

Kaplan, Ben and Oliver, 18
Karolya, Kaitlin, 229
Karolya, Tom, 229
Keirstead, Connor, 406
Keirstead, Hans, 173, 403
Keirstead, Nikki, 403, 408
Keirstead, Sandra and Ken, 407

Keller, Helen, 161, 162, 165
Kennedy, Edward, 160
King, Melissa, 114
Kirk Douglas, 223, 225
Klein, Bob, 1, 2, 4, 10
Knoepfler, Paul, 351
Koh, Charles and David, 218
Kohn, Donald, B., 361, 394

Laird, John, 357
LaLande, Marc, 208
Lamberth, Royce, 239
Lane, Nancy, 361
Langer, Robert S., 331
Lathrop, Steve, 289, 290
Lee, Hsien Loong, 397
Leonhardt, Howard, 265
Lewis, Amy, 11, 12
Liu, Edison, 398
Lipinski, Daniel, 141
Litvack, Frank, 198, 199
Loh, Jonathan, 316, 317
Loring, Jeanne, 298
Lubin, Bert, 189, 190

MacLeod, J.J.R., 351
Maddox, Sam, 345
Malloy, Governor Dannel Patrick, 208, 209
Maxon, Mary, 220, 221
McClurg, Jim, 291, 293
McCulloch, Ernest, 351, 354
McDonald, John, 118
McGuinty, Premier Dalton, 328
McKayle, Donald, 246
McLeery, Beckie, 336–338
McMahon, Linda, 205, 207
Miko, Toshio, 265
Mills, Randy, 269, 272
Miner, Karen, 255
Minger, Stephen, 384
Missouricures, 102
Moffett, Alex, 296
Morrison, Sean, 240
Muotri, Alysson, 115
Murphy, Chris, 206, 209

Name Index (partial)

Nakatsuji, Norio, 385
Newsom, Gavin, 61
Ng, Huck Hui, 314, 318
Nolta, Jan, 247, 249
Norquist, Grover, 215

Obama, President Barack, 86
Okarma, Thomas, 211
Olson, Patricia, 145, 146
O'Malley, Martin, 117
Ortiz, Senator Deborah, 23, 24
Ouyang, H. W., 343

Pachter, Joel, 301
Pachter, Tamar, 45, 46
Paul, Rand, 128
Pereira, Lygia, 341, 342
Piccot, Tom, 349, 350
Pomeroy, Claire, 247, 249
Poon, Suzanne, 346
Powers, Linda, 297
Pranav, 83–86
Prentice, David, 67–69
Prieto, Francisco, 220

Rab, Harriet, 241
Rao, Mahendra, 369, 371
Reed, Charles, 377
Reed, Desiree, 147, 149
Reed, Gloria, 167, 169
Reed, John, 91, 95
Reed, Katherine, 392
Reed, Patty, 129, 130
Reed, Roman, 5, 9, 10, 383
Reed, Roman Jr., and the Family Reed, 391, 395
Reeve, Brock, 381
Reeve, Christopher, 37, 38
Reeve Irvine Research Center, 213
Rehen, Stevens, 341
Reid, Charles, 299, 300
Reijo-Pera, Renee, 252, 253
Rell, Governor Jody, 205, 208
Reynolds, Jesse, 90, 276, 277
Richmond, Alex, 384

Rickman, Catherine Bowes, 163, 164
Rickman, Dennis, 163, 165
Robin Hood, 219, 222
Romney, Mitt, 129
Roosevelt, Franklin Delano, 197
Rose, Melinda and Wayne, 336
Roth, Duane, 263, 265
Roth, Renee, 264
Rove, Karl, 68
Ryan, Paul, 127

Sabraw, Judge Bonnie Lewman, 41, 42, 44, 47
Saldana, Frances and Margie, 243
Samuelson, Joan, 31
Sanford, Denny, 383
Schmit-Albin, Julie, 289, 292
Schuele, Birgit, 324
Schwarzenegger, Arnold, 47
Sebelius, Kathleen, 239, 241
Shapiro, Harold, 278, 280
Sheehy, Jeff, 303, 306
Sherley, James, 239
Shestack, Jonathan, 114
Shiley, Darlene and Donald, 263
Shizuru, Judith, 357
Siegel, Bernie, 24, 25, 27
Simpson, John, 90, 93, 274, 275
Slamon, Dennis, 357
Smith, Richard, 358
Snyder, Evan, 77
Solomon, Susan, 328
Sornberger, Joe, 350
Spink, Katy, 159
Stark, Pete, 295
Steinberg, Gary, 224
Steward, Osward, 385
Stiller, Calvin, 353
Strong, Gwendolyn, 87
Studer, Lorenz, 330
Swee, Hoh Keng, 308, 309

Tan, Tony, 397
Tanzi, Rudolph, 237
Taubman, A. Alfred, 101

Thal, Donna, 263
Thal, Leon, 261–264
Thomas, Jonathan, 198, 199
Thompson, Leslie, 245, 246
Thomson, Jamie, 147
Tian, Xiuchun "Cindy", 262
Till, James, 351, 352, 354
Tong, Dizhou, 283
Torres, Art, 191, 193
Trounson, Alan, 193, 194, 270, 271
Tuszynski, Mark, 171
Twain, Mark, 162

Valdez, Juan J. Parcero, 265
Vertez, Alain, 384
Vetter, David: "Bubble Boy", 394

Wagers, Amy, 381
Walsh, Craig, 301
Wasson, Ann and Greg, 321
Weissman, Irv, 122
Werner, Michael, 120
West, Mike, 212
Wheelock, Vicki, 247–249

Willenbring, Holger, 203
Williams, Robin, 322, 323
Winchell, Paul, 168
Witte, Owen, 149
Wu, Joseph, 168

Xiaoming, Jin, 287
Xu, Ren-He, 208

Yamanaka, Shinya, 51
Yang, Jerry, 261, 262, 264
Yeo, Philip, 309–311
Yew, Lee Kuan, 309, 311
Ying, Jackie Yi-Ru, 318, 319
Young, Wise, 122, 345–347
Yu, Junying, 235

Zatz, Mayana, 341, 342
Zeng, Fanyi, 119, 284–286
Zeng, Xianmin, 324
Zern, Mark, 202, 203
Zhang, Joy Yuehue, 343, 345
Zhang, Wen Cai, 319
Zucker, Jerry, 29

SUBJECT INDEX

abortion, 76, 77, 125, 126
Academy of Sciences of the State of Sao Paulo, 341
AIDS, 303–306
Alliance Defense Fund, 241
Alpha clinics, 270, 271
Alzheimer's disease, 79
ALS (amyotrophic lateral sclerosis, Lou Gehrig's disease), 79
Americans for Cures Foundation, 370
A Night for No Mexican Tears, 391
anti-retroviral therapy (ART), 303
arthritis, 207, 373, 374
A*STAR, 397–401
Asterias Biotherapeutics, 212
attacks on stem cell research, federal, 175, 177
attacks on stem cell research, state, 176, 177, 178
autism, 113, 114, 115, 116

bankruptcies and foreclosures, medical, 139
Beike, Inc., 296, 297, 298
Bible, 65, 66
Biopolis, 309, 310, 311
birth control, 125, 126
bladder control, 253
blastocysts, 16, 17
blindness, 80
breast cancer, 143
Bridges program, CIRM, 153, 154, 155

Campaign, Proposition 71, 19
Canada, 119

Canadian Stem Cell Foundation, 350, 351
cancer, 80
cancer diagnosis, 387, 389
Capricor, Inc., 358
Catholic, 33, 34, 35
Center for Genetics and Society, 276
Center for Regenerative Medicine (CRM), 369
Chiron, Inc, 220
California Institute for Regenerative Medicine (CIRM), 2, 51, 223, 224, 388
   ICOC and author, 393
   location, 145, 146
   stem cell research facilities, 64
cloning, 22–27
Coalition for the Advancement of Medical Research (CAMR), 25, 26
conflict of interest, 46
Connecticut Stem Cell Investment Act, 205
Connecticut United for Research, 208
Consumer Watchdog, 276
core facility, 63
Cancer Prevention and Research Institute of Texas (CPRIT), 335, 336
Cures Action Now (CAN), 2

deafness, 163
definitions, importance of, 75
Democrats, 32
diabetes, 80, 225–229
disability population, U.S. and world, 140

Eli and Edith Broad Center for Regenerative Medicine, 265
embryonic stem cells, 16

Subject Index

Family Research Council, 66–69
FDA (Food and Drug Administration), 234, 236
Fighter, a laboratory rat which walked again after being paralyzed, 213

Genetics Policy Institute (GPI), 117, 118
Geron Stem Cell Department losing their jobs, 158
Geron Trials with Roman Reed program-funded invention, 158

Harris Poll support for embryonic stem cell research, 139
heart attack, 167, 168
Human Life Amendments, 141
Huntington's disease, 243

I-FLY, Inc., 299
ImStem Biotechnology, Inc., 300
incontinence, urinary, 251, 252, 253
Independent Citizens' Oversight Committee (ICOC), 47
induced pluripotent stem cells (iPSCs), 233, 235, 237
Institute of Medicine, 277–280
International Society of Stem Cell Research, 147
in vitro fertilization procedure, 16

Juvenile Diabetes Research Foundation (JDRF), 2

Legislative Bill, 608, 289, 290
leukemia, 130, 131, 132
Little Hoover Commission (LHC), 275–278
liver, described, 201–203
Lupron, 182, 185

Maryland, 117, 118, 122, 123
MELD (model for end-stage liver disease, 202
Monash University, 301
Ministry of Science and Technology (MOST), 286, 287

motor neurons, 87
multiple sclerosis, 300

National Alzheimer's Project Act, 136
National Center for Advancing Translational Sciences (NCATS), 370
Neostem, Inc., 404, 408
New Jersey, 102
New York Stem Cell Foundation, 121
Nightlight, Inc, 239, 240

opposition, ideological, 67

paralysis, 81, 174
paralysis, population, 171–173
Parkinson's, 81
Part Two, 183, 411–413
patient advocacy, need for, 150
patient advocates, 77
personhood, 78, 125–128
polio, 364
poverty as worsening sickle-cell disease, 187
Proposal, 2, 98, 100–103
Proposition, 13, 42
Proposition, 71, 412
prostate cancer, author's, 179, 180, 182
prostatectomy, 180

radiation, 180–184
Republican, 22, 25, 26
Reversal of Bush Stem Cell Policy, 106
right-to-life, 32
Roman Reed Lab, 211
Roman Reed Spinal Cord Injury Research Act of 1999, 10, 255
Roman's Law, 256, 257

Senate Constitutional Amendment 13, 53
sickle cell anemia, described, 187, 188
Singapore, 307–311
spinal muscular atrophy (SMA), 83
Splash (movie with implied Alzheimer's), 135

Stanford University, 114
Stem Cell Research Enhancement Act, 72, 73
stroke, 223, 224
Student Senate of University of Nebraska, 292
Sygen, 6, 7

Texans for the Advancement of Medical Research (TAMR), 336–338
Trop 2 as cancer fighter, 182

United Nations, 27
University of California at Irvine, 138, 404, 409
University of California at San Diego, 131
University of California at San Francisco (UCSF), 50
United Network for Organ Sharing (UNOS), 202

Valley of Death (biomedical term), 269
Values Voters convention, 69
ViaCyte, Inc., 228, 230
voter fraud, 217
voter ID, 216, 217, 218
Voter's Guide, 22

Wisconsin Alumni Research Foundation (WARF), 221
wolverines, 97

MAY 0 4 2016